New and Improved Nanomaterials and Approaches for Optical Bio- and Immunosensors

New and Improved Nanomaterials and Approaches for Optical Bio- and Immunosensors

Editor

Boris B. Dzantiev

MDPI • Basel • Beijing • Wuhan • Barcelona • Belgrade • Manchester • Tokyo • Cluj • Tianjin

Editor
Boris B. Dzantiev
A.N. Bach Institute of Biochemistry,
Research Center of Biotechnology Russian Academy of Sciences,
Moscow, Russia

Editorial Office
MDPI
St. Alban-Anlage 66
4052 Basel, Switzerland

This is a reprint of articles from the Special Issue published online in the open access journal *Biosensors* (ISSN 2079-6374) (available at: https://www.mdpi.com/journal/biosensors/special_issues/optical_immusens).

For citation purposes, cite each article independently as indicated on the article page online and as indicated below:

LastName, A.A.; LastName, B.B.; LastName, C.C. Article Title. *Journal Name* **Year**, *Volume Number*, Page Range.

ISBN 978-3-0365-8316-7 (Hbk)
ISBN 978-3-0365-8317-4 (PDF)

Contents

biosensors

MDPI

Editorial

New and Improved Nanomaterials and Approaches for Optical Bio- and Immunosensors

Boris B. Dzantiev

A.N. Bach Institute of Biochemistry, Research Center of Biotechnology of the Russian Academy of Sciences, Leninsky Prospect 33, 119071 Moscow, Russia; dzantiev@inbi.ras.ru

The current state in the development of biosensors is largely associated with the search for new approaches to simplify measurements and lower detection limits. Nanotechnologies have provided various tools to achieve these goals; however, their comparative assessment and the reasonable choice of the most effective nanomaterials are still in demand. The use of nanoparticles as carriers and labels has good practical prospects due to the availability of simple tools for their optical registration.

The above-listed reasons have driven our interest in preparing this Special Issue. It is focused on new approaches for highly sensitive bio- and immunosensors with nanodispersed labels and their optical registration. The 11 articles of this Special Issue demonstrate the current state of developments and their most promising directions.

Yuliya A. Podkolodnaya et al. characterize composites of carbon nanostructures and silica particles in their review [1]. This combination integrates specific features of both compounds and allows for different assay formats and detection techniques. The publication reports successful applications of the given composites for bioimaging, chromatography, immunoassays, and other important analytical tasks.

Jia Zhang et al. studied the transformation of mycotoxin sterigmatocystin upon processing contaminated rice wine [2]. They used HPLC–MS/MS to investigate changes in sterigmatocystin content and profiles of its metabolites for different rice leaven levels and fermentation times. The obtained data help enable the evaluation of health risks and finding methods for their prevention.

Galina V. Presnova et al. considered the digital detection of nucleic acids on silicon microchips based on the counting of gold nanoparticles in DNA duplexes by scanning electron microscopy [3]. This approach demonstrated high sensitivities (down to 0.04 pM for short oligonucleotides) and was effective for productive testing of genetic variability. The practical potential of the developed technique was confirmed by multiplex quantification of bacterial genes responsible for resistance against β-lactam antibiotics.

An original way to detect low quantities of pathogen-specific immunoglobulins in serum samples for efficient serodiagnosis of infection deceases is proposed in the paper by Dmitriy V. Sotnikov et al. [4]. The authors improved traditional lateral flow test strips via the two-stage incorporation of nanoparticles into immune complexes. This enables increasing the number of the detected markers per one formed immune complex (by more than two orders of magnitude), thus providing the detection of specific antibodies at low concentrations. The proposed approach was successfully applied for the serodiagnosis of COVID-19.

Oleg L. Bodulev et al. combined a chemiluminescent enzyme-linked oligonucleotide assay with mismatched catalytic hairpin assembly amplification [5]. The case study of this approach for microRNA-155 assay demonstrated a low detection limit of 400 fM. The efficiency of the assay and its perspectives for medical purposes was demonstrated by measurements of microRNA-155 content in human cancer lines.

The paper of Zhaodong Li et al. demonstrates the successful application of hapten design techniques aimed at the modulation of immunodetection specificity [6]. Due to

Citation: Dzantiev, B.B. New and Improved Nanomaterials and Approaches for Optical Bio- and Immunosensors. *Biosensors* **2023**, *13*, 443. https://doi.org/10.3390/bios13040443

Received: 23 March 2023
Accepted: 29 March 2023
Published: 31 March 2023

the variety of structurally similar sulfonylurea-based anti-diabetic drugs, broad-specific immunodetection is in demand. The authors present newly produced antibodies and their use in a multiplexed enzyme-linked immunosorbent assay of sulfonylurea adulterants in functional pills.

Larisa V. Sigolaeva et al. considered new approaches for highly sensitive surface-enhanced Raman spectroscopy (SERS) [7]. Their study presents a laser-induced aggregation of silver nanoparticles and the thermoresponsive micelle-forming diblock copolymer of 1,2-butadiene and N,N-dimethylaminoethyl methacrylate. Using 4-mercaptophenylboronic acid as a reporting probe led to the laser-induced enhancement of a SERS signal. The proposed mechanism of this effect was confirmed by theoretical simulation.

Simone Cavalera et al. investigated cases of nonmonotonic dependence of the recorded signal on the analyte concentration ("hook effect") in a lateral flow sandwich immunoassay [8]. The authors explained this effect as due to antigen saturation caused by the excess of the labeled antibody and studied the systematic variation in the immunoassay parameters that affect rapid interactions in lateral flow tests. The obtained results allowed for the development of an effective assay for lumpy skin disease virus detection.

Mariia V. Samodelova et al. developed a SERS-based sensor for detecting the SARS-CoV-2 virus using aptamers as recognizing molecules [9]. The formed sandwich complexes of the primary aptamer at the plasmonic surface, antigen-coated silver nanoparticles, and fluorophore-labeled secondary aptamer allowed an investigation of the impact of the distance between the SERS-active compound and the quencher and proposal of a protocol for high-sensitive detection.

One more SERS-based assay was developed in the study of Artem Tabarov et al. [10]. Influenza viruses have no characteristic spectral peaks and, therefore, are difficult to identify by SERS-based sensors. The authors successfully applied machine learning technologies and provided differentiation of samples containing influenza A and B viruses with high accuracy. The given development contributes to more effective medical diagnostics and assessment of epidemiological situations.

In the study by Olga E. Eremina et al., the development of molecular immobilization and resonant Raman amplification by a complex-loaded enhancers (MIRRACLE) sensor to measure intense Raman signals on a nanostructured silver-based substrate was reported [11]. The authors synthesized plier-like ligands with different functional groups as tools for shifting the maximum absorbance of catecholamines and generating a new absorbance band. The concentration range reached for the determination of dopamine was 3.2 pM–10 nM, which confirmed its prospects in the evaluation of neurotransmitter metabolism disorders.

Funding: This research received no external funding.

Acknowledgments: I would like to acknowledge all the authors who submitted their manuscripts and made this Special Issue successful. I am grateful to the staff of the Editorial Office and the reviewers involved in this Special Issue.

Conflicts of Interest: The author declares no conflict of interest.

References

1. Podkolodnaya, Y.A.; Kokorina, A.A.; Ponomaryova, T.S.; Goryacheva, O.A.; Drozd, D.D.; Khitrov, M.S.; Huang, L.; Yu, Z.; Tang, D.; Goryacheva, I.Y. Luminescent composite carbon/SiO2 structures: Synthesis and applications. *Biosensors* **2022**, *12*, 392. [CrossRef] [PubMed]
2. Zhang, J.; Xu, L.; Xu, X.; Wu, X.; Kuang, H.; Xu, C. Profiles of sterigmatocystin and its metabolites during traditional chinese rice wine processing. *Biosensors* **2022**, *12*, 212. [CrossRef] [PubMed]
3. Presnova, G.V.; Presnov, D.E.; Filippova, A.A.; Tsiniaikin, I.I.; Ulyashova, M.M.; Rubtsova, M.Y. Multiplex digital quantification of β-lactamase genes in antibiotic-resistant bacteria by counting gold nanoparticle labels on silicon microchips. *Biosensors* **2022**, *12*, 226. [CrossRef] [PubMed]
4. Sotnikov, D.V.; Byzova, N.A.; Zherdev, A.V.; Xu, Y.; Dzantiev, B.B. Silent antibodies start talking: Enhanced lateral flow serodiagnosis with two-stage incorporation of labels into immune complexes. *Biosensors* **2022**, *12*, 434. [CrossRef] [PubMed]

5. Bodulev, O.L.; Galkin, I.I.; Zhao, S.; Pletyushkina, O.Y.; Sakharov, I.Y. Quantitation of microRNA-155 in human cells by heterogeneous enzyme-linked oligonucleotide assay coupled with mismatched catalytic hairpin assembly reaction. *Biosensors* **2022**, *12*, 570. [CrossRef] [PubMed]
6. Li, Z.; Xie, H.; Fu, T.; Li, Y.; Shen, X.; Li, X.; Lei, Y.; Yao, X.; Koidis, A.; Liu, Y.; et al. Complementary strategy enhancing broad-specificity for multiplexed immunoassay of adulterant sulfonylureas in functional food. *Biosensors* **2022**, *12*, 591. [CrossRef] [PubMed]
7. Sigolaeva, L.V.; Nechaeva, N.L.; Ignatov, A.I.; Filatova, L.Y.; Sharifullin, T.Z.; Eichhorn, J.; Schacher, F.H.; Pergushov, D.V.; Merzlikin, A.M.; Kurochkin, I.N. In situ SERS sensing by a laser-induced aggregation of silver nanoparticles templated on a thermoresponsive polymer. *Biosensors* **2022**, *12*, 628. [CrossRef] [PubMed]
8. Cavalera, S.; Pezzon, G.; Grazioli, S.; Brocchi, E.; Baselli, S.; Lelli, D.; Colitti, B.; Serra, T.; Di Nardo, F.; Chiarello, M.; et al. Investigation of the "Antigen hook effect" in lateral flow sandwich immunoassay: The case of lumpy skin disease virus detection. *Biosensors* **2022**, *12*, 739. [CrossRef]
9. Samodelova, M.V.; Kapitanova, O.O.; Meshcheryakova, N.F.; Novikov, S.M.; Yarenkov, N.R.; Streletskii, O.A.; Yakubovsky, D.I.; Grabovenko, F.I.; Zhdanov, G.A.; Arsenin, A.V.; et al. Model of the SARS-CoV-2 virus for development of a DNA-modified, surface-enhanced Raman spectroscopy sensor with a novel hybrid plasmonic platform in sandwich mode. *Biosensors* **2022**, *12*, 768. [CrossRef]
10. Tabarov, A.; Vitkin, V.; Andreeva, O.; Shemanaeva, A.; Popov, E.; Dobroslavin, A.; Kurikova, V.; Kuznetsova, O.; Grigorenko, K.; Tzibizov, I.; et al. Detection of A and B influenza viruses by surface-enhanced Raman scattering spectroscopy and machine learning. *Biosensors* **2022**, *12*, 1065. [CrossRef] [PubMed]
11. Eremina, O.E.; Kapitanova, O.O.; Medved'ko, A.V.; Zelenetskaya, A.S.; Egorova, B.V.; Shekhovtsova, T.N.; Vatsadze, S.Z.; Veselova, I.A. Plier ligands for trapping neurotransmitters into complexes for sensitive analysis by SERS spectroscopy. *Biosensors* **2023**, *13*, 124. [CrossRef] [PubMed]

biosensors

MDPI

Review

Luminescent Composite Carbon/SiO_2 Structures: Synthesis and Applications

Yuliya A. Podkolodnaya [1], Alina A. Kokorina [1,*], Tatiana S. Ponomaryova [1], Olga A. Goryacheva [1], Daniil D. Drozd [1], Mikhail S. Khitrov [1], Lingting Huang [2], Zhichao Yu [2], Dianping Tang [2] and Irina Yu. Goryacheva [1]

1 Department of Inorganic Chemistry, Chemical Institute, Saratov State University, Astrakhanskaya Street 83, 410012 Saratov, Russia; podkolodnaya00@mail.ru (Y.A.P.); tatyanka.ponomareva.97@mail.ru (T.S.P.); olga.goryacheva.93@mail.ru (O.A.G.); drozddd@sgu.ru (D.D.D.); mikkhitrov@yandex.ru (M.S.K.); goryachevaiy@mail.ru (I.Y.G.)

2 Key Laboratory for Analytical Science of Food Safety and Biology, Department of Chemistry, Fuzhou University, Fuzhou 350108, China; 201310011@fzu.edu.cn (L.H.); 201320082@fzu.edu.cn (Z.Y.); dianping.tang@fzu.edu.cn (D.T.)

* Correspondence: kokorinaaa@sgu.ru; Tel.: +7-(951)-8861027

Abstract: Luminescent carbon nanostructures (CNSs) have attracted great interest from the scientific community due to their photoluminescent properties, structural features, low toxicity, and a great variety of possible applications. Unfortunately, a few problems hinder their further development. These include the difficulties of separating a mixture of nanostructures after synthesis and the dependence of their properties on the environment and the aggregate state. The application of a silica matrix to obtain luminescent composite particles minimizes these problems and improves optical properties, reduces photoluminescence quenching, and leads to wider applications. We describe two methods for the formation of silica composites containing CNSs: inclusion of CNSs into silica particles and their grafting onto the silica surface. Moreover, we present approaches to the synthesis of multifunctional particles. They combine the unique properties of silica and fluorescent CNSs, as well as magnetic, photosensitizing, and luminescent properties via the combination of functional nanoparticles such as iron oxide nanoparticles, titanium dioxide nanoparticles, quantum dots (QDs), and gold nanoclusters (AuNCs). Lastly, we discuss the advantages and challenges of these structures and their applications. The novelty of this review involves the detailed description of the approaches for the silica application as a matrix for the CNSs. This will support researchers in solving fundamental and applied problems of this type of carbon-based nanoobjects.

Keywords: luminescent composite particles; carbon nanostructures; luminescent carbon-based nanomaterials; silica nanoparticles; luminescence

Citation: Podkolodnaya, Y.A.; Kokorina, A.A.; Ponomaryova, T.S.; Goryacheva, O.A.; Drozd, D.D.; Khitrov, M.S.; Huang, L.; Yu, Z.; Tang, D.; Goryacheva, I.Y. Luminescent Composite Carbon/SiO_2 Structures: Synthesis and Applications. *Biosensors* **2022**, *12*, 392. https://doi.org/10.3390/bios12060392

Received: 18 April 2022
Accepted: 2 June 2022
Published: 6 June 2022

1. Introduction

Luminescent CNSs are colloidal quasi-spherical nanoparticles consisting of amorphous and/or nanocrystalline carbon structure, oxygen/nitrogen groups, and bright and tunable emission [1,2].

Currently, CNSs are relatively new luminescent labels for biological and analytical applications [3]. Their optical properties are comparable or even superior to luminescent semiconductor QDs [3,4], upconversion nanoparticles [5], and organic dyes [6]. Moreover, CNSs have excellent solubility in water, high resistance to photobleaching, absence of toxic components, and biocompatibility [1,3–5]. CNSs can be synthesized from a wide range of start materials without expensive and time-consuming steps [7].

The structure of CNSs has not been fully explored. There are several theories about the structure of CNSs. One of them describes the structure of CNSs as amorphous carbon [8], another reports the crystal structure [9], and a third assumes the presence of a mixture of crystalline and amorphous components [10]. Moreover, there is a disproportion across the structural information provided by various methods, such as X-ray diffraction, high-resolution transmission electron microscopy (HRTEM), and Raman spectroscopy [3].

The nanostructures described in the literature can be called carbon dots [5,6,11], carbon nanoparticles [12,13], carbon quantum dots [14,15], etc. A uniform terminology has not yet been formed. In this review, we use the term "carbon nanostructure" due to their understudied structure and morphology.

Various start materials, synthetic approaches, and conditions allow obtaining CNSs with a wide luminescence range. Despite thousands of studies, there are several scientific open questions about the relationship between structure and emission, optical behavior, and the nature of electronic states [3,4].

Different synthetic methods lead to mixtures of CNSs with various properties. The actual purification and fractionation techniques are mostly multistage, costly, or low-scale productive [12]. The properties of CNSs can depend on the environment, which limits their applications [16]. Moreover, solid-state CNSs usually have low luminescence due to the quenching effect caused by the aggregation of nanoparticles [17].

Some of these problems can be solved by using a "fixation" matrix for CNSs. There is continuous exploration for a suitable matrix for composite formation with CNSs to preserve and even improve their properties. The literature reports various examples of matrices such as silica [18–23], polymers [24–26], starch [27], zeolites [28–30], potassium aluminum sulfate [31], trisodium citrate [32], and sodium chloride crystals [33]. Currently, silica is a powerful and effective matrix thanks to its optical transparency [34,35], biocompatibility [34], low toxicity [34,36], tunable dielectric properties [37], and potential for size and surface customization [34,36].

In this review, we summarize and provide a critical analysis of the synthesis of luminescent composite particles based on CNSs and silica spheres. We demonstrate two dominant methods for obtaining composites: (i) inclusion of CNSs into silica particles (Figure 1A) and (ii) grafting of CNSs onto the silica surface (Figure 1B). Additionally, we demonstrate the obtainment of bifunctional composites consisting of CNSs/silica and fluorescent, magnetic, or photosensitizing particles (Figure 1C). We analyze the advantages and disadvantages of these approaches and complexes, as well as describe their properties and the possibilities of their application (Table 1).

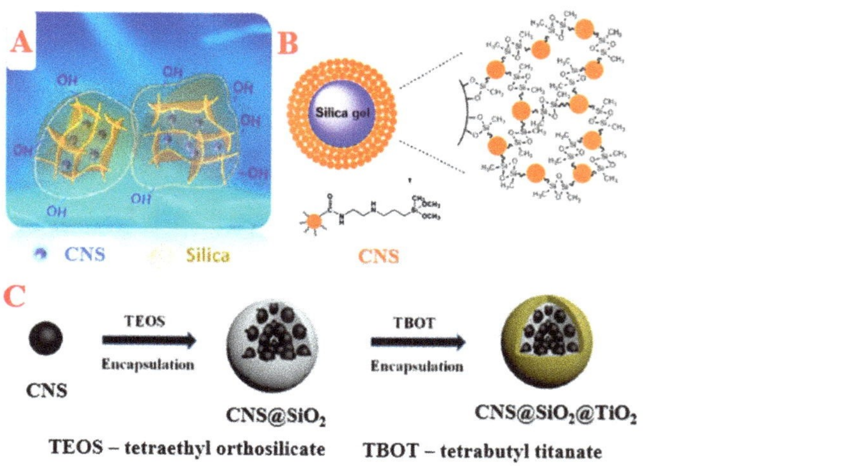

Figure 1. Schematic illustration of the composite formation: inclusion of CNSs into the silica matrix (**A**) [38], grafting of CNSs onto the silica surface (**B**) [22], and synthesis of bifunctional complexes (**C**) [39]. Adapted with permission from [38], ACS 2019, [22], Springer 2017, and [39], Elsevier 2020.

Table 1. Examples of CNSs/SiO_2 composites: synthesis, properties, and applications.

Synthesis CNS		Synthesis Composite		Composite		Application	References
Precursor	**Method**	**Precursor**	**Method**	**Size, nm**	**Property**		
CA, AEAPTMS	Pyrolysis	CNSs, TEOS	Hydrolysis by reverse microemulsion	~150	**PL**: $\lambda_{ex.}$ = 380 nm $\lambda_{em.}$ = 460 nm QY = 56%	Labels for immunochemical analysis	[18]
CA, AEAPTMS				84–190	**PL**: $\lambda_{ex.}$ = 380 nm $\lambda_{em.}$ = 460 nm	Detecting fingermarks	[19]
CA, urea	HT	CNSs, TEOS, APTES		18–159	**PL**: $\lambda_{ex.}$ = 365 nm $\lambda_{em.}$ = 450–650 nm	LEDs	[17]
Polyacrylic acid, EDA		CNSs, TEOS	Hydrolysis by Stober	-	**Phosphorescence**: $\lambda_{em.}$ = 520 nm	Labels in tissues	[38]
CA, urea	Solvothermic	CNSs, ARTMS, and silica nanoparticles	Chemical grafting	25–40	**PL**: $\lambda_{ex.}$ = 400 nm $\lambda_{em.}$ = 455–650 nm QY = 30–60%.	LEDs	[40]
1. Nylon 6, sulfuric acid 2. Obtained CNSs, nitric acid	1. Pyrolysis 2. Solvothermic	Two methods: (a) silica modified with APTES/GOPTMS, CNSs (b) Silica, CNSs	(a) Chemical grafting (b) Thermal annealing	-	**PL**: $\lambda_{ex.}$ = 360 nm $\lambda_{em.}$ = 460 nm QY = 2.5%.	Determination of ions	[23]
CA, AEAPTMS	Pyrolysis	CNSs, silica microsized particles	Sonication, mechanical mixing	>5000	-	Stationary-phase chromatography	[22]
Urea, folic acid	Microwave	Fe_3O_4, TEOS, APTES, CNSs	Hydrolysis by Stober and chemical grafting	155	**PL**: $\lambda_{ex.}$ = 370 nm $\lambda_{em.}$ = 455 nm **A magnetic saturation intensity**: 31.2 emu/g	Synergistic medicine	[41]
CA, EDA		CdTe, TEOS, APTES, CNSs		50	**PL**: $\lambda_{ex.}$ = 350 nm $\lambda_{em.}$ = 455 and 658 nm	Ratiometric optical labels	[21]
CA, EDA, silica spheres	HT	Silica spheres@CNSs, AuNCs, APTES	Chemical grafting	57	**PL**: $\lambda_{ex.}$ = 360 nm $\lambda_{em.}$ = 448 and 610 nm		[20]
CA, urea		CNSs, APTES, TEOS, TiO_2	Hydrolysis by Stober	150	-	Photothermal and photodynamic therapy	[39]

2. Carbon Nanostructures: Features, Structure, and Properties

An unknown highly luminescent material was found in 2004 during the purification and separation of single-walled carbon nanotubes (SWNTs) synthesized from arc-discharge soot [42]. The separation of SWNTs from the species of arc soot was achieved by electrophoresis in agarose gel. The soot was previously oxidized by nitric acid and then extracted with basic water (pH 8.4). After the gel-electrophoretic separation, slow-moving bands corresponded with SWNTs, while fast-moving bands contained the highly luminescent carbon material. This material was separated into several fractions with different emissions in the green-blue, yellow, and orange regions. Since then, similar materials have been called carbon dots, carbon nanoparticles, etc., without a common terminology. In this review, we generally refer to these types of PL materials as CNSs. It is universally acknowledged that the above-described study [42] discovered CNSs for the first time [1,2,7].

These types of nanoparticles have been synthesized and studied by numerous scientific groups to determine the causes of their photophysical properties, improve their synthesis, and identify applications [3,11]. CNSs are nanoparticles with a unit size of nanometers, usually consisting of carbon, hydrogen, oxygen, nitrogen, or sulfur [43]. Their most important property is bright luminescence in the range from blue to near-infrared [44].

Luminescent CNSs are a large class of carbon-based nanosystems of various structural types. Messina's research group classified CNSs according to their differences in structure and morphology. They proposed four groups of CNSs: graphene quantum dots, graphitic carbon dots, g-C_3N_4 carbon dots, and amorphous carbon dots (Figure 2) [4].

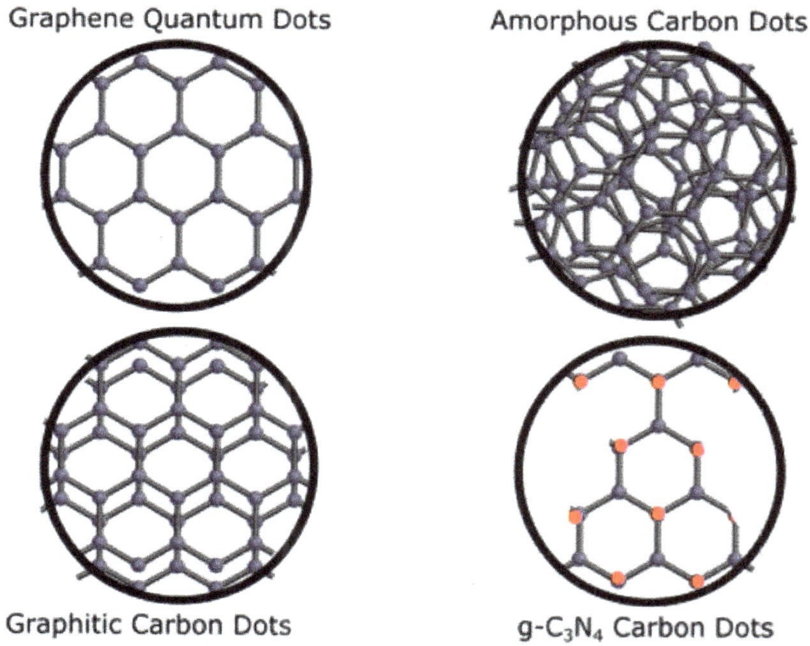

Figure 2. Schematic illustration of reported four types of CNSs. Black and red dots represent carbon and nitrogen atoms, respectively. Reprinted from ref. [4].

Graphitic carbon dots are spherical and the most popular type of nanostructures in the scientific literature [45–49]. However, a detailed study of their structural features by HRTEM was obtained in a few studies [47,50,51]. Their structure consists of several layers of sp^2-hybridized carbon with a diameter of units of nanometers. Graphene quantum dots have a non-spherical structure with 1–3 graphene layers [52,53]. The optical properties of these CNSs are very similar. In fact, their assignment to different types is a subject of discussion.

The third type of CNSs is C_3N_4 carbon dots. Some research groups [54–56] reported on carbon nitride structures with a graphitic- or β-crystalline arrangement (g-C_3N_4 or β-C_3N_4). CNSs of the g-C_3N_4 type have a layered structure similar to graphite with sp^2-hybridized carbon and nitrogen atoms. In contrast, β-C_3N_4 nanostructures have a hexagonal network of sp^3-hybridized carbon atoms connected with sp^2-hybridized nitrogen atoms.

Many studies have described a fourth type of CNSs—amorphous carbon dots [8,57,58]. They are a mixture of differently hybridized carbon (sp^2 and sp^3) in various proportions without a specific core. However, they also exhibit bright luminescence.

Moreover, Zhu and et al. [59] presented polymer dots consisting of aggregated or crosslinked polymers derived from linear polymers or monomers.

Several research groups have presented fluorophore-connected CNSs. These fluorophores are formed during the synthesis of CNSs and can be free or connected to the surface of nanostructures [60,61]. The luminescence of these CNSs has been specified.

Similar sizes ($\leq$10 nm) and surface functional layers characterize all described types of CNSs. These features determine the optical properties of the CNSs [3,4]. CNSs have absorbance in the far-ultraviolet (UV), visible, and near-infrared regions. Usually, intense absorption bands of CNSs have been observed in the range of 190 to 500 nm, corresponding to $\pi \rightarrow \pi^*$ or $n \rightarrow \pi^*$ transitions [4]. It is assumed that these absorption bands are determined by electronic transitions of functional groups located on the surface, as well as defects of the carbon core [4]. The position and intensity of optical bands depend on the temperature, pH, solvent, etc. [4,11].

The most attractive optical characteristic of CNSs is photoluminescence (PL). The spectral range and intensity of the PL bands of CNSs depend on their composition, type, and density of functional groups, and size, as well as energy of the excitation and properties of the environment [3,62]. Typically, CNSs have luminescence bands in the range of 350–700 nm [62]. A distinctive feature of the PL of CNSs is a strong dependence of emission on excitation; an increase in the excitation wavelength leads to a redshift of the spectrum band [62,63]. Researchers have correlated this fact with the presence of different emitting centers [64] or the emission of different chromophore groups in the CNS structure [65]. It should be noted that the optical properties of CNSs depend on the initial components and the synthetic technology [3]. In some cases, it is possible to improve the optical properties by varying the temperature, synthetic time, solvents, concentration, and ratio of starting substances [63]. However, there is no fundamental correlation among the synthetic method, structure, and properties of CNSs. The application of the matrix allows obtaining a better understanding of CNS structure and increases their applicability.

3. Silica Matrix: Synthetic Methods and Properties

Silica structures are often used as a matrix for nanoparticles [35,66–69] due to their thermodynamic and chemical stability in various chemical environments, low toxicity, biodegradability, optical transparency, and high surface area [34,69,70]. The presence of hydroxyl groups on the silica surface leads to high hydrophilicity and contributes to further functionalization [34].

Silica particles can mainly be synthesized by the Stober or the reverse microemulsion methods [36]. The Stober method can be applied to produce colloidal silica particles ranging from 200 to 2000 nm. This method is based on the hydrolysis and condensation of silanes in an aqueous alcohol solution in the presence of an ammonium hydroxide catalyst. The shape and size of the formed silica particles depend on the synthetic parameters: the concentrations of silane, ammonium hydroxide, water, and alcohol, their ratios, and the rate of their interaction [34,70,71]. However, researchers have described difficulties when synthesizing particles with target dimensions and shapes using the Stober approach because of the unlimited hydrolysis of silanes [71]. The reverse microemulsion method can solve these issues.

The reverse microemulsion method involves the formation of a thermodynamically stable dispersion of water in a nonpolar solvent stabilized by surfactant molecules called reverse micelles. Reverse micelles serve as nanoreactors for the hydrolysis and condensation of silanes to form silica nanoparticles. This method allows obtaining a narrow particle size and shape distribution because of similar micelle reactor sizes. In this case, the size and shape of the resulting silica nanoparticles depend on several factors: the nature of the organic solvent and surfactant, the molar ratio of reverse micelle components, the concentration of ammonium hydroxide, the synthesis time, and the amount of tetraethylorthosilicate (TEOS) and/or other organosilanes for the surface modification of nanoparticles [36,69,72].

4. Formation of Composite Carbon/SiO_2

The following problems limit the application of CNSs:

I. Obtainment of a polydisperse product;
II. The dependence of their optical properties and stability on the chemical environment;
III. PL dependence on properties of the microenvironment and the quenching of luminescence in the lyophilized samples;
IV. Nonuniform distribution of surface functional groups;
V. Weak PL intensity.

The above-listed problems can be solved by the application of a silica matrix. CNSs can be associated with silica via two approaches: inclusion of CNSs into the matrix or grafting of CNSs onto the silica surface. Moreover, these approaches can be applied to the synthesis of bifunctional complexes.

4.1. Inclusion of CNSs into the Silica Matrix

The inclusion of CNSs into silica spheres allows solving the above-stated problems. Thus, Xu's group [18] synthesized luminescent composite nanoparticles via simple co-hydrolysis of CNSs with TEOS. The authors synthesized amino-modified CNSs via pyrolysis of citric acid (CA) and *N*-(β-aminoethyl)-γ-aminopropylmethyldimethoxysilane (AEAPTMS) at 240 °C for 5 min under nitrogen atmosphere. The hydrolysis of TEOS and the obtained CNSs was carried out by the reverse microemulsion method. However, the authors highlighted the polydispersity of the luminescent composite nanoparticles. They applied differential centrifugation to obtain uniform-sized nanoparticles. Figure 3A shows a transmission electron microscope (TEM) image of the composites after separation. The PL maximum of the obtained particles was in a blue region at 460 nm and depended on the excitation wavelength (Figure 3B). The proposed approach allowed increasing the composite PL quantum yield (QY) by 9% compared to the initial CNSs (56% and 47%, respectively). The authors described the homogeneous distribution of CNSs in the SiO_2 matrix, as well as the good reproducibility achieved by this approach. This composite was used in an ultrasensitive method for the detection of the thrombocytopenia syndrome virus (SFTSV). The synthesized composite nanoparticles were used as labels for immunochemical analysis. The composite was conjugated with an anti-SFTSV monoclonal antibody for the registration of the optical signal. The detection limit of SFTSV was 10 pg/mL. The sensitivity of the developed assay was two orders of magnitude higher than that of the colloidal gold-based test method. The authors claimed that this method can be used for other viruses, protein biomarkers, nucleic acids, and bacteria in clinical diagnostics.

The Zhao group [19] used a similar technique for the synthesis of composite nanoparticles at 234 °C for 5 min. The obtained particles had a spherical morphology and sizes of 84 to 190 nm. These composite nanoparticles were successfully used for detecting latent fingermarks. The porous structure of the composite increased the adhesion of the nanoparticle to fingermarks. Moreover, these nanocomposites were effective and sensitive for both fresh and aged fingermarks. The composites were sensitive for the detection of latent fingermarks in a range of substrates such as glass, aluminum foil, plastic bags, drug packing, and leather (the use of CNSs to improve the visualization of hidden fingerprints is only applicable to a smooth nonporous surface). The optical signal was obtained at 415 nm irradiation. The luminescence of the found fingermarks was bright and gave a fair contrast, which suggests that the composite selectively targets the latent fingermarks.

Song's research group [17] obtained luminescent composite nanoparticles with multicolor emission via reverse-phase microemulsion. Firstly, the researchers synthesized multicolored CNSs using the one-step hydrothermal (HT) method from CA and urea. The authors varied the ratio of the initial components, reaction time (6/10 h), synthetic temperature (160/180 °C), and solvents (water, ethyl alcohol, and dimethylformamide (DMF)). The PL maxima of the CNSs were in the regions from 450 to 650 nm. The reagent ratio allowed achieving multicolor luminescence, and the HT temperature and time were adjusted to obtain more intensive optical properties. Secondly, a mixture of CNSs, TEOS, and 3-aminopropyltriethoxysilane (APTES) was added to the reverse-phase microemulsion. This synthetic method is effective for preventing CNS quenching caused by the aggregation of nanoparticles. The range of sizes from 18 to 159 nm of the obtained composite particles depended on the TEOS concentrations. The authors described the production of multicolor light-emitting diodes (LED) with the developed luminescent composite nanoparticles. A composite-based LED was successfully fabricated by varying the content ratio of blue-, green-, and red-emitting CNS/SiO_2 with a Commission Internationale de L'Eclairage (CIE) of (0.3497, 0.3045) and color rendering index (CRI) of 85.2.

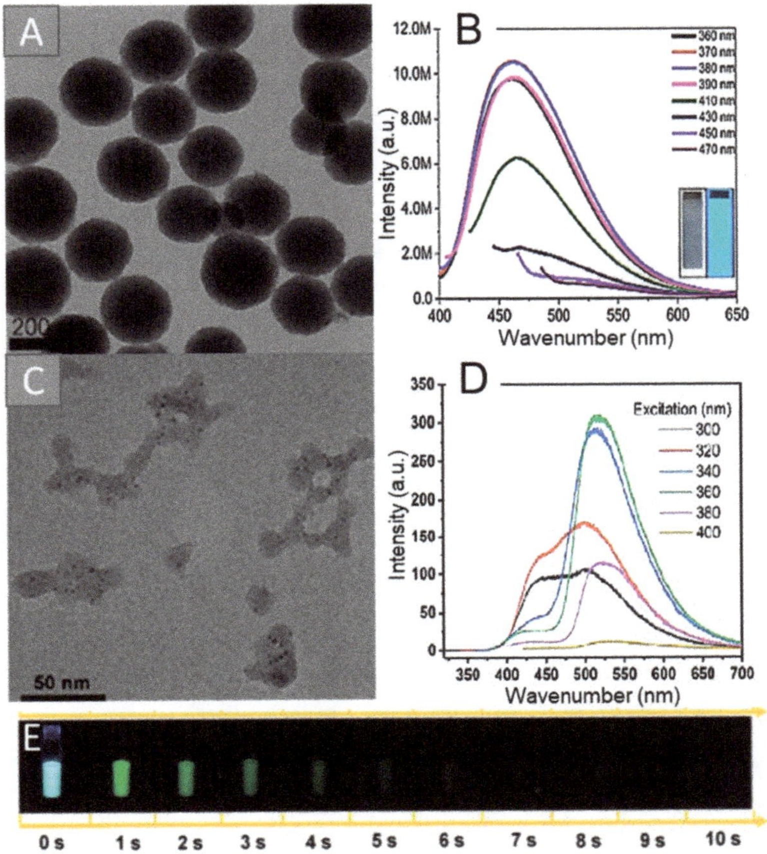

Figure 3. TEM images and emission spectra of the CNSs/SiO_2 composite obtained by reverse microemulsion method after separation (**A**,**B**) [18] and synthesized by Stober method (**C**,**D**) [38]. Images of solid-state composite after 365 nm irradiation from 0 to 10 s (**E**) [38]. Adapted with permission from [18], ACS 2019, and [38], American Chemical Society 2019.

Li's scientific group [38] used the Stober method for the synthesis of composite nanoparticles with CNSs from polyacrylic acid and ethylenediamine (EDA), obtained using the HT method at 200 °C for 8 h. For the composite synthesis, TEOS and CNSs were mixed in the presence of ammonia and heated at 100 °C with vigorous stirring. This method led to the covalent bonding of the CNSs to the matrix. Figure 3C shows HRTEM images of the obtained composites. CNSs were located inside amorphous silica structures. The obtained composites had a nonuniformity of size and shape. CNSs did not have phosphorescent properties, while the composites demonstrated phosphorescence at 520 nm (Figure 3D). The phosphorescence lifetime of the obtained composites was 1.64 s (Figure 3E). The silica matrix was a protective shield preventing quenching and enhancing solubility in water. The authors reported successful phosphorescence imaging of CNS/SiO_2, both in vivo and in vitro, highlighting the advantage of long-lived phosphorescence in bioimaging by eliminating the autofluorescence interference, especially under short-wavelength excitation. The biocompatibility of CNSs@SiO_2 composites was demonstrated on the mouse breast carcinoma EM-6 cell line. The MTT colorimetric assay verified the very low cytotoxicity of composites up to 150 μg/mL.

Thus, the inclusion of CNSs in the silica matrix increases the PL signal because the PL intensity of several CNSs is higher than the intensity of a single one. Moreover, the application of the matrix can reduce the influence of the chemical environment on the optical properties. Silica spheres allow avoiding the PL quenching associated with the aggregation of CNSs in the solid state. However, both reverse microemulsion and Stober methods generally do not provide a synthesis of monodisperse structures. The synthesis of luminescent composites, as a rule, consists of several stages and is quite time-consuming. Many researchers have used differential centrifugation as an additional step to obtain uniform size fractions of composites.

4.2. Grafting of CNSs onto the Silica Surface

Grafting is also a very convenient method since the PL properties of composite nanoparticles can be controlled by the concentration of CNSs during the synthesis. Moreover, this method allows introducing specific functionality to the composite surface. The grafting process proceeds through the formation of covalent bonds and increases the stability of the obtained composites.

Sun et al. [40] synthesized luminescent composite nanoparticles by grafting CNSs onto the surface of a silica matrix via an amide bond formation reaction. They used CA and urea (mass ratio = 1:2) in DMF via the solvothermal method in an autoclave at 160 °C for 8 h for CNS synthesis. An increase in the concentration of reagents (from 6 to 240 mg/mL) led to a gradual shift of the PL maximum to longer wavelengths (Figure 4A–H). The solvothermal reaction of citric acid and urea was demonstrated to be an available route to acquire full-color-emitting CNSs at controlled reactant concentrations in DMF. They mixed different concentrations of a solution of CNSs, 3-aminopropyltrimethoxysilane, and silica nanoparticles in DMF with stirring at room temperature for 3 h (Figure 4I). The reaction mixtures were centrifuged at 8000 rpm for 10 min, and the precipitated solids were solidified in a vacuum oven at 50 °C for 24 h. The size of the obtained composites was 25–40 nm. The use of a matrix prevented the quenching of the CNS luminescence in a solid state. The combined application of full-color-emitting CNS/SiO_2 and InGaN chips with different peak wavelengths led to design flexible packaging schemes for white LEDs (WLEDs), obtaining a pure white light at the CIE coordinates of (0.33, 0.33) with CRI of 80.4 and a high color-rendering white light with CIE coordinates of (0.34, 0.36) and CRI of 97.4. This fact indicated the significant application potential of the SiO_2/CNSs composite in the LED field.

The Thongsai group [23] used other methods for chemical grafting of CNSs onto the silica substrate. The authors synthesized CNSs from nylon-6 by pyrolysis at 250 °C for 6 h in the presence of sulfuric acid. The resulting product was HT treated with nitric acid at 200 °C for 6 h. The PL QY of the product was 2.53%. The obtained CNSs were grafted onto the silica surface primarily modified with APTES or 3-glycidyloxypropyltrimethoxysilane (GOPTMS). Grafting of CNSs was carried out via a carbodiimide reaction with EDC. The functionalized substrates were immersed and stirred in a solution with CNSs at room temperature for 24 h. The product was washed several times with deionized and sonicated to remove unreacted CNSs. The composites were dried with N_2 and kept in a vacuum oven before use. The authors also described the preparation of composite nanoparticles via thermal annealing. A water solution of CNSs was added dropwise onto the purified silica substrate and subjected to thermal annealing in a vacuum chamber at a high temperature (160 °C). The data of the root-mean-square roughness and contact angle of the silica surface after grafting indicated successful CNS attachment. The authors noted that the functionalization of both APTES and GOPTMS was effective for introducing CNSs to the silica surface, but a uniform dense monolayer was not obtained. In addition, the two-stage methods required a long reaction time and several purification stages. Therefore, the one-stage method of thermal annealing is the most promising. Moreover, the successful grafting of CNSs onto APTES- and GOPTMS-terminated surfaces showed the presence of carboxyl and amino groups. The obtained composites were used for the determination of heavy-metal ions. The composite demonstrated selective detection and high adsorption of Cu^{2+} ions, suggesting practical applications as two-in-one sensors and adsorbents.

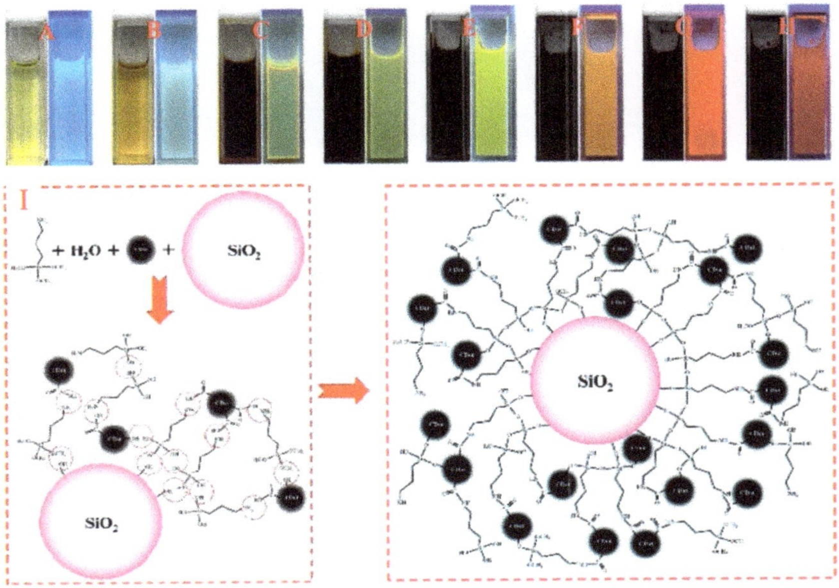

Figure 4. Images of CNSs solutions (**A–H**) with concentrations (6–240 mg/mL) under daily and UV light (λex. = 365 nm). Scheme of CNSs/SiO_2 formation mechanism (**I**). Reprinted with permission from [40], Royal Society of Chemistry 2020.

The Cai group [22] successfully grafted CNSs onto the matrix surface without linkers. The authors synthesized silanized CNSs via pyrolysis of CA in hot AEAPTMS at 240 °C for 5 min, which were grafted on the silica micro-sized particles. A mixture of silanized CNSs and silica microparticles was sonicated, heated to 110 °C, mechanically stirred for 24 h, and mixed until the formation of a homogeneous dispersion. After that, the resulting substance was washed sequentially with toluene and ethanol and dried at 60 °C in a vacuum oven. This method allowed obtaining a uniform distribution on the silica surface. The resulting luminescent composite particles had a large surface area and a variety of functional groups, high adsorption, and thermal and mechanical stability. These composites were used as a stationary phase in hydrophilic interaction chromatography for the separation of sulfonamides, flavones, amino acids, nucleosides, and bases. The composite stationary phase was packed into stainless-steel columns (150 mm × 4.6 mm i.d.) with carbon tetrachloride as the slurry solvent and acetonitrile as the mobile solvent. This work revealed a new way to enhance the chromatographic selectivity by CNSs, which increases the density of the interaction sites on the stationary phase.

The synthesized composite nanoparticles from CNSs and silica helped to better understand the structure of CNSs, to consider surface and interfacial phenomena with their participation (such as adsorption, energy, and charge transfer), and to apply CNSs in important applications (sensing, chromatography, LEDs). However, the chemical grafting of CNSs onto the silica surface is a complex procedure and requires appropriate chemical reactions. The authors note the difficulties in obtaining a monolayer of CNSs on the surface of silica. The resulting composite requires several stages of purification to separate uncombined CNSs.

4.3. *Formation of Bifunctional Structures*

Composite nanoparticles can be used to obtain bifunctional complexes with a simultaneous attachment of structures both inside and on the surface of silica. The application of a silica matrix allows solving the previously described problems, as well as combining the unique fluorescent, magnetic, and photosensitizing properties of particles of different nature.

Guan et al. [41] fabricated bifunctional composites from Fe_3O_4, SiO_2, and CNSs. A single-stage microwave method was applied for CNS synthesis from urea and folic acid at 800 W for 8 min. Magnetic nanoparticles Fe_3O_4 were synthesized using the solvothermal method from iron chloride hexahydrate and anhydrous sodium acetate in a mixture of ethylene glycol and diethylene glycol at 200 °C for 12 h. Then, Fe_3O_4 particles were silanized using the classical Stober method with TEOS and coated with APTES for amino modification of the SiO_2 surface. The carbodiimide method was used for grafting of CNSs onto the amino-modified Fe_3O_4/SiO_2 surface with EDC and NHS linkers for functional group activation. The magnetic–fluorescent composite had a size of 155 nm and PL maximum at 455 nm. The PL intensity of the composite was lower than that for CNSs. This fact can be explained by the possible interaction of functional groups on the surface of the CNSs with iron ions. The resulting composites had a magnetic saturation intensity of 31.2 emu/g. In vitro experiments for Fe_3O_4/SiO_2/CNSs to load and release gambogic acid in PBS (pH = 7.4 or 5.7) were provided. The cell uptake experiments were performed by incubating the cell line with bifunctional nanoparticles. Blue luminescence from CNSs in Fe_3O_4/SiO_2/CNSs nanoparticles was observed near the nucleus, indicating that these nanoparticles penetrate the cells via endocytosis. The authors reported the successful application of the obtained composite in synergistic therapy, including the release of gambogic acid and magnetic targeting. The release of gambogic acid led to the inhibition of tumor cells, and their survival rate was less than 20% at a concentration of 100 μg/mL.

A similar synthetic procedure was applied by Xu's research group [21]. They obtained luminescent composites via covalent linking of blue-emissive CNSs with the surface of silica nanoparticles containing red-emissive ($\lambda_{em.}$ = 658 nm) quantum dots. The QD-embedded silica nanoparticles were synthesized with TEOS hydrolysis and condensation using the Stober method, and their surface was modified with APTES. CNSs were obtained from CA and EDA using the microwave approach at 750 W for 10 min and contained carboxyl groups on their surface. These functional groups reacted with amino-modified silica nanoparticles. The size of the composites was ~50 nm. The obtained composite nanoparticles had PL maxima at 453 and 658 nm under the excitation of 350 nm. The obtained composites were used for the determination of mercury ions. The intensity of the PL maximum at 453 nm had a linear dependence on the concentration of mercury ions, while the luminescence at 658 nm was unchanged. The detection limit of mercury ions was 0.47 nM, corresponding to modern analytical standards.

A different dual-emission ratiometric optical probe was developed by An's group [20] (Figure 5). The authors HT treated CA, EDA, and silica spheres at 200 °C for 5 h. They suggested the simultaneous formation of CNSs and their fixation inside the silica nanoparticles during HT synthesis. The composite nanoparticles were covalently bonded with AuNCs. Nanoclusters were prepared thermally at 70 °C for 24 h from chloroauric acid and glutathione. AuNCs were pre-functionalized by APTES for covalent binding to the surface of composite particles. The size of the obtained composites was approximately 57 nm. The composites simultaneously displayed two emission maxima at 448 nm and 610 nm under a single excitation wavelength of 380 nm. The obtained composites were dispersed in buffer solutions with different pH values (from 2 to 12), while the PL intensity was the same, which indicates that the composites were stable in highly alkaline and highly acidic conditions. The developed composite was used to determine the level of silver ions; with an increase in their concentration, the PL in the region of 610 nm increased noticeably, while the PL in the region of 448 nm remained unchanged. The detection limit of Ag^+ for this system was 1.6 nM. Thus, this method has advantages in Ag^+ detection sensitivity compared to other analytical methods.

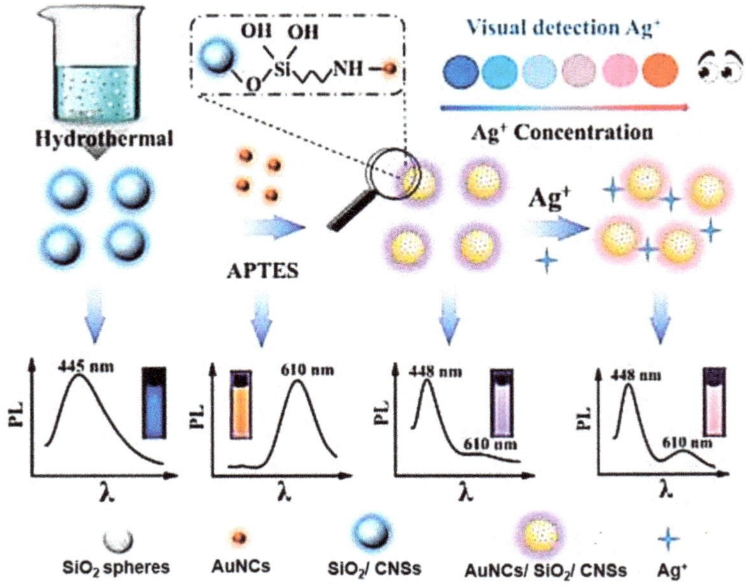

Figure 5. Schematic illustration of the composite synthesis and Ag^{+} detection. Reprinted with permission from [20], Elsevier 2021.

Bing et al. [39] also used CA and an amine-containing agent (urea) as precursors for the HT (150 °C, 5 h) production of CNSs. Composite CNSs/SiO_2 particles were synthesized by mixing of CNSs with polyvinylpyrrolidone in ethanol and water followed by the addition of APTES and TEOS. The mixture was stirred for 3 h at 20 °C. The polyvinylpyrrolidone acted as a dispersant, structurant, and linker during the reaction. The synthesized composite particles CNSs/SiO_2 were additionally wrapped with the photosensitizing agent of titanium dioxide as a shell. The CNSs inside the silica matrix had a multifunctional core for photothermal therapy and photothermal imaging. These composite nanoparticles had good biocompatibility and a uniform size of 150 nm. CNSs/SiO_2 composites can be used for photothermal therapy and photothermal imaging due to the induction apoptosis in cancer cells. These structures could absorb long-wavelength light for the activation of photothermal effects and provide cancer therapy in deep tissues. Moreover, these composites responded to ultrasonic stimulation with a generation of the oxygen active forms for cancer sonodynamic therapy.

Thus, we demonstrated examples of silica application as a matrix for the synthesis of bifunctional composites. Researchers could control and adjust the concentration of the particles, as well as the size of the composites, at each stage of synthesis. However, the preparation of composites is a multistage process. Moreover, a few authors noted the quenching of PL due to the interaction of CNSs with the other components. These bifunctional systems can be used in sensing (detection of heavy-metal ions), drug delivery systems, bioimaging, and antitumor therapy.

5. Conclusions

CNSs have attractive characteristics such as unique optical properties, excellent biocompatibility and photostability, the possibility of surface functionalization, high colloidal stability, and low toxicity. To this day, the scientific community is looking for ways to solve fundamental and applied issues for the wider application of these nanoobjects. The issues include the unclear structure of CNSs and/or their luminescence mechanism, the dependence of properties on the environment and/or state of aggregation, and difficulties in their purification and separation. The application of a silica matrix for CNSs allows unifying the composite CNSs/SiO_2 shape, size, and optical properties. Silica is a proper matrix due to its optical transparency, variety of functional groups for surface modifications, and controllable dimensions during synthesis.

We demonstrated examples of the inclusion of CNSs into the silica matrix or grafting onto its surface. Composite formation solves the problems associated with luminescence quenching in the solid state and purification of the product, as well as improves the luminescent properties via the emission of several CNSs. Moreover, silica and CNS composites led to a better understanding of the structure of CNSs, as well as surface and interfacial phenomena, such as adsorption, energy, and charge transfer. CNSs/SiO_2 composites have improved brightness and chemical stability compared to CNSs.

Furthermore, the silica matrix makes it possible to obtain multifunctional nanocomposites due to the combination of several nanoparticles with different properties such as fluorescent, magnetic, or photosensitizing properties. This review reported examples of the successful application of composites in bioimaging, medicine, chemical, and immunoassays, in LED production, and as a stationary phase in hydrophilic interaction chromatography.

It is important to note that the approaches for composite synthesis require further improvement of the obtained structures with the necessary dimensions, while avoiding separation and purification steps. The development of effective and well-controlled strategies for the synthesis of these composites can lead to their wider application in medical, biological, analytical, or technical areas.

Author Contributions: Writing—original draft preparation, Y.A.P., T.S.P., D.D.D. and M.S.K.; writing—review and editing, A.A.K., O.A.G., L.H. and Z.Y.; supervision, I.Y.G. and D.T. All authors have read and agreed to the published version of the manuscript.

Funding: This work was financially supported by the Ministry of Science and Higher Education of the Russian Federation in the framework of the state task (project no. FSRR-2020-0002).

Institutional Review Board Statement: Not applicable.

Informed Consent Statement: Not applicable.

Conflicts of Interest: The authors declare no conflict of interest.

Abbreviations

AEAPTMS	*N*-(β-aminoethyl)-γ-aminopropylmethyldimethoxysilane
APTES	3-aminopropyltriethoxysilane
AuNC	gold nanocluster
CA	citric acid
CIE	Commission Internationale de L'Eclairage
CNS	carbon nanostructure
CRI	color rendering index
DMF	dimethylformamide
EDA	ethylenediamine
GOPTMS	3-glycidyloxypropyltrimethoxysilane
HRTEM	high-resolution transmission electron microscopy
HT	hydrothermal
LED	light-emitting diode
PL	photoluminescence
QD	quantum dot
QY	quantum yield
SFTSV	thrombocytopenia syndrome virus
SWNT	single-wall carbon nanotube
TEM	transmission electron microscopy
TEOS	tetraethyl orthosilicate
UV	ultraviolet

References

1. Sun, X.; Lei, Y. Fluorescent Carbon Dots and Their Sensing Applications. *TrAC-Trends Anal. Chem.* **2017**, *89*, 163–180. [CrossRef]
2. Xiao, L.; Sun, H. Novel Properties and Applications of Carbon Nanodots. *Nanoscale Horiz.* **2018**, *3*, 565–597. [CrossRef] [PubMed]
3. Goryacheva, I.Y.; Sapelkin, A.V.; Sukhorukov, G.B. Carbon Nanodots: Mechanisms of Photoluminescence and Principles of Application. *TrAC-Trends Anal. Chem.* **2017**, *90*, 27–37. [CrossRef]
4. Sciortino, A.; Cannizzo, A.; Messina, F. Carbon Nanodots: A Review—From the Current Understanding of the Fundamental Photophysics to the Full Control of the Optical Response. *C* **2018**, *4*, 67. [CrossRef]
5. Himmelstoß, S.F.; Hirsch, T. A Critical Comparison of Lanthanide Based Upconversion Nanoparticles to Fluorescent Proteins, Semiconductor Quantum Dots, and Carbon Dots for Use in Optical Sensing and Imaging. *Methods Appl. Fluoresc.* **2019**, *7*, 022002. [CrossRef]
6. Li, F.; Yang, D.; Xu, H. Non-Metal-Heteroatom-Doped Carbon Dots: Synthesis and Properties. *Chem.-A Eur. J.* **2019**, *25*, 1165–1176. [CrossRef] [PubMed]
7. He, C.; Xu, P.; Zhang, X.; Long, W. The Synthetic Strategies, Photoluminescence Mechanisms and Promising Applications of Carbon Dots: Current State and Future Perspective. *Carbon* **2022**, *186*, 91–127. [CrossRef]
8. Margraf, J.T.; Strauss, V.; Guldi, D.M.; Clark, T. The electronic structure of amorphous carbon nanodots. *J. Phys. Chem. B* **2015**, *119*, 7258–7265. [CrossRef] [PubMed]
9. Yang, P.; Zhu, Z.; Chen, M.; Zhou, X.; Chen, W. Microwave-Assisted Synthesis of Polyamine-Functionalized Carbon Dots from Xylan and Their Use for the Detection of Tannic Acid. *Spectrochim. Acta-Part A: Mol. Biomol. Spectrosc.* **2019**, *213*, 301–308. [CrossRef]
10. Guo, L.; Ge, J.; Liu, W.; Niu, G.; Jia, Q.; Wang, H.; Wang, P. Tunable Multicolor Carbon Dots Prepared from Well-Defined Polythiophene Derivatives and Their Emission Mechanism. *Nanoscale* **2016**, *8*, 729–734. [CrossRef]
11. Jorns, M.; Pappas, D. A Review of Fluorescent Carbon Dots, Their Synthesis, Physical and Chemical Characteristics, and Applications. *Nanomaterials* **2021**, *11*, 1448. [CrossRef] [PubMed]
12. Kokorina, A.A.; Sapelkin, A.V.; Sukhorukov, G.B.; Goryacheva, I.Y. Luminescent Carbon Nanoparticles Separation and Purification. *Adv. Colloid Interface Sci.* **2019**, *274*, 102043. [CrossRef]
13. Mazrad, Z.A.I.; Lee, K.; Chae, A.; In, I.; Lee, H.; Park, S.Y. Progress in Internal/External Stimuli Responsive Fluorescent Carbon Nanoparticles for Theranostic and Sensing Applications. *J. Mater. Chem. B* **2018**, *6*, 1149–1178. [CrossRef]
14. Wang, X.; Feng, Y.; Dong, P.; Huang, J. A Mini Review on Carbon Quantum Dots: Preparation, Properties, and Electrocatalytic Application. *Front. Chem.* **2019**, *7*, 671. [CrossRef] [PubMed]
15. Chahal, S.; Yousefi, N.; Tufenkji, N. Green Synthesis of High Quantum Yield Carbon Dots from Phenylalanine and Citric Acid: Role of Stoichiometry and Nitrogen Doping. *ACS Sustain. Chem. Eng.* **2020**, *8*, 5566–5575. [CrossRef]
16. Long, C.; Jiang, Z.; Shangguan, J.; Qing, T.; Zhang, P.; Feng, B. Applications of Carbon Dots in Environmental Pollution Control: A Review. *Chem. Eng. J.* **2021**, *406*, 126848. [CrossRef]
17. Song, X.; Guo, Q.; Cai, Z.; Qiu, J.; Dong, G. Synthesis of Multi-Color Fluorescent Carbon Quantum Dots and Solid State CQDs@SiO_2 Nanophosphors for Light-Emitting Devices. *Ceram. Int.* **2019**, *45*, 17387–17394. [CrossRef]
18. Xu, L.D.; Zhang, Q.; Ding, S.N.; Xu, J.J.; Chen, H.Y. Ultrasensitive Detection of Severe Fever with Thrombocytopenia Syndrome Virus Based on Immunofluorescent Carbon Dots/SiO_2 Nanosphere-Based Lateral Flow Assay. *ACS Omega* **2019**, *4*, 21431–21438. [CrossRef] [PubMed]
19. Zhao, Y.B.; Ma, Y.J.; Song, D.; Liu, Y.; Luo, Y.; Lin, S.; Liu, C.Y. New Luminescent Nanoparticles Based on Carbon Dots/SiO_2 for the Detection of Latent Fingermarks. *Anal. Methods* **2017**, *9*, 4770–4775. [CrossRef]
20. An, J.; Chen, R.; Chen, M.; Hu, Y.; Lyu, Y.; Liu, Y. An Ultrasensitive Turn-On Ratiometric Fluorescent Probes for Detection of Ag^+ Based On Carbon Dots/SiO_2 and Gold Nanoclusters. *Sens. Actuators B Chem.* **2021**, *329*, 129097. [CrossRef]
21. Xu, H.; Zhang, K.; Liu, Q.; Liu, Y.; Xie, M. Visual and Fluorescent Detection of Mercury Ions by Using a Dually Emissive Ratiometric Nanohybrid Containing Carbon Dots and CdTe Quantum Dots. *Microchim. Acta* **2017**, *184*, 1199–1206. [CrossRef]
22. Cai, T.; Zhang, H.; Rahman, A.F.M.M.; Shi, Y.P.; Qiu, H. Silica Grafted with Silanized Carbon Dots as a Nano-on-Micro Packing Material with Enhanced Hydrophilic Selectivity. *Microchim. Acta* **2017**, *184*, 2629–2636. [CrossRef]
23. Thongsai, N.; Supchocksoonthorn, P.; Dwyer, J.H.; Wei, W.; Sun, J.; Gopalan, P.; Paoprasert, P. High-Capacity Adsorbent/Sensor from Nylon 6 Derived Carbon Dots on SiO_2 Substrate via One-Step Surface Grafting. *Mater. Sci. Eng. B Solid-State Mater. Adv. Technol.* **2020**, *262*, 114692. [CrossRef]
24. Li, Q.; Zhou, M.; Yang, Q.; Wu, Q.; Shi, J.; Gong, A.; Yang, M. Efficient Room-Temperature Phosphorescence from Nitrogen-Doped Carbon Dots in Composite Matrices. *Chem. Mater.* **2016**, *28*, 8221–8227. [CrossRef]
25. Chen, B.; Feng, J. White-Light-Emitting Polymer Composite Film Based on Carbon Dots and Lanthanide Complexes. *J. Phys. Chem. C* **2015**, *119*, 7865–7872. [CrossRef]
26. Tian, Z.; Li, D.; Ushakova, E.V.; Maslov, V.G.; Zhou, D.; Jing, P.; Shen, D.; Qu, S.; Rogach, A.L. Multilevel Data Encryption Using Thermal-Treatment Controlled Room Temperature Phosphorescence of Carbon Dot/Polyvinylalcohol Composites. *Adv. Sci.* **2018**, *5*, 1800795. [CrossRef] [PubMed]
27. Sun, M.; Qu, S.; Hao, Z.; Ji, W.; Jing, P.; Zhang, H.; Zhang, L.; Zhao, J.; Shen, D. Towards Efficient Solid-State Photoluminescence Based on Carbon-Nanodots and Starch Composites. *Nanoscale* **2014**, *6*, 13076–13081. [CrossRef] [PubMed]

28. Liu, J.; Wang, N.; Yu, Y.; Yan, Y.; Zhang, H.; Li, J.; Yu, J. Carbon dots in zeolites: A new class of thermally activated delayed fluorescence materials with ultralong lifetimes. *Sci. Adv.* **2017**, *3*, e1603171. [CrossRef]
29. Wang, Y.; Li, Y.; Yan, Y.; Xu, J.; Guan, B.; Wang, Q.; Li, J.; Yu, J. Luminescent Carbon Dots in a New Magnesium Aluminophosphate Zeolite. *Chem. Commun.* **2013**, *49*, 9006–9008. [CrossRef]
30. Zhang, H.; Liu, J.; Wang, B.; Liu, K.; Chen, G.; Yu, X.; Li, J.; Yu, J. Zeolite-Confined Carbon Dots: Tuning Thermally Activated Delayed Fluorescence Emission: Via Energy Transfer. *Mater. Chem. Front.* **2020**, *4*, 1404–1410. [CrossRef]
31. Dong, X.; Wei, L.; Su, Y.; Li, Z.; Geng, H.; Yang, C.; Zhang, Y. Efficient Long Lifetime Room Temperature Phosphorescence of Carbon Dots in a Potash Alum Matrix. *J. Mater. Chem. C* **2015**, *3*, 2798–2801. [CrossRef]
32. Shen, C.L.; Zang, J.H.; Lou, Q.; Su, L.X.; Li, Z.; Liu, Z.Y.; Dong, L.; Shan, C.X. In-Situ Embedding Carbon Dots in Trisodium Citrate Crystal Matrix for Tunable Solid-State Fluorescence. *Carbon* **2018**, *136*, 359–368. [CrossRef]
33. Zhai, Y.; Zhou, D.; Jing, P.; Li, D.; Zeng, H.; Qu, S. Preparation and Application of Carbon-Nanodot@NaCl Composite Phosphors with Strong Green Emission. *J. Colloid Interface Sci.* **2017**, *497*, 165–171. [CrossRef] [PubMed]
34. Li, L.; Wang, W.; Tang, J.; Wang, Y.; Liu, J.; Huang, L.; Wang, Y.; Guo, F.; Wang, J.; Shen, W.; et al. Classification, Synthesis, and Application of Luminescent Silica Nanoparticles: A Review. *Nanoscale Res. Lett.* **2019**, *14*, 190. [CrossRef]
35. Fedorenko, S.; Stepanov, A.; Sibgatullina, G.; Samigullin, D.; Mukhitov, A.; Petrov, K.; Mendes, R.; Rümmeli, M.; Giebeler, L.; Weise, B.; et al. Fluorescent Magnetic Nanoparticles for Modulating the Level of Intracellular Ca^{2+} in Motoneurons. *Nanoscale* **2019**, *11*, 16103–16113. [CrossRef]
36. Narayan, R.; Nayak, U.Y.; Raichur, A.M.; Garg, S. Mesoporous Silica Nanoparticles: A Comprehensive Review on Synthesis and Recent Advances. *Pharmaceutics* **2018**, *10*, 118. [CrossRef]
37. Ren, G.; Su, H.; Wang, S. The Combined Method to Synthesis Silica Nanoparticle by Stöber Process. *J. Sol-Gel Sci. Technol.* **2020**, *96*, 108–120. [CrossRef]
38. Li, W.; Wu, S.; Xu, X.; Zhuang, J.; Zhang, H.; Zhang, X.; Hu, C.; Lei, B.; Kaminski, C.F.; Liu, Y. Carbon Dot-Silica Nanoparticle Composites for Ultralong Lifetime Phosphorescence Imaging in Tissue and Cells at Room Temperature. *Chem. Mater.* **2019**, *31*, 9887–9894. [CrossRef]
39. Wei, B.; Dong, F.; Yang, W.; Luo, C.; Dong, Q.; Zhou, Z.; Yang, Z.; Sheng, L. Synthesis of Carbon-Dots@SiO_2@TiO_2 Nanoplatform for Photothermal Imaging Induced Multimodal Synergistic Antitumor. *J. Adv. Res.* **2020**, *23*, 13–23. [CrossRef]
40. Sun, M.; Han, Y.; Yuan, X.; Jing, P.; Zhang, L.; Zhao, J.; Zheng, Y. Efficient Full-Color Emitting Carbon-Dot-Based Composite Phosphors by Chemical Dispersion. *Nanoscale* **2020**, *12*, 15823–15831. [CrossRef]
41. Guan, Y.; Yang, Y.; Wang, X.; Yuan, H.; Yang, Y.; Li, N.; Ni, C. Multifunctional Fe_3O_4@SiO_2-CDs Magnetic Fluorescent Nanoparticles as Effective Carrier of Gambogic Acid for Inhibiting VX_2 Tumor Cells. *J. Mol. Liq.* **2021**, *327*, 114783. [CrossRef]
42. Xu, X.; Ray, R.; Gu, Y.; Ploehn, H.J.; Gearheart, L.; Raker, K.; Scrivens, W.A. Electrophoretic Analysis and Purification of Fluorescent Single-Walled Carbon Nanotube Fragments. *J. Am. Chem. Soc.* **2004**, *126*, 12736–12737. [CrossRef] [PubMed]
43. Kokorina, A.A.; Prikhozhdenko, E.S.; Sukhorukov, G.B.; Sapelkin, A.V.; Goryacheva, I.Y. Luminescent Carbon Nanoparticles: Synthesis, Methods of Investigation, Applications. *Russ. Chem. Rev.* **2017**, *86*, 1157–1171. [CrossRef]
44. Song, Y.; Zhu, S.; Yang, B. Bioimaging Based on Fluorescent Carbon Dots. *RSC Adv.* **2014**, *4*, 27184–27200. [CrossRef]
45. Sciortino, A.; Cayuela, A.; Soriano, M.L.; Gelardi, F.M.; Cannas, M.; Valcárcel, M.; Messina, F. Different Natures of Surface Electronic Transitions of Carbon Nanoparticles. *Phys. Chem. Chem. Phys.* **2017**, *19*, 22670–22677. [CrossRef]
46. Cayuela, A.; Soriano, M.L.; Valcárcel, M. Reusable Sensor Based on Functionalized Carbon Dots for the Detection of Silver Nanoparticles in Cosmetics via Inner Filter Effect. *Anal. Chim. Acta* **2015**, *872*, 70–76. [CrossRef]
47. Ding, H.; Zhang, P.; Wang, T.Y.; Kong, J.L.; Xiong, H.M. Nitrogen-Doped Carbon Dots Derived from Polyvinyl Pyrrolidone and Their Multicolor Cell Imaging. *Nanotechnology* **2014**, *25*, 205604. [CrossRef]
48. Das, A.; Gude, V.; Roy, D.; Chatterjee, T.; De, C.K.; Mandal, P.K. On the Molecular Origin of Photoluminescence of Nonblinking Carbon Dot. *J. Phys. Chem. C* **2017**, *121*, 9634–9641. [CrossRef]
49. Cayuela, A.; Soriano, M.L.; Carrión, M.C.; Valcárcel, M. Functionalized Carbon Dots as Sensors for Gold Nanoparticles in Spiked Samples: Formation of Nanohybrids. *Anal. Chim. Acta* **2014**, *820*, 133–138. [CrossRef]
50. Nguyen, V.; Si, J.; Yan, L.; Hou, X. Direct Demonstration of Photoluminescence Originated from Surface Functional Groups in Carbon Nanodots. *Carbon* **2016**, *108*, 268–273. [CrossRef]
51. Sciortino, L.; Sciortino, A.; Popescu, R.; Schneider, R.; Gerthsen, D.; Agnello, S.; Cannas, M.; Messina, F. Tailoring the Emission Color of Carbon Dots through Nitrogen-Induced Changes of Their Crystalline Structure. *J. Phys. Chem. C* **2018**, *122*, 19897–19903. [CrossRef]
52. Fuyuno, N.; Kozawa, D.; Miyauchi, Y.; Mouri, S.; Kitaura, R.; Shinohara, H.; Yasuda, T.; Komatsu, N.; Matsuda, K. Drastic Change in Photoluminescence Properties of Graphene Quantum Dots by Chromatographic Separation. *Adv. Opt. Mater.* **2014**, *2*, 983–989. [CrossRef]
53. Shinde, D.B.; Pillai, V.K. Electrochemical Preparation of Luminescent Graphene Quantum Dots from Multiwalled Carbon Nanotubes. *Chem. Eur. J.* **2012**, *18*, 12522–12528. [CrossRef] [PubMed]
54. Zhou, J.; Yang, Y.; Zhang, C.Y. A Low-Temperature Solid-Phase Method to Synthesize Highly Fluorescent Carbon Nitride Dots with Tunable Emission. *Chem. Commun.* **2013**, *49*, 8605–8607. [CrossRef]

55. Messina, F.; Sciortino, L.; Popescu, R.; Venezia, A.M.; Sciortino, A.; Buscarino, G.; Agnello, S.; Schneider, R.; Gerthsen, D.; Cannas, M.; et al. Fluorescent Nitrogen-Rich Carbon Nanodots with an Unexpected β-C_3N_4 Nanocrystalline Structure. *J. Mater. Chem. C* **2016**, *4*, 2598–2605. [CrossRef]
56. Sciortino, A.; Mauro, N.; Buscarino, G.; Sciortino, L.; Popescu, R.; Schneider, R.; Giammona, G.; Gerthsen, D.; Cannas, M.; Messina, F. β-C_3N_4 Nanocrystals: Carbon Dots with Extraordinary Morphological, Structural, and Optical Homogeneity. *Chem. Mater.* **2018**, *30*, 1695–1700. [CrossRef]
57. Serhan, M.; Sprowls, M.; Jackemeyer, D.; Long, M.; Perez, I.D.; Maret, W.; Tao, N.; Forzani, E. Total Iron Measurement in Human Serum with a Smartphone. In *Proceedings of the 2019 AIChE Annual Meeting, Orlando, FL, USA, 10–15 November 2019*; American Institute of Chemical Engineers: New York, NY, USA, 2019. [CrossRef]
58. Cayuela, A.; Soriano, M.L.; Valcárcel, M. Photoluminescent Carbon Dot Sensor for Carboxylated Multiwalled Carbon Nanotube Detection in River Water. *Sens. Actuators B: Chem.* **2015**, *207*, 596–601. [CrossRef]
59. Zhu, S.; Song, Y.; Zhao, X.; Shao, J.; Zhang, J.; Yang, B. The Photoluminescence Mechanism in Carbon Dots (Graphene Quantum Dots, Carbon Nanodots, and Polymer Dots): Current State and Future Perspective. *Nano Res.* **2015**, *8*, 355–381. [CrossRef]
60. Yan, F.; Sun, Z.; Zhang, H.; Sun, X.; Jiang, Y.; Bai, Z. The Fluorescence Mechanism of Carbon Dots, and Methods for Tuning Their Emission Color: A Review. *Microchim. Acta* **2019**, *186*, 583. [CrossRef]
61. Schneider, J.; Reckmeier, C.J.; Xiong, Y.; von Seckendorff, M.; Susha, A.S.; Kasak, P.; Rogach, A.L. Molecular Fluorescence in Citric Acid-Based Carbon Dots. *J. Phys. Chem. C* **2017**, *121*, 2014–2022. [CrossRef]
62. Zhou, Y.; Zahran, E.M.; Quiroga, B.A.; Perez, J.; Mintz, K.J.; Peng, Z.; Liyanage, P.Y.; Pandey, R.R.; Chusuei, C.C.; Leblanc, R.M. Size-Dependent Photocatalytic Activity of Carbon Dots with Surface-State Determined Photoluminescence. *Appl. Catal. B: Environ.* **2019**, *248*, 157–166. [CrossRef] [PubMed]
63. Zuo, P.; Lu, X.; Sun, Z.; Guo, Y.; He, H. A Review on Syntheses, Properties, Characterization and Bioanalytical Applications of Fluorescent Carbon Dots. *Microchim. Acta* **2016**, *183*, 519–542. [CrossRef]
64. Tao, S.; Zhu, S.; Feng, T.; Xia, C.; Song, Y.; Yang, B. The Polymeric Characteristics and Photoluminescence Mechanism in Polymer Carbon Dots: A Review. *Mater. Today Chem.* **2017**, *6*, 13–25. [CrossRef]
65. Pirsaheb, M.; Mohammadi, S.; Salimi, A.; Payandeh, M. Functionalized Fluorescent Carbon Nanostructures for Targeted Imaging of Cancer Cells: A Review. *Microchim. Acta* **2019**, *186*, 231. [CrossRef] [PubMed]
66. Althagafi, I.I.; Ahmed, S.A.; El-Said, W.A. Colorimetric Aflatoxins Immunoassay by Using Silica Nanoparticles Decorated with Gold Nanoparticles. *Spectrochim. Acta-Part A: Mol. Biomol. Spectrosc.* **2021**, *246*, 118999. [CrossRef] [PubMed]
67. Prajapati, J.P.; Das, D.; Katlakunta, S.; Maramu, N.; Ranjan, V.; Mallick, S. Synthesis and Characterization of Ultrasmall Cu_2O Nanoparticles on Silica Nanoparticles Surface. *Inorg. Chim. Acta* **2021**, *515*, 120069. [CrossRef]
68. Drozd, D.; Zhang, H.; Goryacheva, I.; de Saeger, S.; Beloglazova, N.V. Silanization of Quantum Dots: Challenges and Perspectives. *Talanta* **2019**, *205*, 120164. [CrossRef] [PubMed]
69. Asgari, M.; Soleymani, M.; Miri, T.; Barati, A. A Robust Method for Fabrication of Monodisperse Magnetic Mesoporous Silica Nanoparticles with Core-Shell Structure as Anticancer Drug Carriers. *J. Mol. Liq.* **2019**, *292*, 111367. [CrossRef]
70. Fernandes, R.S.; Raimundo, I.M.; Pimentel, M.F. Revising the Synthesis of Stöber Silica Nanoparticles: A Multivariate Assessment Study on the Effects of Reaction Parameters on the Particle Size. *Colloids Surf. A: Physicochem. Eng. Asp.* **2019**, *577*, 1–7. [CrossRef]
71. Ghimire, P.P.; Jaroniec, M. Renaissance of Stöber Method for Synthesis of Colloidal Particles: New Developments and Opportunities. *J. Colloid Interface Sci.* **2021**, *584*, 838–865. [CrossRef] [PubMed]
72. Zhang, Y.; Zhen, B.; Hu, Y.; Liang, G.; Feng, Y. A Reverse Micellar System with Triton X-100: Effect of Surfactant Polydispersity and Preparation of Monodisperse Silica Nanoparticles. *Soft Matter* **2020**, *16*, 383–389. [CrossRef] [PubMed]

biosensors

MDPI

Article

Profiles of Sterigmatocystin and Its Metabolites during Traditional Chinese Rice Wine Processing

Jia Zhang [1,2,3], Liwei Xu [1,2,3], Xinxin Xu [1,2,3], Xiaoling Wu [1,2,3,*], Hua Kuang [1,2,3] and Chuanlai Xu [1,2,3,*]

1 State Key Laboratory of Food Science and Technology, Jiangnan University, Wuxi 214122, China; 7200112107@stu.jiangnan.edu.cn (J.Z.); 7170112080@stu.jiangnan.edu.cn (L.X.); xuxinxin@jiangnan.edu.cn (X.X.); kuangh@jiangnan.edu.cn (H.K.)
2 International Joint Research Laboratory for Biointerface and Biodetection and School of Food Science and Technology, Jiangnan University, Wuxi 214122, China
3 Collaborative Innovation center of Food Safety and Quality Control in Jiangsu Province, Jiangnan University, Wuxi 214122, China
* Correspondence: wuxiaoling@jiangnan.edu.cn (X.W.); xcl@jiangnan.edu.cn (C.X.); Tel./Fax: +86-510-85329077 (X.W. & C.X.)

Abstract: Mycotoxin pollution is widespread in cereal, which greatly threatens food security and human health. In this study, the migration and transformation of sterigmatocystin (STG) mycotoxin during the contaminated rice wine processing was systematically assessed. QuEChERS (Quick, Easy, Cheap, Effective, Rugged, and Safe) coupled with ultrahigh-performance liquid chromatography coupled with tandem mass spectrometry (UPLC−MS/MS) method was firstly established for STG analysis in rice wine. It was found that high levels of rice leaven caused a significant reduction in STG in the fermented rice and wine, which was mainly due to the adsorption of yeast cells and *Rhizopus* biological degradation. However, compared with rice, the levels of STG in separated fermented wine was significantly decreased by 88.6%, possibly attributed to its high log Kow (3.81) and low water solubility (1.44 mg/L). The metabolites of STG (i.e., monohydroxy STG) were identified in rice wine fermentation for the first time. Moreover, STG disturbed the metabolic profile rice wine composition mainly by glycine, serine and threonine metabolism, alanine, aspartate and glutamate metabolism, purine metabolism pathway, particularly with regard to eight amino acids and sixteen lipids. This study elucidated the STG migration and transformation mechanism during the rice wine processing. The finding provided new analytical method for mycotoxin exposure and pollutant in food production, which may support agricultural production and food security.

Keywords: mycotoxin; food processing; migration and transformation; metabolites; UPLC–MS/MS

Citation: Zhang, J.; Xu, L.; Xu, X.; Wu, X.; Kuang, H.; Xu, C. Profiles of Sterigmatocystin and Its Metabolites during Traditional Chinese Rice Wine Processing. *Biosensors* **2022**, *12*, 212. https://doi.org/10.3390/bios12040212

Received: 7 March 2022
Accepted: 31 March 2022
Published: 1 April 2022

1. Introduction

Food security has been severely affected by plant diseases and pollutants such as pesticides, mycotoxins and heavy metals, particularly mycotoxins, which threaten the precarious food supply of millions of people on the planet [1,2]. Mycotoxins are secondary metabolites produced by several fungal species and are known to frequently contaminate small grain crops in the world [3–5], which may pose severe threats to human and animal bodies because of their toxicity [6]. Sterigmatocystin (STG) is a polyketide mycotoxin that is structurally related to aflatoxin B_1 and produced by *Aspergillus flavus*, *A. parasiticus*, *A. versicolor* and *A. nidulans*; the most common source of STG is *A. versicolor* [7,8]. STG can arise due to fungal infestation at the post-harvesting stage in a range of small grain cereals and grain-based products, including maize, rice, rye, wheat, oats, and barley [8], especially rice and oats [9], the concentration levels ranging from 0 to 83 μg/kg [10,11]. Over the past 10 years, the severe contamination problem in cereal grains caused by mycotoxin has drawn increased attention with regards to food safety [3]. Furthermore, the contamination of mycotoxin in food and environment may cause potential toxic effects in human and livestock, including liver cancer [8]. Like aflatoxin B_1 (AFB_1), STG resulted as genotoxic, also able to induce DNA damage and form DNA adducts [8]. However, due to the lack of data related to the occurrence of STG, the EFSA Panel on Contaminants in the Food Chain (CONTAM Panel) were not able to carry out a reliable assessment of human and animal dietary exposure. Therefore, it is essential to collect more data relating to STG in food and animal feed to accomplish the assessment of dietary exposure [8].

Rice (*Oryza sativa L.*) is consumed by more than 50% of the world's population, functions as a vital source to produce foods for a growing population and forage for livestock, especially in Asian countries. Global rice consumption is predicted to increase 61 Mt to reach 582 Mt by 2020–2029 [12]. Furthermore, rice provides 20% of the world's dietary energy supply and represents a major source of nutrients due to its daily consumption. Rice is consumed in various forms and can be processed into different foods, including wholegrain flour (brown, milled, or parboiled) and fermented products (e.g., rice wine). Chinese rice wine, also named sweet rice wine, is a traditional fermented food in China. Sweet rice wine has high nutritional value and is rich in amino acids and vitamins [13].

Food processing (e.g., roasting, fermentation, bread and cheese making, milling, heating, or enzymes) can potentially lead to alterations in the levels of STG; the extent of these modifications depends on the type of food involved and the food processing conditions [14,15]. The migration, transformation and degradation of contaminants in food processing, such as mycotoxins, is closely related to its physicochemical properties, especially the melting point, the Octanol-Water Partition Coefficient (Kow), water solubility, and vapor pressure [16–18]. STG has a high melting point (246 °C) and is relatively insoluble in water (1.44 mg/L at 25 °C); these characteristics make it stable in the food and surroundings. Veršilovskis et al. [18] reported that STG remained stable during the bread-making process; the levels of STG were determined in 5 of the 29 bread samples analyzed at concentrations ranging from 2.4 to 7.1 μg/kg in Riga, Latvia. Metwally et al. [17] reported high levels of STG (80%) in the curd and much lower levels (20%) in the whey during the cheese making process, thus demonstrating the low solubility of STG in aqueous media.

To our knowledge, only a limited amount of data is available for STG and its metabolites during food processing. However, there are many studies relating to the behavior of mycotoxin during food processing [14,15,18]. Most previous studies were carried out in cereals that were naturally or artificially contaminated. However, the behavior of target mycotoxins in contaminated food is often difficult to predict and characterize. This is because of the wide levels of variation exhibited by mycotoxins, including different derivatives, bonding forms, isomers, and precursors; often, these variations are observed synchronously in grains [15]. Therefore, to investigate the real behavior of target mycotoxins during food processing, free of contamination cereals were spiked with the targeted mycotoxins. Although STG are similar to AFB_1 in terms of chemical structure, literatures on the conversion profile of STG were scarce in food processing [8], particularly with regards to the process used to make rice wine. However, with the increasing levels of attention targeted to the risk assessment of mycotoxins, the need for data related to the behavior of mycotoxins in food processing has become increasingly urgent. Thus, it is necessary to investigate changes in the levels of STG in cereal-based products that are naturally or artificially infected with *Aspergillus versicolor* during food processing. Additionally, the metabolomic profiles of rice were demonstrated to undergo changes during Chinese rice wine fermentation [13]. Currently, most studies on STG metabolism focus on the transformation in animals [19], but there are few studies on the effects of STG on metabolism in fermentation and other processing technology. Thus, it is essential to investigate whether the level of STG has any effect on the contents of various components based on the metabolic profiles, including organic acids, amino acids, and lipids, which in turn leads to changes in the composition and quality in rice wine during fermentation.

It is worth noting that, the complexity of the rice wine product matrix creates potential difficulties for the analysis of STG levels and its metabolites; thus, researchers have begun to focus on the development of propitious analytical methods [20]. We first performed the analysis of STG levels in rice wine samples using UPLC–MS/MS coupled with the modified QuEChERS ("Quick, Easy, Cheap, Effective, Rugged, and Safe") method. A non-targeted metabolomics methodology was carried out to identify and compare the metabolic differences between the rice wine products treated with and without STG. However, the main purpose was to enforce the scheme for the evaluation of the fate of STG during the rice wine production and identify the key procedures conducing its possible decrease and to identify the different STG levels' exposure effect on metabolite profiles during rice wine production. This research enhances our knowledge of the effects of food processing on STG levels. Furthermore, data relating to changes in STG levels during rice wine production might provide further insight into the assessment of chronic dietary risk, as determined by the risk quotients (RQs) method and based on Chinese dietary habits. In addition, it was first observed that the transformation of STG was converted into metabolite (monohydroxy STG) during food fermentation.

2. Materials and Methods

2.1. Rice Pre-Treatment and Rice Wine Preparation

Rice contaminated with STG were acquired by soaking in an aqueous solution referring to the treatment procedures described in previous studies [15,21].

Spiked sample 1: Rice (2 kg) was individually soaked in STG aqueous solution (2 mg/L) for 8 h in a glass beaker (6 L). Subsequently, the rice was allowed to air dry naturally at room temperature (25 °C) for 84 h in order to restore the original state. The treated rice was stored in a freezer at −20 °C until further use.

Spiked sample 2: Rice (0.5 kg) was individually soaked in STG aqueous solution at six spiked levels (0.3, 0.5, 0.8, 1, 1.5, and 2 mg/L) for 8 h in a glass beaker (2 L). The rice was stirred every 0.5 h to ensure the uniform absorption of STG. Then, the soaked rice was separated from the aqueous solution for the next step as a raw material for rice wine processing.

Generally, rice wine processing includes the following consecutive steps, as shown in Figure 1. The treated rice was washed with tap water for 1 min and soaked in water at room temperature (approximately 25 °C) for 8 h. Then, the soaked rice was individually steamed for 30 min. The steamed rice was cooled to about 35 °C and was transferred to a glass (2.5 L). The rice leaven 3 g (ANGEL YEAST CO., LTD., Hubei, China) was dissolved in 100 mL of water; this was then poured into the steamed rice in batches while stirring. Then, the rice was sealed and fermented at room temperature in the dark for 84 h, respectively.

The STG levels of fermented rice and wine were determined separately. The concentration of STG in rice wine was calculated by weighing the content of STG in fermented rice and wine, as follows:

$$C = \sum C1 \times \frac{m1}{m1 + m2} + C2 \times \frac{m2}{m1 + m2} \quad (1)$$

In Equation (1), C represents the concentration of STG in rice wine (μg/kg), C1 represents the content of STG in fermented rice (μg/kg), C2 represents the content of STG in fermented wine (μg/kg), $m1$ represents the weight of fermented rice (g), and $m2$ represents the weight of fermented wine (g).

2.2. Instrument Conditions

2.2.1. LC–MS/MS Method for STG Analysis

The UPLC System was coupled to a tandem mass spectrometry QTRAP 5500 (AB SCIEX; Toronto, ON, Canada) with electrospray ionization (ESI) in positive mode. This was equipped with a ACQUITY UPLC HSS T3 column (2.1 mm × 100 mm, 1.8 μm; Waters) maintained at 40 °C, and the injection volume was 2 μL. The mobile phase included A: water (0.1% formic acid, 2 mM ammonium formate) and B: acetonitrile. The flow rate was 0.3 mL/min, and the gradient elution is given in Table S1. The mass detection parameters for STG were optimized, including declustering potential (DP), entrance potential (EP), collision energy (CE), and collision cell exit potential (CXP) (Table S1).

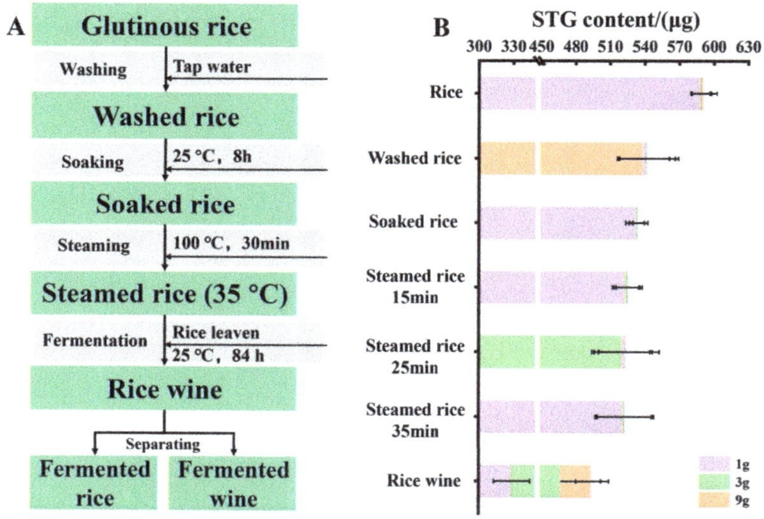

Figure 1. (**A**) The processed step of rice wine. The samples highlighted in green were used for mycotoxin analysis. (**B**) Changes of STG absolute content (μg) in each procedure during rice wine production.

2.2.2. LC–HRMS/MS Method for the Identification of STG Degradation Products and Non-Targeted Metabonomics

Chromatographic analysis and the identification of STG metabolites was performed by the Vanquish UHPLC system equipped with a reverse-phase C18 Acquity UPLC HSS T3 column (2.1 mm × 100 mm, 1.8 μm; Waters) heated to 35 °C, coupled with high-resolution tandem mass spectrometer (HRMS/MS) Q-Exactive (Thermo Scientific, Bremen, Germany) equipped with a heated ESI probe. The injection volume was 5 μL and sample determination was carried out over a 22-minute run time. The mass spectrometer was operated in full MS-data-dependent MS/MS (full MS-dd MS/MS) mode [22]. Detailed instrumental and chromatographic conditions are shown in Table S2.

2.3. *Statistical Analysis*

Differences were considered to be statistically significant if $p < 0.05$. Comparisons were carried out by paired-samples *t*-tests using SPSS version 19.0 software. A nonlinear curve fitting equation ($Y = aX^2 + bX + c$) was performed using Origin Pro version 9.0 software. The coefficient of determination (R^2) was used to evaluate the nonlinear curve fitting results and evaluate whether equations had a satisfying goodness of fit and good predictive capability.

The partial least squares–discriminant analysis (PLS–DA) model [23] was accomplished using SIMCA-P (V14.1) for discriminate between different groups. The variable importance in projection (VIP) were calculated by this model. Permutation tests ($n = 200$) were used to evaluate the quality of each PLS–DA model (Figure S4) [24]. Permutation tests were also used to assess whether a particular classification of individuals in either of the designed groups was prominently better than any other random classification in two arbitrary groups. *p* values were calculated by one-way analysis of variance (ANOVA) and pathway analysis was based on metabolites identified by MetaboAnalyst 5.0.

3. Results and Discussion

3.1. Method Validation

The method validation was assessed according to guidelines and standards from the European Commission [25,26].

The overview of the acquired validation parameters are summarized in Table 1. The product ion chromatograms of STG in fermented rice and wine are shown in Figure S1. The linearity of the calibration curve was evaluated with a standard solution of blank matrix extracts and acetonitrile; this was performed by preparing five matrix-matched calibration standards (5, 10, 20, 100 and 200 μg/L) for STG in each matrix (acetonitrile, soaked rice, steamed rice, fermented rice and fermented wine). Outstanding linearity ($R^2 \geq 0.9910$) was observed in each matrix (Table 1). Matrix effects (ME) were calculated by the slope ratios of the matrix and solvent calibration curves. As shown in Table 1, the results suggested no significant enhancement or suppression effects for STG in soaked rice, steamed rice, and fermented rice, within 10% of the slope ratio, ranging from 0.93 to 1.09. The slope ratio (1.35) of fermented wine showed the matrix enhancement effect. Recovery was performed by spiked experimental samples at three different levels (20, 100, and 200 μg/kg) of STG. Mean recovery of STG ranged from 102–119% for soaked rice, 77–112% for steamed rice, 116–118% for fermented rice, 73–119% for fermented wine, with RSDs ranging from 2.1–8.7% (Table 1). The limits of detection (LODs, signal-to-noise ratio = 3) defined as the minimum detection level, and limits of quantification (LOQs, signal-to-noise ratio = 10) defined as the minimum quantitation level for STG, were 0.01 and 0.03 μg/kg for all matrices (0.07 and 0.25 μg/kg for fermented wine) (Table 1).

Table 1. Linear range (μg/L), regression equation, calibration curve coefficients (R^2), Matrix effects (ME), Limit of detection (LOD) and Limit of quantitation (LOQ) for STG in rice wine products. Recoveries and RSDs of sterigmatocystin in rice wine products at different spiked levels ($n = 5$).

Mycotoxin	Matrix	Linear Range	Regression Equation	R^2	ME/%	LOD (μg/kg)	LOQ (μg/kg)
STG	Solvent	5–200	y = 190,442x + 268,234	0.9953			
	Soaked rice	5–200	y = 176,167x + 236,281	0.9979	−7.0	0.01	0.03
	Steamed rice	5–200	y = 179,674x + 238,395	0.9969	−6.0	0.01	0.03
	Fermented rice	5–200	y = 208,146x + 392,325	0.9959	+9.0	0.01	0.03
	Fermented wine	5–200	y = 256,280x + 565,254	0.9910	+35.0	0.07	0.25

Sample	20 μg/kg		100 μg/kg		200 μg/kg	
	Recoveries (%)	RSD (%)	Recoveries (%)	RSD (%)	Recoveries (%)	RSD (%)
Soaked rice	102	2.3	107	2.3	119	2.4
Steamed rice	77	6.2	85	8.7	112	3.5
Fermented rice	118	2.1	118	7.6	116	6.9
Fermented wine	73	3.6	105	5.0	119	4.3

3.2. The Fate of STG within the Chinese Rice Wine Process

The process used to make traditional Chinese rice wine involves four key steps: washing, soaking, steaming, and fermenting [13]. Although mycotoxins are stable, the levels and structure of STG may change as a result of the complex physicochemical modifications that occur during the processing of raw materials into a processed product [14,27].

3.2.1. Washing and Soaking

The occurrence and concentration of STG in rice wine products are shown in Figure 1, Figure S2 and Table 2. In the present study, the treated rice samples were washed with water for 3 min. During the washing process, the initial level of STG (986.1 μg/kg) in rice had a significant decrease of 16.6% ($p < 0.05$) in STG levels; this may be due to the STG dilution with the water absorption of rice or STG partially dissolved into water during the washing process. Rice could sufficiently absorb amounts of water during soaking, which is conducive to the next effective steaming. Compared to the level of STG in washed rice, 8.7% ($p < 0.05$) of STG was removed by soaking (Figure S2 and Table 1). When washed rice absorbs water and expands, it follows that some STG is redistributed into the soaking water, thus resulting in a decrease of STG in soaked rice.

Table 2. Changes of STG level in spiked samples in different steam time, fermentation time and rice leaven addition level during the rice wine production (mean, n = 3).

Sample	Rice	Washed Rice	Soaked Rice	Steam Rice			Fermented Rice-1 g			Fermented Rice-3 g		
				15 min	25 min	35 min	12 h	36 h	84 h	12 h	36 h	84 h
Level/ (μg/kg)	986.1	822.3 *	750.8 *	738.8 ^{a}	733.2 ^{a}	737.7 ^{a}	594.8 ^{*,a}	711.6 ^{*,b}	913.1 ^{*,c}	592.8 ^{*,a}	712.3 ^{*,b}	858.1 ^{*,c}
SD	16.4	39.8	14.4	16.2	46.4	37.1	60.8	32.3	23.6	36.8	78.4	17.7
Sample	Fermented rice-9g			Separated fermented wine			Separated fermented rice			Total rice wine		
	12 h	36 h	84 h	1 g	3 g	9 g	1 g	3 g	9 g	1 g	3 g	9 g
Level/ (μg/kg)	586.6 ^{*,a}	736.7 ^{*,b}	818.5 ^{*,c}	169 ^{a}	126.2 ^{b}	112.2 ^{c}	1214.1 ^{a}	1164.9 ^{a}	956.0 ^{b}	925.4	850.6	613.4 *
SD	74.9	19.4	58.3	2	5.5	3.9	43.1	108.4	39.0			

Note: * Indicates a significant difference of STG in rice wine product of the step versus the prior step ($p < 0.05$), as determined by Student's t-test. ^{a,b,c} The different letters show a remarkable difference ($p < 0.05$) between the effects of the different factors in same processing; conversely, the same letter shows no significant difference observed. Fermented rice-1 g: the 1 g level of rice leaven during rice wine production, all else follows.

3.2.2. Influence of Steaming Time on STG Levels

In order to investigate the impact of steaming time on the level of STG during rice wine production, steaming times of 15, 25 and 35 min were used, respectively. The results of different steaming conditions on STG concentration in soaked rice are demonstrated in Table 2 and Figure S2. The levels of STG in soaked rice decreased after steaming ($p > 0.05$) by 1.6%, 2.3%, and 1.7%, after 15, 25, and 35 min, respectively (Table 2). These decreases of STG levels did not differ significantly ($p > 0.05$) when compared across different steaming times (Figure 1 and Figure S2). Our results were in line with the previous studies, veršilovskis et al. [18] found that the levels of STG were stable during bread production (17 min, 200–220 °C); Wu et al. [15] found that the deoxynivalenol (DON) was stable during a Chinese steamed bread making process (20 min, 100 °C). This may be related to the high melting point of STG at 245–246 °C [8].

Steaming is the most widely used method for rice processing and the treatment time is an essential processing factor for food production. Although some studies have reported that different processing times exerted influence on the levels of mycotoxin in food material [15], this is the first detailed investigation reporting changes in the profile of STG in response to different thermal treatments.

3.2.3. Influence of Rice Leaven Levels on the Concentration of STG

Prior to the fermentation process, rice leaven was added at low, medium, and high levels (1, 3, and 9 g, respectively). Then, 100 mL of water was added to the steamed rice, mixed thoroughly, and then fermented in a sealed container. As is shown in Table 2 and Figure 2A, the fermentation step (12 h and 36 h) caused no significant change in STG concentrations in fermented rice samples when treated with different levels of rice leaven. However, after 84 h of fermentation, the STG concentration (913.1 μg/kg) in the fermented rice that was mixed with 1 g of rice leaven was higher than that (858.1 μg/kg) in the fermented rice that was mixed with 3 g of rice leaven; the lowest level of STG was found in the fermented rice that was mixed with 9 g of rice leaven (Table 2 and Figure 2A). There were significant ($p < 0.05$) reduction (10.4%) in STG levels when compared between fermented rice samples containing 1 g and 9 g of rice leaven (Figure 2A and Table 2). Therefore, the addition of a larger amount of rice leaven (consisting of yeast and *Rhizopus*) may more easily lead to a reduction in the concentration of STG in rice wine products by binding mycotoxins [28,29]. The relevant study showed lactic acid bacteria and yeast composed of a rather complex biological ecosystem was contributed to dough fermentation [30]. It is possible that this biological ecosystem, containing yeast and *Rhizopus*, plays a key role in the fermentation of rice wine that results in the adsorption, biotransformation, or degradation of STG. Whereas, no other studies have confirmed the prediction. The results of this work were similar to previous studies. Previous research demonstrated that the beta-1,3/1,6-glucan moieties play an essential role during the Saccharomyces cerevisiae cell wall adsorption of the mycotoxin [31]. The glucomannans and mannan-oligosaccharides have been proposed to be the most crucial elements responsible for AFB_1 binding in yeast [32]. The cell wall of bakery yeast can adsorb 29% of AFB_1, 68% of zearalenone (ZEA), and 62% of Ochratoxin A (OTA) [33]. Cole et al. found that *Rhizopus* had an effect on biological degradation of the AFB_1 [34] and Aflatoxin G_1 (AFG_1) [35]. Since STG and AFB_1 are similar in structure, the degradation of STG by *Rhizopus* is similar to that of AFB_1. Based on the previous literature, and the results of our current research, the weight of rice leaven likely affected the levels of STG via yeast adsorption and *Rhizopus* biological degradation during the rice wine fermentation process. Hence, the addition of high weight rice leaven led to the decrease of STG level.

3.2.4. Influence of Fermentation Time on STG Levels

To monitor the effect of fermentation time on STG levels during rice wine production, three different fermentation times (12, 36, and 84 h) were set as sampling points. The results are shown in Table 2. Compared to the levels of STG in steamed rice, the levels of STG in fermented rice that was mixed with 1 g of rice leaven decreased by 19.4% (12 h) and 3.5% (36 h) and increased by 24% (84 h) of fermentation. The STG level of fermented rice that was mixed with 3 g of rice leaven decreased by 19.6% (12 h) and 3.4% (36 h) and increased by 16.3% (84 h); and the STG levels of fermented rice that was mixed with 9 g of rice leaven decreased by 20.5% (12 h) and 0.1% (36 h) and increased by 11.0% (84 h) (Table 2). Compared to the steamed rice, the decrease in STG in fermented rice at 12 and 36 h was mainly due to the dilution effect of adding water prior to fermentation (Table 2); therefore, after 12 and 36 h of fermentation, the STG concentration had not increased relative to the original STG concentration in steamed rice. As fermentation time increased, the level of STG in fermented rice showed a gradually increasing trend. Generally, the longer the fermentation time, the higher the level of STG in fermented rice (Table 2). Furthermore, the level of STG in separated fermented wine from the three groups supplemented with rice leaven was significantly lower than that in separated fermented rice when fermented rice and fermented wine were completely separated (Table 2). Compared to the level of STG in steamed rice, STG increases of 39.2%, 36.7%, and 22.8% were observed in the separated fermented rice in the 1, 3, and 9g rice leaven groups; STG decreases of 77.1%, 82.9%, and 84.8% were observed in the separated fermented wine in the 1, 3, and 9g rice leaven groups (Table 2). This finding was similar to the results reported by [14], who reported that the levels of fumonisin B_1 (FB_1) prominently increased by 166% during the second part of DDGS fermentation in naturally contaminated maize. This is because the rice leaven consisted of complex biological ecosystem [30] converts rice into rice wine via biotransformation during the fermentation process, and STG is a mycotoxin with high fat solubility (log Kow = 3.81) and low water solubility (1.44 mg/L) which tends to be distributed and accumulated in fermented rice [21]. Another reason for the increase in STG level may be yeast extracellular enzymes which might be responsible for STG release from covalent bonds with rice constituents, such as starch or proteins [14].

After the separation of fermented rice and fermented wine, the level of STG in fermented wine were significantly lower than that in fermented rice (Figure 2B and Table 2), this is mainly due to its high log Kow (3.81) and low water-solubility (1.44 mg/L), which caused it to accumulate in fermented rice rather than fermented wine. According to the weight of separated fermented rice and separated fermented wine, the final concentration of STG in rice wine was calculated as 925.4, 850.0 and 613.4 μg/kg, respectively (Table 2). Compared to the STG level in rice, the level of STG decreased by 6.2%, 13.8%, and 37.8%

in the final rice wine (1, 3 and 9 g rice leaven, respectively) during rice wine production. Considering the information mentioned previously in Section 3.2.3, it might be assumed that the reduction in STG in the whole rice wine process stage most likely resulted from adsorption by yeasts cells [14,33,36] and *Rhizopus* biological degradation [34,35], or to some extent, via biotransformation.

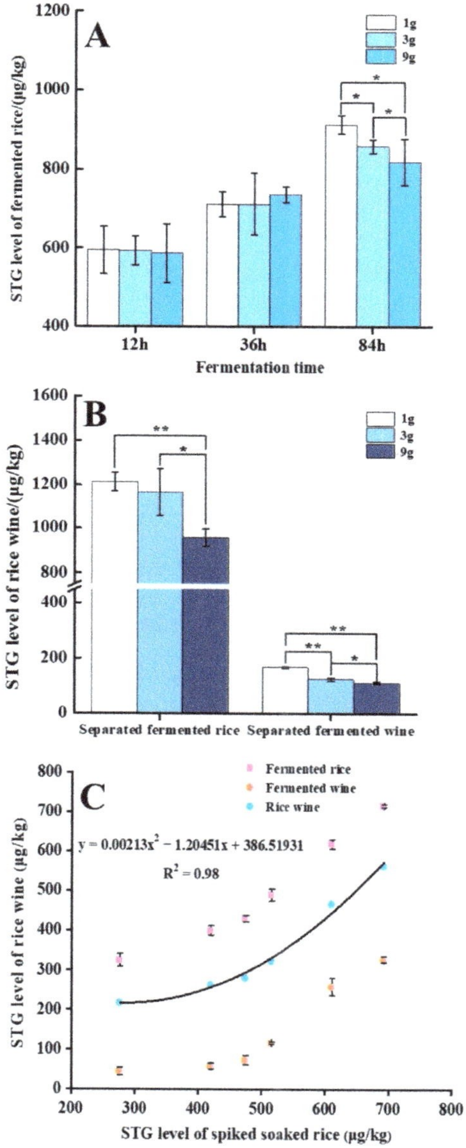

Figure 2. (**A**) The STG level in fermented rice of different fermentation time (12 h, 36 h, 84 h) during rice wine production (1g, 3g, and 9g mean different rice leaven levels); (**B**) the STG level of fermented wine and fermented rice after complete separation; (**C**) correlation of STG level between the original soaked rice and final rice wine product. Data are expressed as means ± standard error of means (n = 3). * Error bars represent the standard deviation. * Indicates a significant difference of STG content in rice wine product of the step versus the prior step, (*: $p < 0.05$, **: $p < 0.01$), as determined by Student's t-test.

3.3. Correlation of STG Levels between Soaked Rice and Rice Wine Final Product

Fermented wine was generated from fermented rice during fermentation; the fermented rice and fermented wine represented two food matrices with entirely disparate physicochemical properties.

The levels of STG in fermented wine were significantly ($p < 0.05$) lower than the levels in fermented rice. The final levels of STG in rice wine were calculated by Equation (1) and depended on the STG levels in fermented rice and fermented wine (Table S4). A linear relationship was built using the STG levels in the soaked rice and the rice wine products, with a non-linear fitting curve of $Y = 0.00213x^2 - 1.20451x + 386.51931$, $R^2 = 0.98$ (Figure 2C).

The developed mathematical model demonstrated a satisfying goodness-of-fit and prediction at concentrations ranging from 280 to 700 μg/kg (Figure S3) and showed that the STG level in the final product (such as rice wine) could be evaluated according to the original concentration in raw food material (the STG levels in rice or soaked rice, for instance). Based on the established model, the linear relationship between STG residue levels in rice wine products and rice or soaked rice were also predicted. This clearly demonstrated the potential of this approach to investigate other behaviors of other toxins in the residue of rice wine products and other processed food products.

3.4. Identification of Fermentation Degradation Products of STG

The calculated data clearly showed that the fermentation procedure (9 g) had a noteworthy effect on the final levels of STG in rice wine. There was a 37.8% decrease in the level of STG during the rice wine production (Table 2). In order to investigate whether the changes in STG level was associated with biotransformation products generated by the catalysis of yeast enzymes during the fermentation process. While quantifying STG levels in rice wine products, the biotransformation products of STG were determined by UPLC–HRMS/MS. Two STG biotransformation products, monohydroxy STG A (M1) and monohydroxy STG B (M2), were identified for the first time in the food processing field (Figure 3). The result showed the transformation behavior of STG existed in food processing. The monohydroxy of STG were major metabolite (phase I metabolism) formed by human and rat hepatic microsomes, via hydroxylation of the aromatic ring [8]. Previous studies showed that phase I metabolism of STG consists of a cytochrome P450 (CYP)-mediated formation of mono-hydroxylation reactions [8]. Saccharomyces cerevisiae also contains the P450 enzyme system, which could be the reason of the transformation of STG during fermentation. Unfortunately, it has not been able to quantify for it because there are no standards.

3.5. Effects of STG on the Metabolite Profiles of Rice Wine

Changes in the metabolite profiles during the preparation of rice wine using different groups of rice (with low and high levels of STG) were analyzed by UPLC–HRMS/MS. Partial least-squares discriminant analysis (PLS-DA) of nontargeted metabolites in rice wine samples are shown in Figure 4A. These data demonstrated the quality parameters of the loading plot with an appropriate explanation, including goodness of fit (R2Y), and accuracy (Q2). The permutation test ($n = 200$) was performed on all samples; this confirmed that the model was good quality and that there was no overfitting (Figure S4) [37]. UPLC–HRMS analysis identified a total of 54 highly differential metabolites with a VIP > 1.0 and $p < 0.05$ (Table S5). The PLS-DA models acquired from HRMS analysis showed that the metabolites varied depending on the addition of STG, and each group was obviously discriminated (Figure 4A).

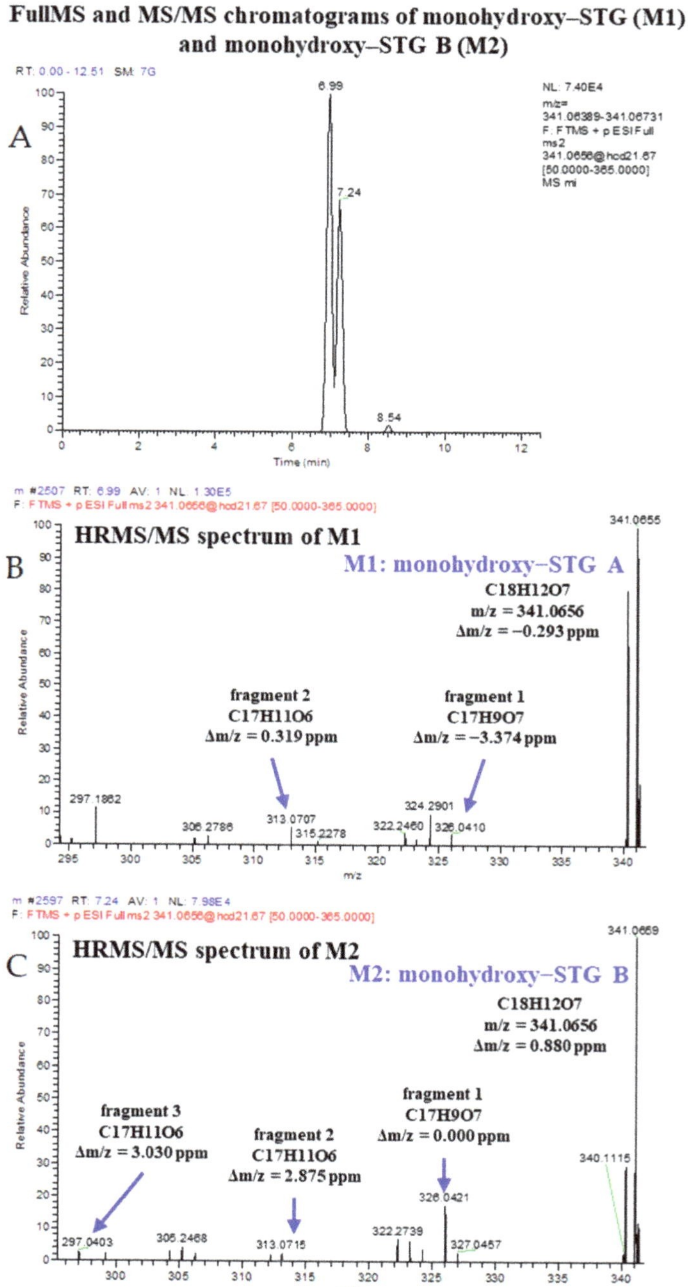

Figure 3. (**A**) Chromatograms of metabolite of STG found in rice wine product; (**B**,**C**) HRMS/MS spectra of metabolites of STG found in rice wine product.

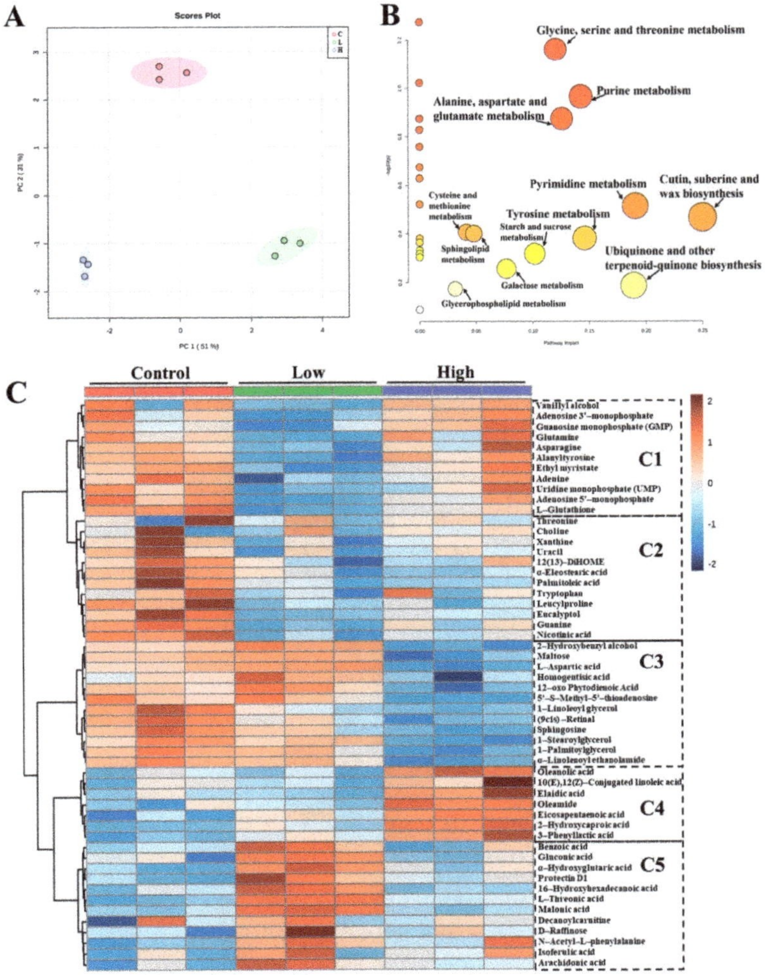

Figure 4. (**A**) PLS-DA score plot of STG exposure rice wine in positive and negative mode results (C, L, and H mean control group, low, and high level STG group); (**B**) pathway impact analysis showing changing metabolism in rice wine treated with STG compared to normal rice wine; (**C**) based on UPLC–HRMS/MS system identified, a heat map of identified metabolites in rice wine with varied STG levels exposure by hierarchical clustering of the most significantly differential metabolites in rice wine ($p < 0.05$ and VIP > 1.0).

To visualize the changes in metabolite levels based on the STG addition level, a heat map showed that 35 metabolites (Figure 4: C1, C2, and C3) were significantly down-regulated and 19 metabolites (Figure 4: C4, and C5) were significantly up-regulated in a dose-dependent manner when compared with the controls.

Generally, the breakdown of macronutrients can produce micronutrients, such as monosaccharides, amino acids, and lipid metabolites, which can then partake in glycolysis and the TCA cycle to make energy. All metabolites were classified into five groups (C1, C2, C3, C4, and C5) (Figure 4C). Group C1, mainly included amino acids and nucleotides; these decreased when exposed to STG. When compared to the control group, other metabolites in group C2 were also significantly decreased at low and high STG levels. These metabolites were mainly amino acids, nucleic acids, and lipids (Table S5 and Figure 4C). Two saturated lipid acids (α-eleostearic acid and palmitoleic acid) and one oxylipin (DiHOME) were significantly reduced after exposure to STG; these were down-regulated in the low and high STG treatment groups, thus indicating abnormal lipid metabolism. In group C3, these metabolites were significantly reduced in the high STG treatment groups. Although the flavor of long-chain saturated fatty acids is generally regarded as unpleasant, some of them were involved in esterification to form esters during fermentation and aging [13]. Three esters of long-chain saturated fatty acids (1-linoleoyl glycerol, 1-stearoylglycerol, and 1-palmitoylglycerol) were significantly down-regulated in the high STG treatment groups. The metabolites in group C4, including 4 lipids (10(E),12(Z)-Conjugated linoleic acid, Elaidic acid, Oleamide, Eicosapentaenoic acid) and 3 organic acids (Oleanolic acid, 2-Hydroxycaproic acid, 3-Phenyllactic acid) (Table S5 and Figure 4C), were significantly increased after exposure to high levels of STG. Group C5 mainly included six organic acids and others (Table S5 and Figure 4C) which showed a significant increase after exposure to low levels of STG. The distinct changes of these biomarkers revealed a clear metabolite profile for rice wine, and clear indicators of promotion or suppression in the relevant metabolism pathways (Figure 4B). As shown in Figure 4B, the STG exposure may have a negative effect on secretion of yeast protease, lipase and other relevant enzyme using for the hydrolysis of sugars and the breakdown of substances such as proteins and fats [38], which resulted in twelve differential metabolic pathways of rice wine production, especially for 6 pathways of glycine, serine and threonine metabolism, alanine, aspartate and glutamate metabolism, purine metabolism, pyrimidine metabolism, tyrosine metabolism and cutin, suberine and wax biosynthesis. As shown in Figure 4C, the relative contents of amino acids and saturated fatty acids showed a downward trend, while organic acids showed an upward trend, which had a certain influence on the quality of rice wine.

4. Conclusions

In this research, we used UPLC–MS/MS to investigate changes in STG levels, and its metabolites, during rice wine production. We found that the levels of STG first decreased but then increased during processing. When compared to the prior initial processing, the levels of STG decreased after washing, soaking, and steaming treatment. High levels of rice leaven may also cause STG levels of fermented rice to fall during fermentation. However, the opposite trend was observed after different fermentation times. Our analysis showed that rice leaven levels and fermentation times were both critical factors in food production. To the best of our knowledge, this study is the first to demonstrate that rice leaven levels and fermentation times can have significant ($p < 0.05$) effects on the levels of STG during rice wine production. Furthermore, for the first time, we also identified the presence of metabolites of STG (monohydroxy STG) during food processing. We also investigated relative changes in the characteristic metabolism of amino acids, lipids, and organic acids. This analysis showed that metabolite profiles changed due to exposure to STG during rice wine fermentation. In conclusion, the results of this study provide a preliminary investigation of STG residues behavior and exposure effect on composition of rice wine during fermentation production and constitute a reference for formulating future rice wine process and risk-assessment programs. In addition, more studies would be essential to replenish the conditions of STG during food production.

Supplementary Materials: The following supporting information can be downloaded at: https://www.mdpi.com/article/10.3390/bios12040212/s1, Figure S1: TIC of STG in fermented rice (A1) and wine (B1); product ion chromatograms of STG in fermented rice (A2, A3) and wine (B2, B3). Figure S2. Changes of STG level in fermented rice during rice wine production. Figure S3. Changes of STG level in spiked samples during the rice wine process. Note: Group 1–6, spiking STG levels of 276.7, 420.3, 474.9, 515.7, 611.2, 692.7 μg/kg, respectively (Table S4). Data are expressed as means ± standard error of means (n = 3). Error bars represent the standard deviation. * Indicates a significant difference of STG in rice wine product of the step versus the prior step (* $p < 0.05$, *** $p < 0.001$), as determined by Student's t-test. Figure S4. Permutation test on fermented rice (FR) and fermented wine (FW) of exposure groups to control group on PLS-DA model. Spectra are randomly assigned to a class by 200 permutations. (D1, D3, E1, E3) Low level treatment; (D2, D4, E2, E4) High level treatment.

Table S1. Chromatography gradient elution procedure and mass parameters Table S2. Instrumental and chromatographic conditions for the analysis of metabolomics in rice wine samples. Table S3. Changes of STG absolute content (μg) in each procedure during rice wine production (mean ± SD, $n = 3$). Table S4. Changes of STG level in spiked samples during rice wine production (mean ± SD, $n = 3$). Table S5. Differential metabolites of rice wine by untargeted metabolomics ($p < 0.05$ and VIP > 1).

Author Contributions: Conceptualization, H.K. and C.X.; methodology, J.Z.; software, X.W. and X.X.; validation, J.Z. and L.X.; formal analysis, J.Z.; investigation, J.Z.; resources, J.Z.; data curation, J.Z.; writing—original draft preparation, J.Z.; writing—review and editing, X.W. and C.X.; visualization, L.X.; supervision, X.W., H.K. and C.X; project administration, H.K. and C.X; funding acquisition, X.W., H.K. and C.X. All authors have read and agreed to the published version of the manuscript.

Funding: This project was supported by the National Key Research and Development Program of China (2017YFC1601102) and Jiangsu Association for Science and Technology Youth Science and Technology Talent Support Project (TJ-2021-049).

Institutional Review Board Statement: Not applicable.

Informed Consent Statement: Not applicable.

Data Availability Statement: Not applicable.

Conflicts of Interest: The authors declare no conflict of interest.

Abbreviations

UPLC–MS/MS, ultra-performance liquid chromatography coupled with tandem mass spectrometry; HRMS, high resolution mass spectrometer; Kow, Octanol-Water Partition Coefficient; Sw, solubility water; STG, sterigmatocystin; DON, deoxynivalenol; AFB_1, Aflatoxin B_1; AFG_1, Aflatoxin G_1; QuEChERS (Quick, Easy, Cheap, Effective, Rugged, and Safe); ZEA, zearalenone. DDGS, distiller's dried grains with solubles; OTA, Ochratoxin A.

References

1. Gao, Q.; Ma, J.J.; Liu, Q.; Liao, M.; Xiao, J.J.; Jiang, M.H. Effect of application method and formulation on prothioconazole residue behavior and mycotoxin contamination in wheat. *Sci. Total Environ.* **2020**, *729*, 139019. [CrossRef] [PubMed]
2. Ristaino, J.B.; Anderson, P.K.; Bebber, D.P.; Brauman, K.A.; Cunniffe, N.J.; Fedoroff, N.V.; Finegold, C.; Garrett, K.A.; Gilligan, C.A.; Jones, C.M.; et al. The persistent threat of emerging plant disease pandemics to global food security. *Proc. Natl. Acad. Sci. USA* **2021**, *118*, e2022239118. [CrossRef] [PubMed]
3. Lee, H.J.; Ryu, D. Worldwide occurrence of mycotoxins in cereals and cereal derived food products: Public health perspectives of their co-occurrence. *J. Agric. Food Chem.* **2017**, *65*, 7034–7051. [CrossRef] [PubMed]
4. Yang, C.W.; Song, G.; Lim, W. Effects of mycotoxin-contaminated feed on farm animals. *J. Hazard. Mater.* **2020**, *389*, 122087. [CrossRef] [PubMed]
5. Jallow, A.; Xie, H.; Tang, X.; Qi, Z.; Li, P. Worldwide aflatoxin contamination of agricultural products and foods: From occurrence to control. *Compr. Rev. Food Sci. Food Saf.* **2021**, *20*, 2332–2381. [CrossRef]
6. Tian, Y.; Yu, D.Z.; Liu, N.; Tang, Y.; Wu, A. Confrontation assays and mycotoxin treatment reveal antagonistic activities of trichoderma and the fate of fusarium mycotoxins in microbial interaction. *Environ. Pollut.* **2020**, *267*, 115559. [CrossRef]
7. Kong, D.Z.; Xie, Z.J.; Liu, L.Q.; Song, S.S.; Kuang, H.; Cui, G.; Xu, C.L. Development of indirect competitive ELISA and lateral-flow immunochromatographic assay strip for the detection of sterigmatocystin in cereal products. *Food Agric. Immunol.* **2017**, *28*, 260–273. [CrossRef]
8. (EFSA) European Food Safety Authority. Scientific Opinion on the risk for public and animal health related to the presence of sterigmatocystin in food and feed. *EFSA J.* **2013**, *11*, 3254. [CrossRef]
9. Mol, J.; Pietri, A.; Macdonald, S.J.; Anagnostopoulos, C.; Spanjer, M. Survey on sterigmatocystin in food. *EFSA Support. Publ.* **2015**, *12*, 774E.
10. Noel, L.; Chafey, C.; Testu, C. Sterigmatocystin presence in typical Latvian grains. *Food Chem.* **2008**, *109*, 243–248.
11. Zhao, Y.R.; Liu, X.X.; Zhao, J.; Wang, F.H. Research progress of sterigmatocystin pollution in food. *Chin. J. Food Hyg.* **2016**, *28*, 680–682.
12. OECD/FAO. OECD-FAO Agricultural Outlook 2020–2029. 2020. Available online: https://www.oecd-ilibrary.org/agriculture-and-food/oecd-fao-agricultural-outlook-2020-2029_1112c23b-en (accessed on 16 July 2020).
13. Jin, Z.; Cai, G.; Wu, C.; Hu, Z.; Lu, J. Profiling the key metabolites produced during the modern brewing process of Chinese rice wine. *Food Res. Int.* **2021**, *139*, 109955. [CrossRef] [PubMed]

14. Dzuman, Z.; Stranska-Zachariasova, M.; Vaclavikova, M.; Tomaniova, M.; Veprikova, Z.; Slavikova, P.; Hajslova, J. Fate of free and conjugated mycotoxins within the production of distiller's dried grains with solubles (DDGS). *J. Agric. Food Chem.* **2016**, *64*, 5085–5092. [CrossRef]
15. Wu, L.; Wang, B. Transformation of deoxynivalenol and its acetylated derivatives in Chinese steamed bread making, as affected by pH, yeast, and steaming time. *Food Chem.* **2016**, *202*, 149–155. [CrossRef] [PubMed]
16. Rahimi, E.; Sadeghi, E.; Bohlouli, S.; Karami, F. Fates of deoxynivalenol and deoxynivalenol-3-glucoside from wheat flour to iranian traditional breads. *Food Control* **2018**, *91*, 339–343. [CrossRef]
17. Metwally, M.M.; El-Sayed, A.M.; Abd, A.M.E.; Mehriz, A.; Abu, S.Y.H. Sterigmatocystin: Incidence, fate and production by *A. versicolor* in Ras cheese. *Mycotoxin Res.* **1997**, *13*, 61–66. [CrossRef]
18. Veršilovskis, A.; Bartkevičs, V. Stability of sterigmatocystin during the bread making process and its occurrence in bread from the Latvian market. *Mycotoxin Res.* **2012**, *28*, 123–129. [CrossRef] [PubMed]
19. Vidal, A.; Mengelers, M.; Yang, S.P.; Saeger, S.D.; Boevre, M.D. Mycotoxin biomarkers of exposure: A comprehensive review. *Compr. Rev. Food Sci. Food Saf.* **2018**, *17*, 1127–1155. [CrossRef] [PubMed]
20. Zhao, Y.; Huang, J.; Ma, L.; Wang, F. Development and validation of a simple and fast method for simultaneous determination of aflatoxin B_1 and sterigmatocystin in grains. *Food Chem.* **2017**, *221*, 11–17. [CrossRef] [PubMed]
21. Zhang, J.; Li, M.-M.; Zhang, R.; Jin, N.; Quan, R.; Chen, D.-Y.; Francis, F.; Wang, F.Z.; Kong, Z.Q.; Fan, B. Effect of processing on herbicide residues and metabolite formation during traditional Chinese tofu production. *LWT Food Sci. Technol.* **2020**, *131*, 109707. [CrossRef]
22. Xu, L.W.; Xu, X.X.; Guo, L.L.; Wang, Z.X.; Wu, X.L.; Kuang, H.; Xu, C.L. Potential Environmental Health Risk Analysis of Neonicotinoids and a Synergist. *Environ. Sci. Technol.* **2021**, *55*, 7541–7550. [CrossRef] [PubMed]
23. Xu, L.W.; Guo, L.L.; Wang, Z.X.; Xu, X.X.; Zhang, S.; Wu, X.L.; Kuang, H.; Xu, C.L. Profiling and identification of biocatalyzed transformation of sulfoxaflor in vivo. *Angew. Chem. Int. Ed.* **2020**, *59*, 16218–16224. [CrossRef] [PubMed]
24. Mao, L.; Fang, S.; Zhao, M.; Liu, W.; Jin, H. Effects of bisphenol a and bisphenol s exposure at low doses on the metabolome of adolescent male sprague-dawley rats. *Chem. Res. Toxicol.* **2021**, *34*, 1578–1587. [CrossRef]
25. EC. Commission Regulation No. 657/2002 of 12 August 2002 implementing Council Directive 96/23/EC concerning the performance of analytical methods and the interpretation of results. *Off. J. Eur. Union* **2002**, *L221*, 8–36.
26. EC. Commission Regulation No. 401/2006 of 23 February 2006 laying down the methods of sampling and analysis for the official control of the levels of mycotoxins in foodstuffs. *Off. J. Eur. Union* **2006**, *L70*, 12–34.
27. Zhong, L.; Carere, J.; Mats, L.; Lu, Z.X.; Lu, F.X.; Zhou, T. Formation of glutathione patulin conjugates associated with yeast fermentation contributes to patulin reduction-sciencedirect. *Food Control* **2021**, *123*, 107334. [CrossRef]
28. Sampaolesi, S.; Gamba, R.R.; Antoni, G.; Peláez, Á. Potentiality of yeasts obtained as beer fermentation residue to be used as probiotics. *LWT Food Sci. Technol.* **2019**, *113*, 108251. [CrossRef]
29. Wall-Martínez, H.A.; Pascari, X.; Ramos, A.J.; Marín, S.; Sanchis, V. Fate of the mycotoxins in the wort and yeast during ale and lager fermentation and their evaluation under different technological parameters. *LWT Food Sci. Technol.* **2020**, *132*, 109877. [CrossRef]
30. Huys, G.; Daniel, H.M.; Vuyst, L.D. Taxonomy and Biodiversity of Sourdough Yeasts and Lactic Acid Bacteria. 2013. Available online: https://link.springer.com/chapter/10.1007/978-1-4614-5425-0_5 (accessed on 1 January 2013).
31. Yiannikouris, A.; François, J.; Poughon, L.; Dussap, C.G.; Bertin, G.; Jeminet, G.; Jouany, J.P. Adsorption of Zearalenone by beta-D-glucans in the Saccharomyces cerevisiae cell wall. *J. Food Prot.* **2004**, *67*, 1195–1200. [CrossRef]
32. Kosztik, J.; Mörtl, M.; Székács, A.; Kukolya, J.; Bata-Vidács, I. Aflatoxin b1 and sterigmatocystin binding potential of lactobacilli. *Toxins* **2020**, *12*, 756. [CrossRef]
33. Joannis-Cassan, C.; Tozlovanu, M.; Hadjeba-Medjdoub, K.; Ballet, N.; Pfohl-Leszkowic, A. Binding of zearalenone, aflatoxin B_1, and ochratoxin a by yeast-based products: A method for quantification of adsorption performance. *J. Food Prot.* **2011**, *74*, 1175–1185. [CrossRef] [PubMed]
34. Cole, R.J.; Kirksey, J.W.; Blankenship, B.R. Conversion of aflatoxin B_1 to isomeric hydroxy compounds by Rhizopus subspecies. *J. Agric. Food Chem.* **1972**, *20*, 1100–1102. [CrossRef]
35. Cole, R.J.; Kirksey, J.W. Aflatoxin G_1 metabolism by *Rhizopus* species. *J. Agric. Food Chem.* **1971**, *19*, 222–223. [CrossRef] [PubMed]
36. García-Béjar, B.; Owens, R.A.; Briones, A.; Arévalo-Villena, M. Proteomic profiling and glycomic analysis of the yeast cell wall in strains with Aflatoxin B_1 elimination ability. *Environ. Microbiol.* **2021**, *23*, 5305–5319. [CrossRef] [PubMed]
37. Lu, Y.; Gao, K.; Li, X.N.; Tang, Z.; Xiang, L.; Zhao, H.Z.; Fu, J.J.; Ling Wang, L.; Zhu, N.L.; Cai, Z.W.; et al. Mass Spectrometry-Based Metabolomics Reveals Occupational Exposure to Per- and Polyfluoroalkyl Substances Relates to Oxidative Stress, Fatty Acid beta-Oxidation Disorder, and Kidney Injury in a Manufactory in China. *Environ. Sci. Technol.* **2019**, *53*, 9800–9809. [CrossRef] [PubMed]
38. Lee, D.E.; Lee, S.; Singh, D.; Jang, E.S.; Shin, H.W.; Moon, B.S.; Lee, C.H. Time-resolved comparative metabolomes for koji fermentation with brown-, white-, and giant embryo-rice. *Food Chem.* **2017**, *231*, 258–266. [CrossRef] [PubMed]

biosensors MDPI

Article

Multiplex Digital Quantification of *β*-Lactamase Genes in Antibiotic-Resistant Bacteria by Counting Gold Nanoparticle Labels on Silicon Microchips

Galina V. Presnova [1], Denis E. Presnov [2,3,4], Anna A. Filippova [1], Ilia I. Tsiniaikin [3,4], Mariya M. Ulyashova [1] and Maya Yu. Rubtsova [1,*]

1 Department of Chemistry, Lomonosov Moscow State University, 119991 Moscow, Russia; gkovba@enzyme.chem.msu.ru (G.V.P.); iiffii@mail.ru (A.A.F.); umarusya@mail.ru (M.M.U.)
2 D.V. Skobeltsyn Institute of Nuclear Physics, M.V. Lomonosov Moscow State University, 119991 Moscow, Russia; denis.presnov@physics.msu.ru
3 MSU Quantum Technology Centre, 119991 Moscow, Russia; ii.tcinyaykin@physics.msu.ru
4 Cryoelectronics Lab, Faculty of Physics, M.V. Lomonosov Moscow State University, 119991 Moscow, Russia
* Correspondence: myr@enz.chem.msu.ru

Abstract: Digital quantification based on counting of individual molecules is a promising approach for different biomedical applications due to its enhanced sensitivity. Here, we present a method for the digital detection of nucleic acids (DNA and RNA) on silicon microchips based on the counting of gold nanoparticles (GNPs) in DNA duplexes by scanning electron microscopy (SEM). Biotin-labeled DNA is hybridized with capture oligonucleotide probes immobilized on the microchips. Then biotin is revealed by a streptavidin–GNP conjugate followed by the detection of GNPs. Sharp images of each nanoparticle allow the visualization of hybridization results on a single-molecule level. The technique was shown to provide highly sensitive quantification of both short oligonucleotide and long double-strand DNA sequences up to 800 bp. The lowest limit of detection of 0.04 pM was determined for short 19-mer oligonucleotide. The method's applicability was demonstrated for the multiplex quantification of several β-lactamase genes responsible for the development of bacterial resistance against β-lactam antibiotics. Determination of nucleic acids is effective for both specific DNA in lysates and mRNA in transcripts. The method is also characterized by high selectivity for single-nucleotide polymorphism discrimination. The proposed principle of digital quantification is a perspective for studying the mechanisms of bacterial antibiotic resistance and bacterial response to drugs.

Keywords: high-sensitive digital detection; quantification; functionalized gold nanoparticles; microchips; β-lactamases; antibiotic-resistant bacteria

Citation: Presnova, G.V.; Presnov, D.E.; Filippova, A.A.; Tsiniaikin, I.I.; Ulyashova, M.M.; Rubtsova, M.Y. Multiplex Digital Quantification of β-Lactamase Genes in Antibiotic-Resistant Bacteria by Counting Gold Nanoparticle Labels on Silicon Microchips. *Biosensors* **2022**, *12*, 226. https://doi.org/10.3390/bios12040226

Received: 21 March 2022
Accepted: 7 April 2022
Published: 9 April 2022

1. Introduction

Modern progress in nucleic acids (DNA and RNA) detection techniques has a great impact on biomedical applications in diagnostics of bacterial and viral infections, genetic predispositions to disorders, food safety, anti-bioterrorism, and others [1–3]. Among these tasks, the diagnosis of infections caused by various microorganisms is one of the primary areas in clinical diagnostics. According to the WHO, bacterial infections represent one of the leading causes of death in developing countries [4]. Despite the presence of a variety of antibacterial drugs, the widespread presence of strains resistant to their action limits options for effective therapy [5,6]. Major challenges concern the early diagnosis of the etiology of infections and their susceptibility to antibiotics. Over the last decade, many biosensor platforms have been developed to detect different bacterial and viral markers [7]. Among them, fast and label-free technologies based on opto-photonic, electro-photonic, and Raman spectroscopy have been developed for real-time monitoring and primary typing of bacteria at the level of single cells [8–10]. Novel methods have been proposed for studying biofilms as communities of surface-attached bacterial cells which are more tolerant to antibiotics compared to planktonic cells [11]. Various methods involving arrays of micro- and nanoelectrodes combined with optoelectronic devices; arrays of interdigitated electrodes combined with impedance spectroscopy; and the method of resonant hyperspectral imaging have been reported for monitoring the initial bacterial attachment to control biofilm formation [12–14].

A more detailed characterization of bacteria and their resistance to antibiotics is carried out using molecular typing of nucleic acids. Recent developments involve methods providing multiplexing capability and miniaturization (DNA-based nanobiosensors, biobarcode assays and DNA microchips) [15–17]. The main focus of genetic methods is achieving high sensitivity. For this purpose, single-molecule techniques for detecting molecular interactions are being actively developed [18], and great attention is drawn to digital detection based on single-molecule counting, which is considered one of the most powerful tools for ultrasensitive biomarker identification [19].

Digital counting of single molecules can be carried out using various high-resolution imaging techniques. Among them, scanning electron microscopy (SEM) is being effectively applied for the visualization and characterization of various biological objects from 1 nm in size, involving cells, viruses, bacteriophages, etc. [20,21]. SEM is also actively used to evaluate biosensor surface morphology and characteristics [22–24]. However, its use for the direct analysis of nucleic acids is difficult due to their small size. To overcome this limitation, indirect detection of nanoparticles as molecular tags for biomolecular complexes can be used. Metal nanoparticles, including gold nanoparticles (GNPs), have attracted considerable interest in various molecular nanotechnological applications involving the detection of pathogens [25–27]. Single-particle counting of fluorescent magnetic nanospheres was used for ultrasensitive multiplexed identification of influenza viruses with a detection limit of 0.02 pg/mL [28]. The application of a field-emission SEM was demonstrated for the quantification of single-strand model DNA of 46 bases in the format of a two-step sandwich hybridization assay on plastic microchips by visualization and counting of GNPs [29].

This study aimed to develop multiplex digital quantification of nucleic acids (oligonucleotides, DNA and RNA) by allele-specific hybridization on silicon microchips via the visualization and counting of GNP labels with SEM. To reduce the effect of the label on the formation of DNA duplexes, DNA targets were labeled with biotin, which was then detected in DNA duplexes by a conjugate of streptavidin with GNPs.

β-Lactamase genes were selected as model DNAs, the multiplex determination of which is extremely relevant for clinical diagnosis. β-Lactamases are responsible for the development of bacterial resistance to β-lactam antibiotics (penicillins, cephalosporins, carbapenems, monobactams), which are the most widely used in clinical practice [30,31]. These enzymes are extremely diverse, including serine and metallo-enzymes, and are divided into four molecular classes (A, B, C, and D) according to their structure [32]. Multi- and pan-antibiotic-resistant bacteria carry many resistance genetic determinants, including several types of β-lactamase genes [33], which explain the need to develop multiplex methods to determine them. For this work, we have selected four relevant β-lactamases genes: TEM-1, CTX-M-3, CTX-M-5 (molecular class A), and VIM-1 (molecular class B). Identification was performed in DNA lysates isolated from bacterial cultures as well as in RNA transcripts to explore the possibility of applying this technique to determine β-lactamase gene expression. We also studied the applicability of the developed method for determining single-nucleotide polymorphism (SNP) in genes. In class A β-lactamases, certain SNPs encode substitutions of key amino acids that lead to changes in substrate specificity for cephalosporins. Two of them were selected for identification by digital counting of GNP labels.

We optimized the conditions for DNA hybridization analysis using a highly specific biotin–streptavidin reaction for labeling DNA duplexes with GNPs and the conditions for their visualization and counting by SEM. To determine long DNA molecules, the conditions for biomolecules' unfolding were optimized to increase the efficiency of hybridization. The analytical parameters for the quantification of short oligonucleotides and long DNAs corresponding to the genes of four types of β-lactamases were determined. Total DNA and RNA isolated from bacterial cells were used as samples of nucleic acids. The high specificity of SNP determination in β-lactamase genes was also shown.

2. Materials and Methods

2.1. Materials

All chemicals and organic solvents were of analytical grade. dNTPs, *Taq* DNA polymerase, and dUTP-11-Biotin were obtained from Fermentas (St. Leon-Rot, Germany); 3-glycidopropyl trimethoxysilane (GPTMS), chloroauric acid, mercaptosuccinic acid, dithiothreitol, ethylenediaminetetraacetic acid (EDTA), sodium sulfate, sodium chloride, sodium dodecyl sulfate (SDS), bovine serum albumin (BSA), casein, Tween-20, toluene, methanol, and isopropanol were purchased from Sigma-Aldrich (St. Louis, MO, USA). Sodium citrate dihydrate was purchased from MP Biomedicals (Eschwege, Germany). Streptavidin was obtained from Imtek (Moscow, Russia). All water used in experiments was purified with a Milli-Q system (Millipore, Billerica, MA, USA).

Primers and capture oligonucleotide probes containing a 5′-amino group and 13-mer thymidine spacer were synthesized and purified by Syntol (Moscow, Russia). Their sequences are listed in Table 1.

Table 1. Oligonucleotide probe sequences used in this work.

Probe Name	Sequence, 5′-3′	Length, Bases
Probe A	GATTGGACGAGTCAGGAGC	19
Probe B	TTCTAGACAGCCACTCATA	19
Probe C	Biotin-GCTCCTGACTCGTCCAATC	19
Probe CTX-M	ATATCGCGGTGATCTGGCC	19
Probe TEM	CCAGAAACGCTGGTGAAAGT	20
Probe VIM	GTGGTTGTGCCGTTCAT	17
CTX-M-3_167_A	GACCGTACCGAGACGACGTTAAAC	24
CTX-M-3_167_G	GACCGTACCGAGGCGACGTTAAAC	24
CTX-M-3_167_C	GACCGTACCGAGCCGACGTTAAAC	24
CTX-M-3_167_T	GACCGTACCGAGTCGACGTTAAAC	24
CTX-M-3_240_A	GGCAGCGGTGACTATGGCAC	20
CTX-M-3_240_G	GGCAGCGGTGGCTATGGCAC	20
CTX-M-3_240_C	GGCAGCGGTGCCTATGGCAC	20
CTX-M-3_240_T	GGCAGCGGTGTCTATGGCAC	20

GNPs with a diameter of 25 nm were prepared by the Frens method [34], based on the reduction of chloroauric acid by sodium citrate. A conjugate of streptavidin with GNPs was prepared by covalent binding of streptavidin modified with mercaptosuccinic acid with the colloidal solution of GNPs as described earlier [35].

2.2. Silicon Microchips Fabrication

The surface of the silicon plates was purified with oxygen plasma for 30 min using a reactive-ion etching technique (Alcatel RDE-300, 30 Pa, 25 Wt) and was then treated with 10 mM GPTMS in dry toluene for 12 h at 80 °C, followed by washing for 10 min at 100 °C [36]. Further, samples of modified silicon were stored in highly pure alcohol.

Capture oligonucleotide probes (20 μM) dissolved in 0.25 M Na-phosphate buffer, containing 0.3 M Na_2SO_4, were spotted onto the silicon microchips with XactII Microarrayer

(Lab Next Inc., Glenview, IL, USA). The microchip fabrication scheme and the layout of the capture probes are shown in Figure S1 (Supplementary Materials). Each probe was spotted in triplicate. After the immobilization, the microchips were blocked using a solution of 1% BSA and 1% casein in 10 mM K-phosphate buffer, pH 7.2, containing 0.15 M NaCl.

2.3. Amplification and Labeling of Target DNA

DNA samples isolated from four phenotypically and genotypically characterized control and laboratory strains (*E. coli* EPI300, producing β-lactamase CTX-M-3; *E. coli* TOP10/STY20, producing β-lactamase CTX-M-5; *E. coli* J53, producing β-lactamase TEM-1; *Ps. aeruginosa* VR-143/97, producing β-lactamase VIM-2) were kindly provided by Dr. M.V. Edelstein (Institute of Antimicrobial Chemotherapy of Smolensk State Medical Academy, Smolensk, Russia).

Total bacterial DNA was extracted with the InstaGene Matrix Kit (Bio-Rad, Hercules, CA, USA) and used as a template for PCR amplification. Total RNA was extracted with the «RNA-Extran» Kit (Syntol, Moscow, Russia) by the method of acidic phenol extraction. cDNA synthesis was carried out via the reverse transcription reaction using «OT-1» Kit (Syntol, Moscow, Russia). Samples of cDNA were then used as a template for PCR amplification.

Target DNA was amplified by the PCR and labeled with biotin by introducing dUTP-11-biotin during the reaction. The PCR was carried out in a total volume of 25 μL containing 1.0 μL of bacterial DNA in PCR buffer (2.5 mM $MgCl_2$, 50 mM KCl, 10 mM Tris-HCl, pH 8.3), 0.5 μM of direct and reverse primers, 100 μM (each) dATP, dGTP, and dCTP; 60 μM dTTP; 40 μM dUTP-11-biotin and 2.5 U of Taq DNA polymerase [37]. The amplification was performed in a Mastercycler gradient thermal cycler (Eppendorf AG, Hamburg, Germany) accordingly to the protocol: an initial denaturation step (94 °C for 2 min) was followed by 20 cycles, each consisting of 94 °C for 20 s, 60 °C for 30 s, and 72 °C for 45 s, and a final extension step at 72 °C for 4 min. The quantity and molecular weight of PCR products were estimated by horizontal electrophoresis in 1% agarose gel.

2.4. Hybridization of Target DNA on Silicon Microchips

The microchips were placed in the wells of a 48-well polystyrene plate (Greiner, Germany) and hybridized with biotinylated target oligonucleotide (Probe C) or DNA resuspended in 2xSSPE buffer (0.04 M NaH_2PO_4, 0.6 M NaCl, 4 mM EDTA, 6 mM $MgCl_2$, pH 7.4). Hybridization was carried out in a Thermomixer Comfort (Eppendorf AG, Hamburg, Germany) for 2 h at 45 °C. After the hybridization, the microchips were washed once with 2xSSPE containing 0.2% SDS at 45 °C for 10 min and then twice with PBST (10 mM K-phosphate buffer, pH 7.2, containing 0.15 M NaCl and 0.1% Tween-20) for 10 min at room temperature. To identify SNPs, long target DNA was fragmented prior to the hybridization by DNase I (0.5 mU per 1 ng of DNA) to yield 50 to 150 bp fragment sizes.

2.5. Digital Detection and Data Processing

The microchips were incubated in a solution of the streptavidin–GNPs conjugate in PBST containing 1% BSA at 37 °C for 30 min, washed twice with PBST for 10 min, once with H_2O for 3 min, and dried on air. The GNPs on the silicon surface were detected using a Supra-40 field emission scanning electron microscope (Carl Zeiss AG, Jena, Germany) with an InLens secondary electron detector built in the microscope column. The accelerating voltage and beam current were adopted to achieve the best resolution and contrast for GNP visualization. The number of GNPs on microchip fragments was counted using the Gwyddion software (Czech Metrology Institute, Brno, Czech Republic). The limit of DNA detection (LOD) was calculated as the mean number of GNPs registered for a blank DNA sample (0 pM DNA) plus 2 standard deviations (SD, $n = 10$).

3. Results and Discussion

3.1. Assay Principle, Counting of Gold Nanoparticles on Silicon Microchips

The experimental procedure is shown schematically in Figure 1. Standard silicon wafers with orientation <100> were chosen for the fabrication of microchips. The choice of the material was determined by its compatibility with all microelectronic processes and availability. A semiconductor silicon crystal is characterized by low conductivity, which is sufficient to avoid charge effects during the process of silicon sample imaging in an electron microscope. As a result, images with a resolution up to 1 nm are obtained. Since, under normal conditions, the surface of silicon is covered with a layer of native silicon oxide ~2–3 nm thick, it was treated with oxygen plasma to increase the number of free hydroxyl groups for subsequent chemical modification using GPTMS.

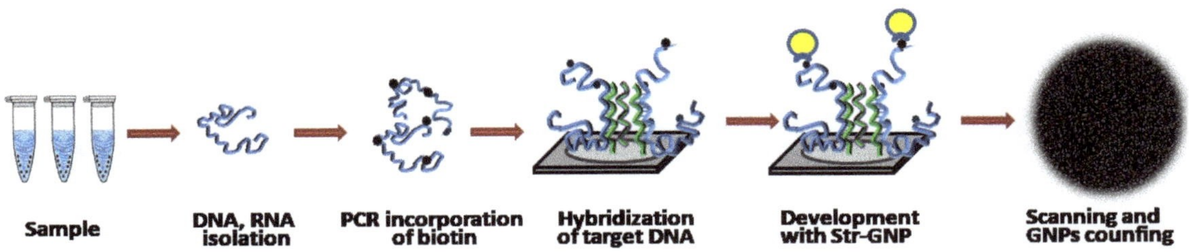

Figure 1. Scheme of successive stages of DNA hybridization analysis on microchips with digital detection of gold nanoparticles.

Capture oligonucleotide probes with 13 T spacer and amino-group at 5′-end were covalently immobilized on modified silicon microchips (Figure S1). Biotin-labeled short oligonucleotides or target DNA were hybridized on the microchips, and then biotin labels in DNA duplexes were developed by a conjugate of streptavidin with GNPs (Figure 2). The conjugate was obtained by covalent attachment of streptavidin modified with mercaptosuccinic acid to GNPs, since this method was shown to improve the signal-to-noise ratio substantially [35]. After washing and drying, SEM images of the microchips were registered and analyzed.

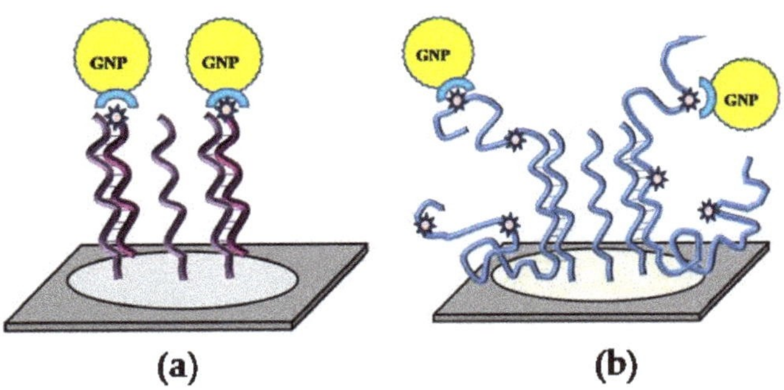

Figure 2. Schematic representation of hybridization of biotin-labeled short target oligonucleotides (**a**) and long target DNA (**b**) on silicon DNA microchips. —capture oligonucleotide probes immobilized on the microchip surface, —a conjugate of streptavidin with gold nanoparticles, —target oligonucleotides labeled with biotin, —target DNA labeled with biotin.

First, we studied the hybridization of the model biotin-labeled 19-mer oligonucleotide (Probe C) with complementary (Probe A) and non-complementary (Probe B) capture oligonucleotide probes immobilized on the microchip. SEM images of specific (Probe A) and control (Probe B) microchip spots after interaction with streptavidin-GNPs conjugate are shown in Figure 3. Nanoparticles were detected clearly by SEM on the spots with complementary Probe A, and their distribution was uniform. Each recorded nanoparticle is referred to as one duplex molecule, as each target oligonucleotide contains one biotin label. So, a sharp image of each nanoparticle allows the visualization of hybridization results on a single-molecule level. A significant difference was observed between the number of GNPs registered on the specific and control microchip spots. Only trace amounts of GNPs were registered on the control spot and the areas outside of the spots with the complementary Probe A as well.

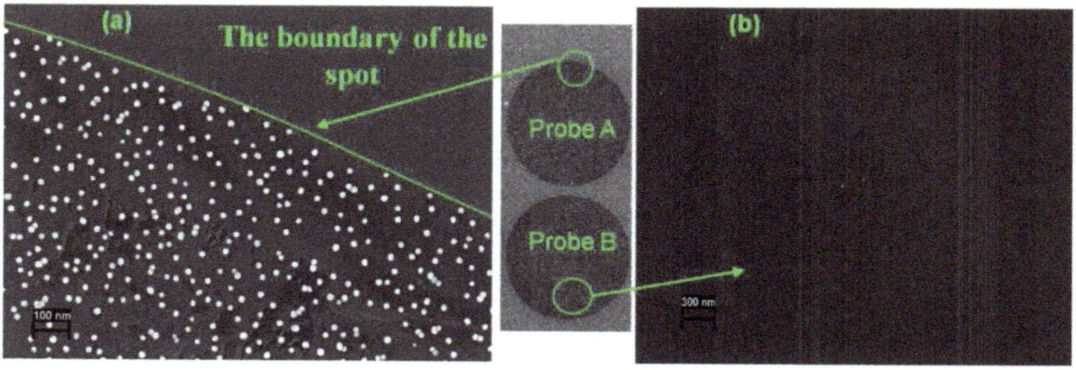

Figure 3. SEM images of the specific microchip spots after hybridization of 300 pM target Probe C with complementary Probe A (**a**) and non-complementary Probe B (**b**). Biotin in DNA duplexes was developed with a streptavidin–GNPs conjugate.

Then, the SEM was applied for the characterization of long DNA hybridization. Two target DNAs were studied: 507 bp DNA, coding a large fragment of bla_{VIM-1}, and 870 bp DNA, coding the whole-length $bla_{CTX-M-3}$. Several capture oligonucleotide probes were spotted on the microchip surface: two specific probes (Probe VIM and Probe CTX-M) and Probe B as a negative control. After hybridization with the target DNA, biotin labels were revealed by the conjugate streptavidin–GNPs as described above, followed by the registration of GNPs with SEM. Similarly to the hybridization of the short oligonucleotide, individual nanoparticles were recorded clearly on SEM images of the spots with complementary duplexes of target DNA and specific immobilized probes. The number of nanoparticles registered on the negative control spots was negligible.

Long DNA biopolymer molecules were present in the hybridization medium in a folded conformation, depending on the ionic strength and salt concentration. Since it may have reduced the hybridization effectiveness, we investigated the addition of magnesium chloride and SDS to promote their unfolding. The effect of $MgCl_2$ and SDS concentrations on the GNP density registered on the specific spots is shown in Figure S2 (Supplementary Materials). The number of particles detected was dependent on the concentration of magnesium ions, and the dependence was bell-shaped (Figure S2a). Although the flatness of the curve maximum depended on the type of gene, we chose a concentration of 6mM magnesium ions, at which the number of nanoparticles on the surface increased as much as possible. The effect of SDS was also characterized by the dependence of a similar form (Figure S2b). The increased number of formed DNA duplexes was observed at a wide range of SDS concentrations, and it did not depend on the DNA type. However, SDS concentrations above 7.5 mM led to an uneven distribution of the detected DNA duplexes; accordingly, the standard deviation of the number of registered particles increased also.

Labeling of long DNA molecules in our method is based on the incorporation of biotin–dUTP during the PCR, and each DNA amplicon contains several biotin labels (data not shown). Their further development with the conjugate of streptavidin with GNPs may result in several nanoparticle labels per DNA strand. Figure 4 shows SEM images obtained after hybridization of the microchips with different concentrations of target DNA, containing $bla_{CTX\text{-}M\text{-}3}$. The image for high DNA concentrations contains closely spaced nanoparticles and their agglomerates (Figure 4a). In contrast the image for low DNA concentration (Figure 4b) is comparable with that for short oligonucleotides (Probe C), containing one biotin molecule (Figure 4c). In this case, we did not observe a large number of neighboring nanoparticles or their conglomerates for comparable concentrations of short and long DNA molecules, and the distribution of GNPs was similar. Most likely, the large size of the nanoparticles represents a steric hindrance to the interaction of several conjugate molecules with one molecule of the target DNA. Thus, it can be assumed that the visualization of GNPs as labels by SEM provides single-molecule analysis for both short and long target DNA molecules.

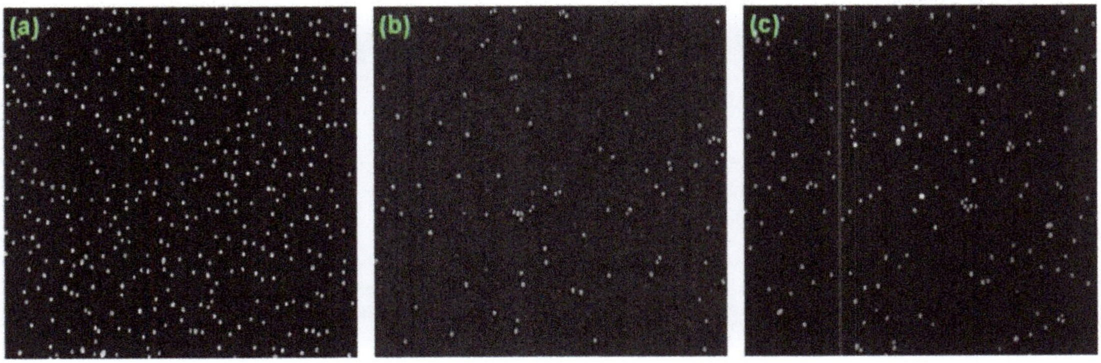

Figure 4. SEM images of the specific spots after hybridization of the microchip with 0.8 nM (**a**) and 60 pM (**b**) target DNA, containing blaCTX-M-3, and 60 pM target oligonucleotide Probe C (**c**).

3.2. Analysis of Cleaved Silicon Microchips

In order to analyze the thickness and structure of biospecific layers formed on the silicon surface after the hybridization and development steps, the microchips spots were split up along the silicon crystallographic axis near the center of the spot, and the sample edge was explored by placing it closer to the objective lens of the microscope. This enabled us to obtain images of the surface at different angles, including the orthogonal edge view. Figure 5 shows the cross-sectional SEM images of the microchips obtained at 80° angles after hybridization of the short oligonucleotide (Probe C) and 870 bp DNA ($bla_{CTX\text{-}M\text{-}3}$). It is clearly seen that the surface location of GNPs is different depending on the DNA length. In the case of short oligonucleotide duplexes, nanoparticles are recessed in the surface layer of the support formed by DNA duplexes, immobilized probes, proteins, and GPTMS (an enlarged image of one nanoparticle is shown in Figure 5b). However, in the case of large DNA duplexes, nanoparticles are located on the silicon surface (an enlarged view of one nanoparticle is shown in Figure 5d).

Differences in the surface localization of nanoparticles when analyzing short and long DNA targets can be explained as follows. The length of the capture probes, including the 13-mer spacer, is about 9–10 nm. The layer of blocking proteins on the microchip surface is non-uniform in height and is about 10–15 nm (this was shown by us earlier in another study). Thus, oligonucleotide duplexes with short capture probes (about 6 nm in size) are located very close to or even inside the protein layer, and the nanoparticles in them appear to be partially immersed in this layer. When the capture probes interact with DNA targets of 500 bp or more, the length of which is more than 170 nm, the size of the duplexes exceeds the protein layer, and GNPs appear on its surface.

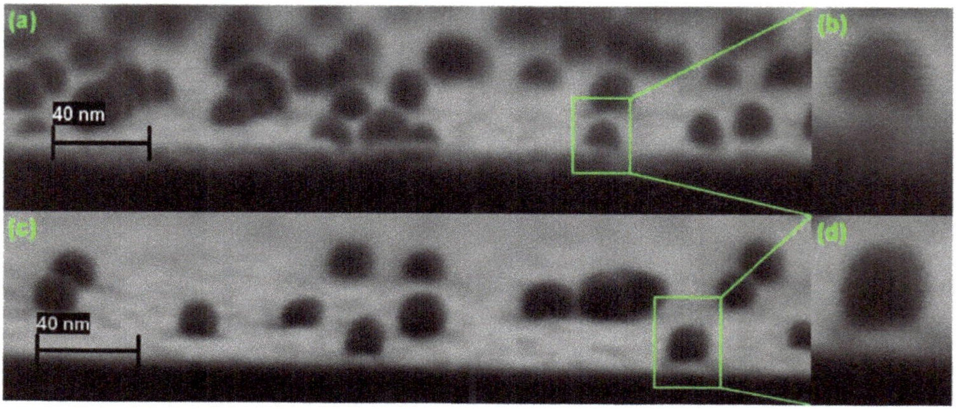

Figure 5. Cross-sectional SEM images of the microchip spots obtained at 80° angle after hybridization of target oligonucleotide or DNA and development of the duplexes with a streptavidin–GNP conjugate: (**a**) 80 pM target Probe C, (**b**) an enlarged image of one GNP of the duplex with short oligonucleotide, (**c**) 50 pM target DNA containing blaCTX-M-3, (**d**) an enlarged image of one GNP of the duplex with blaCTX-M-3.

SEM detects secondary electrons, passing through a thin non-conductive layer; the method is effective in the detection of all GNPs irrespective of their location on the surface of a microchip. In this regard, SEM has certain advantages over atomic force microscopy (AFM), which detects only objects placed above the surface and cannot detect reliably the nanoparticles immersed in the surface layer of the support in whole or in part [38].

3.3. Optimization of Digital Analysis of DNA Duplexes by SEM

Counting GNP labels was used for quantitative analysis of the short target 19-mer oligonucleotide, assuming that one nanoparticle corresponds to one DNA duplex. In order to optimize the technical parameters of SEM for better performance, counting nanoparticles at various microscope magnifications was investigated (Figure S3 in Supplementary Materials). Each magnification corresponds to a fixed frame size: 2.8 μm^2, 11.3 μm^2, and 52 μm^2 for the magnifications of 150 KX, 75 KX, and 35 KX, respectively.

To study the reproducibility, the GNPs were counted on several frames on each spot. Table 2 contains the number of GNPs and variation coefficients (CV) estimated for different quantities of frames, labeled probe concentration, and microscope magnification. The most accurate values of the number of GNPs with low CVs were obtained at frames of larger size; at the same time, the resolution of individual nanoparticles was worse. At low Probe C concentrations, the particles were distributed less uniformly, which led to an increase in the CVs of up to 29% at high magnification. The optimum microscope magnification for the GNP counting was revealed to be 75 KX with CVs not more than 5–8% depending on the probe concentration. An average of three frames is sufficient to count the number of nanoparticles with good reproducibility.

For the characterization of GNP distribution in different spots, it is convenient to estimate the value of the particle density (D), which is defined as the number of nanoparticles registered per area unit:

$$D = Xav/S \tag{1}$$

where X_{av}—the average number of GNPs registered in the frame and S—the area of the frame.

Table 2. Counting average GNP number (X_{av}), coefficients of variation (CV) and particle density (D) on separate frames of specific spots of the microchips by SEM at various microscope magnifications.

MF * / NF **	150 KX			75 KX			35 KX		
	X_{av}	CV%	D	X_{av}	CV%	D	X_{av}	CV%	D
Concentration of Probe C, 5 pM									
3	24	29	8	77	8.9	7	302	4.6	6
6	23	27	8	76	5,9	7	296	3.1	6
9	23	22	8	77	4.9	6	302	2.4	6
Concentration of Probe C, 500 pM									
3	1193	3.5	426	4599	3.7	407	20, 956	3.1	403
6	1190	2.7	425	4567	2.9	412	21,060	2.3	405

*—Microscope magnification; **—Number of frames for GNPs counting.

D values, depending on microscope magnification and probe concentration, are shown in Table 2. This parameter does not depend strongly on microscope magnification and can be used for quantitative analysis of hybridization results.

Optimization of the hybridization protocol and washing procedure has allowed us to significantly reduce non-specific conjugate binding resulting in low values of GNPs recorded on the control spot (noise value). This value did not exceed 1–2 GNPs per frame at a magnification of 75KH.

3.4. *Digital Quantification of Nucleic Acids on Silicon Microchips*

To study the feasibility of GNP digital counting by SEM for the quantitative detection of target DNA, various concentrations of short single-strand oligonucleotide (Probe C) and fragmented long double-strand DNA, containing genes of β-lactamases ($bla_{TEM\text{-}1}$, $bla_{CTX\text{-}M\text{-}3}$, $bla_{CTX\text{-}M\text{-}5}$, $bla_{VIM\text{-}2}$), were hybridized on the microchips. It was found that the density of GNPs increased with the target DNA concentration (Figure 6). The lower LOD of 0.04 pM was determined for the short 19-mer oligonucleotide (Figure 6a,c). It is one order of magnitude higher compared to previously published data for digital detection of the GNPs in sandwich hybridization of short oligonucleotides with the capture probes, which in turn was shown to be three orders of magnitude more sensitive than fluorescence detection [13]. This demonstrates the advantages of incorporating nanoparticles into DNA duplexes using a highly specific biotin–streptavidin interaction, which reduces steric hindrances. The use of streptavidin conjugated with nanoparticles makes it possible to use one universal approach for multi analysis, which was demonstrated by the determination of several β-lactamase genes on a single microchip (Figure 6b,d).

Among long double-strand DNA targets, the lower LOD of 0.3 pM was determined for $bla_{CTX\text{-}M\text{-}5}$ and LODs of 0.5 pM, 5.0 pM, and 16 pM were determined for $bla_{CTX\text{-}M\text{-}5}$, $bla_{VIM\text{-}2}$, and $bla_{TEM\text{-}1}$, respectively. The variation in the LODs for different genes is mainly due to differences in the properties of the capture probes, leading to various hybridization stringencies with the target DNAs.

Table 3 presents the data on the quantitative determination of DNA by various methods proposed recently. Most of them use GNPs as a label for DNA duplexes, which are incorporated either directly into oligonucleotide probes or using the streptavidin–biotin interaction. Several methods are based on the application of other labels (magnetic nanoparticles and enzymes). Among them, there are methods for determining both types of DNA: short (no more than 50 bases) and long (more than 150 bases). Our method with GNP digital detection is more sensitive in determining short oligonucleotide targets when compared to standard target concentration determination. The indirect method of incorporating GNPs into duplexes via the streptavidin-biotin interaction reduces the LOD by one order of magnitude compared to the digital detection of nanoparticles incorporated directly into detecting oligonucleotide probes.

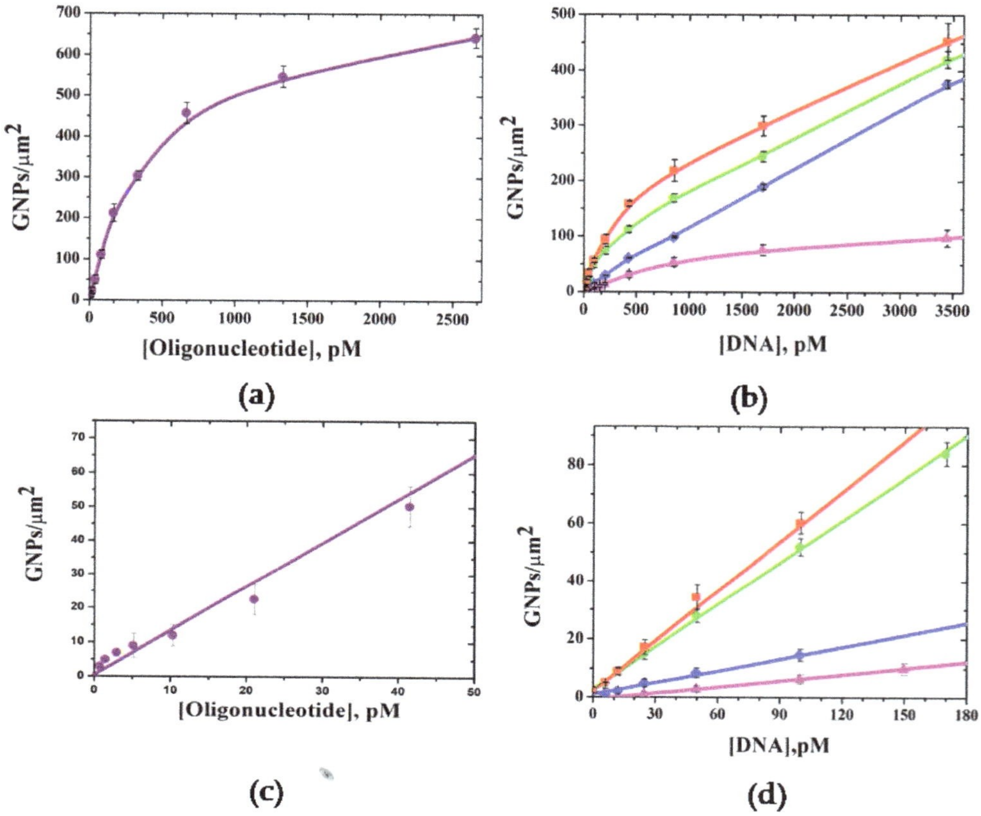

Figure 6. Calibration curves for the determination of short oligonucleotide Probe C (**a**) and long DNAs (**b**) containing blaTEM-1(▲), blaCTX-M-3(•), blaVIM-2 (♦), blaCTX-M-5 (■), on silicon microchips with digital detection. (**c**,**d**)—enlarged sections of the calibration curves (**a**,**b**) for low concentrations of target DNAs.

All methods for determining longer DNA fragments show worse sensitivity. Presumably, this is due to the steric problems of DNA target interaction with capture and detection probes. Our method is characterized by sensitivity, comparable to the best analytical performance of other methods, while it should be noted that the longest DNA fragment was determined in our study. When compared to hybridization analysis of the same genes based on the same capture probes on DNA microchips with colorimetric detection, the LOD of our method is more than three orders of magnitude lower; also, the dynamic range is 1–2 orders of magnitude wider [42]. Comparing the analytical characteristics of the determination of short and long DNA fragments, it can be assumed that the lengthening of the capture probes may contribute to an increase in the sensitivity of the method.

The size of the detected genes is important for increasing the specificity of the analysis and reducing false positives in the diagnosis of infectious diseases. The method of real-time PCR widely used for these purposes is based on the amplification of relatively short fragments of genes. Even though our method is longer in time and more complex, one of the advantages concerns the amplification of long DNA, which was shown by the amplification of full-size β-lactamase genes.

Table 3. Analytical performances of different methods developed for quantitative determination of DNA.

Method/Detection Principle	DNA Target Size	Label	Limit of Detection	Reference
Determination of short oligonucleotides				
Hybridization on silicon microchips/counting of GNP labels	Oligonucleotide (19 b)	Indirect labeling of DNA duplexes with GNPs via streptavidin-biotin interaction	0.04 pM	This work
Sandwich hybridization on plastic microchips/counting of GNP labels	Oligonucleotide (46 b)	Direct labeling of detection oligonucleotide probe with GNPs	1 pM	[18]
Sandwich hybridization on the microelectrodes/detection of conductivity	Oligonucleotide (27 b)	Direct labeling of detection oligonucleotide probe with GNPs, silver enhancement	0.5 pM (500 fM)	[39]
DNA hybridization with PNA probes/colorimetric detection of GNPs	Oligonucleotide (18 b)	Electrostatic interaction of DNA duplexes with GNPs, gold or silver enhancement	10 pM	[40]
Hybridization on DNA microarrays/scanometric detection with optical scanner)	Oligonucleotide (21 b)	Labeling of the ds-DNA with DNA intercalator (daunorubicin) conjugated to GNPs, enhancement of the GNPs	10 pM	[41]
Determination of long DNA				
Hybridization on silicon microchips/counting of GNP labels	Full-size gene of β-lactamase *bla*$_{CTX-M-5}$ (870 bp)	Indirect labeling of DNA duplexes with GNPs via streptavidin-biotin interaction	0.3 pM	This work
Hybridization on membrane microchips/colorimetric detection	Full-size gene of β-lactamase *bla*$_{CTX-M-5}$ (870 bp)	Indirect labeling of DNA duplexes with horseradish peroxidase via streptavidin-biotin interaction	0.71 nM (0.40 ng μL^{-1})	[42]
Hybridization on biosensor array/detection of magnetoresistive ratio	Synthetic ssDNA (167 p) GAPDH gene (the fragment size is not specified)	Indirect labeling of DNA duplexes with magnetic NPs via streptavidin-biotin interaction	39 pM 0.1–1 pM depending on amount of amplification cycles	[43,44]
Sandwich hybridization on glass microchips/optical detection	Fragment of Hepatitis E virus RNA (DNA target of 500 bp)	Direct labeling of detection oligonucleotide probe with nano-gold, silver enhancement	0.1 pM (100 fM)	[45]

The method of digital nucleic acid quantification by SEM was also applied for the detection of expressed β-lactamase genes in RNA transcripts isolated from cultivated bacteria. Sample preparation for the hybridization stage involved total RNA isolation from *E. coli* cells — producers of recombinant β-lactamase TEM-1, reverse transcription of specific mRNA β-lactamase to cDNA followed by PCR similar to the preparation of target DNA from DNA lysates. Several samples of total RNA isolated from bacterial cultures containing different cell concentrations (from 2.5×10^4 to 6.2×10^7 CFU/mL) were prepared. From them, samples of labeled target DNA, containing bla_{TEM-1}, were obtained in sequential reverse transcription and PCR reactions. Calibration curves for the quantitative determination of the bla_{TEM-1} in RNA transcripts isolated from *E. coli* cells by sequential reverse transcription and PCR followed by hybridization and digital detection of GNP tags are shown in Figure 7. All calculated values of GNP density corresponded well with the calibration curve for the quantification of the bla_{TEM-1} gene amplified from DNA lysates. Thus, the developed hybridization analysis of specific mRNA on the DNA microchips with digital detection can be used for the identification of expressed β-lactamase genes in antibiotic-resistant bacteria. Determining the expression levels of beta-lactamase genes is important when studying the mechanisms of multi-resistance and the response of bacteria to the action of drugs.

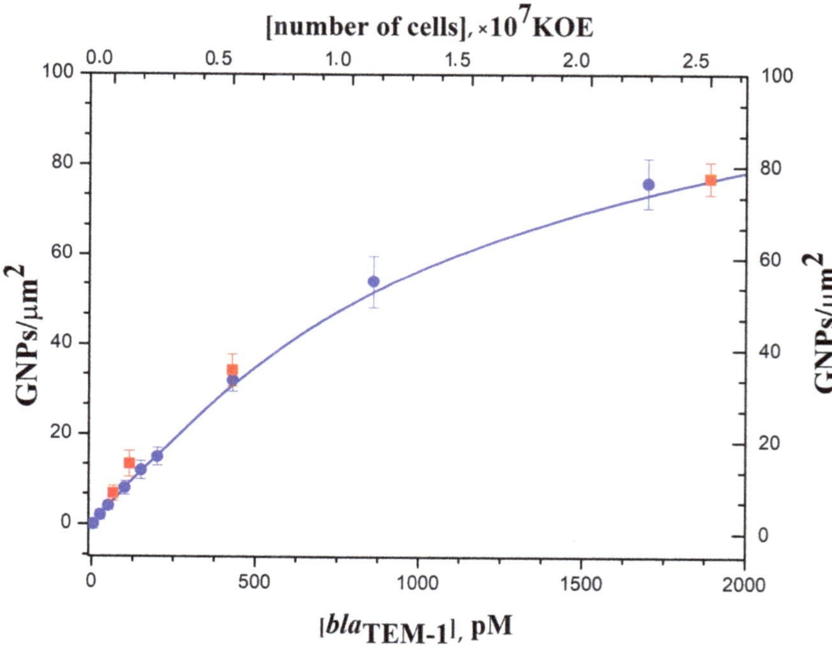

Figure 7. Determination of specific mRNA of the blaTEM-1 in RNA transcripts isolated from *E. coli* cells by sequential reverse transcription and PCR followed by hybridization and digital detection of GNPs (red). Calibration curve for the determination of blaTEM-1 (blue).

3.5. Digital Identification of Single Nucleotide Polymorphism

Determination of nucleic acids for diagnostic purposes often includes the identification of SNPs responsible for mutations in encoded proteins and enzymes [46]. This task is extremely important in the characterization of bacteria resistant to β-lactam antibiotics since the key substitutions in bacterial β-lactamases result in the enhanced substrate specificity of the enzymes (extended-spectrum β-lactamases, ESBLs) [31,32]. We demonstrated the capabilities of the microchips with SEM-based digital detection in identifying the SNPs encoding two key mutations of amino acid residues (P167 and D240) in β-lactamases belonging to the CTX-M-1 subcluster. Identification is based on an allele-specific hybridization approach, which exploits the differences in thermal stability of a fully complementary DNA duplex and duplexes with a single mismatch. For the detection of the SNP, a set of four probes with identical sequences except for the nucleotide at the central (or close to central) base, which was A, T, G, or C, was used (Table 1). After hybridization with target DNA and digital detection, the density of GNPs was determined for each probe of the SNP set. The probe showing the highest signal value in the result of complementary hybridization was considered the perfect match (PM); the remaining ones with lower signal values were regarded as mismatches (MM). To investigate the discriminatory power of the method, the relative GNP densities were defined by normalizing to the net GNP density of the PM probe. Figure 8 represents relative GNP densities determined after the hybridization of 20 ng $bla_{CTX\text{-}M\text{-}3}$ target DNA with the probe sets for identification of SNPs encoding amino acid replacement at positions 167 (a) and 240 (b) in β-lactamases of the CTX-M-1 subcluster. Nucleotide triplets in the antisense strain of $bla_{CTX\text{-}M\text{-}3}$ contain dGTP and dTTP in the positions, corresponding to amino acids P167 and D240. They were identified successfully on the microchips by complementary hybridization with the capture probes CTX-M-3_167_C and CTX-M-3_240_A, which correspond to the sense strand and include dCTP and dATP in the central position (Table 1). In both cases, identification of nucleotide type was characterized by high discriminatory power, and mismatch signals, which show the level of nonspecific hybridization, were no more than 15% of the PM signals.

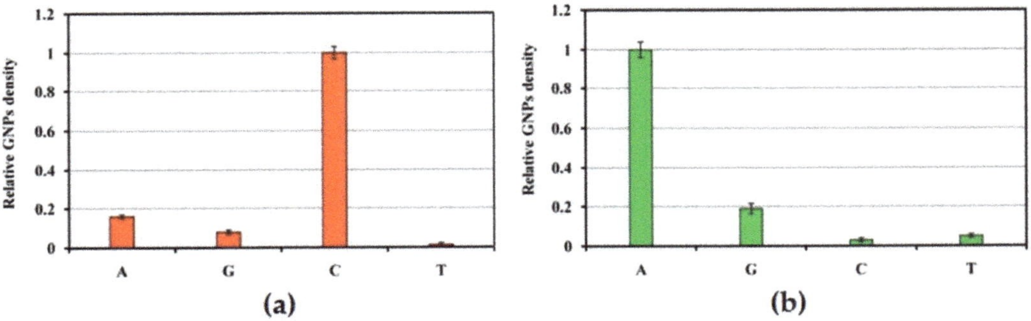

Figure 8. Relative GNP density determined after the hybridization of 20 ng blaCTX-M-3 target DNA with the probe sets for identification of SNPs encoding amino acid replacement at position 167 (**a**) and 240 (**b**) in β-lactamases of the CTX-M-1 subcluster. A, G, C, or T indicates a central nucleotide in the probe.

4. Conclusions

In this study, we developed a new technique for nucleic acids quantification based on digital counting of GNP labels by high-resolution SEM. It combines the advantages of nanoparticles as labels for biospecific interactions and the ability of SEM to produce a sharp image of the microchip surface with distinguishable GNPs. GNP counting is used for the digital detection of single DNA duplexes formed on the surface of a microchip. Immobilization of capture probes of different specificity on a microchip enables multiplex determination. Indirect labeling of DNA duplexes via biotin-streptavidin interaction minimizes steric hindrances and makes it possible to analyze both short oligonucleotides and long DNA molecules of several hundred bases in length. The method is characterized by enhanced sensitivity and high signal-to-noise values.

The method of nucleic acid analysis on silicon microchips is characterized by high stability. It is ensured by the high stability of microchips with covalently immobilized oligonucleotide probes (they retain their properties for more than one year when stored at RT) and high stability of DNA duplexes and duplexes after the introduction of GNPs and washing, since streptavidin conjugate with nanoparticles was obtained by covalent cross-linking.

The method showed high sensitivity in determining both short oligonucleotides and long DNA molecules up to 800 bp. We have shown its application in the identification of full-size genes of β-lactamases, bacterial enzymes responsible for the development of resistance to β-lactam antibiotics. Such an approach also provides highly selective identification of SNPs, which is important for the identification of extended-spectrum β-lactamases. When combined with a reaction of reverse transcription, the method allows the determination of specific mRNAs of β-lactamases in RNA transcripts.

We believe that our technique may be applicable for fundamental studies of antibiotic resistance mechanisms, the development of multi-resistance in bacteria, their response to the action of drugs, and monitoring of β-lactamase genes.

Supplementary Materials: The following supporting information can be downloaded at: https://www.mdpi.com/article/10.3390/bios12040226/s1, Figure S1: The microchip fabrication scheme and the layout of the capture probes; Figure S2: Effect of Mg^{2+} and SDS concentrations on the GNPs' density on specific microchip spots corresponding to immobilized Probe A and Probe CTX-M after the hybridization of $bla_{CTX\text{-}M\text{-}3}$ and $bla_{CTX\text{-}M\text{-}5}$; Figure S3: SEM images of the specific microchip spots after hybridization with 500 pM short target Probe C obtained at different magnifications of the microscope.

Author Contributions: Conceptualization, G.V.P. and M.Y.R.; methodology, G.V.P. and D.E.P.; investigation, validation and formal analysis, G.V.P., A.A.F., I.I.T. and M.M.U.; writing—original draft preparation, G.V.P. and M.Y.R.; writing—review and editing, G.V.P., D.E.P., M.M.U. and M.Y.R.; visualization, G.V.P., D.E.P. and M.M.U.; supervision, M.Y.R. All authors have read and agreed to the published version of the manuscript.

Funding: The work was supported by the Interdisciplinary Scientific and Educational School of Moscow State University "Photonic and Quantum Technologies. Digital Medicine" and by M.V. Lomonosov Moscow State University state task (Government Grant 121041500039-8). The research infrastructure of the "Educational and Methodical Center of Lithography and Microscopy" of M.V. Lomonosov Moscow State University was used.

Institutional Review Board Statement: Not applicable.

Informed Consent Statement: Not applicable.

Data Availability Statement: Research data are not shared.

Conflicts of Interest: The authors declare no conflict of interest. The funders had no role in the design of the study; in the collection, analyses, or interpretation of data; in the writing of the manuscript, or in the decision to publish the results.

Abbreviations

AFM: atomic force microscopy; BSA, bovine serum albumin; GNP, gold nanoparticle; GPTMS, 3-glycidopropyl trimethoxysilane; EDTA, ethylenediaminetetraacetic acid; SEM, scanning electron microscopy; SNP, single nucleotide polymorphism; SDS, sodium dodecyl sulfate.

References

1. Du, Y.; Dong, S. Nucleic Acid Biosensors: Recent Advances and Perspectives. *Anal. Chem.* **2017**, *89*, 189–215. [CrossRef] [PubMed]
2. Zhou, H.; Liu, J.; Xu, J.-J.; Zhang, S.; Chen, H.-Y. Advances in DNA/RNA detection using nanotechnology. *Adv. Clin. Chem.* **2019**, *91*, 31–98. [CrossRef] [PubMed]
3. Scheler, O.; Glynn, B.; Kurg, A. Nucleic acid detection technologies and marker molecules in bacterial diagnostics. *Expert Rev. Mol. Diagn.* **2014**, *14*, 489–500. [CrossRef] [PubMed]
4. World Health Organization. *World Health Statistics 2020*; World Health Organization: Geneva, Switzerland, 2020. Available online: https://apps.who.int/iris/bitstream/handle/10665/332070/9789240005105-eng.pdf (accessed on 2 April 2022).
5. Eichenberger, E.M.; Thaden, J.T. Epidemiology and Mechanisms of Resistance of Extensively Drug Resistant Gram-Negative Bacteria. *Antibiotics* **2019**, *8*, 37. [CrossRef]
6. Chokshi, A.; Sifri, Z.; Cennimo, D.; Horng, H. Global contributors to antibiotic resistance. *J. Glob. Infect. Dis.* **2019**, *11*, 36–42. [CrossRef] [PubMed]
7. Qureshi, A.; Niazi, J.H. Biosensors for detecting viral and bacterial infections using host biomarkers: A review. *Analyst* **2021**, *145*, 7825–7848. [CrossRef] [PubMed]
8. Therisod, R.; Tardif, M.; Marcoux, P.R.; Picard, E.; Jager, J.-B.; Hadji, E.; Peyrade, D.; Houdré, R. Gram-type differentiation of bacteria with 2D hollow photonic crystal cavities. *Appl. Phys. Lett.* **2018**, *113*, 111101. [CrossRef]

9. Conteduca, D.; Brunetti, G.; Dell'Olio, F.; Armenise, M.N.; Krauss, T.F.; Cimenelli, C. Monitoring of individual bacteria using electro-photonic traps. *Biomed. Opt. Express* **2019**, *10*, 3463–3471. [CrossRef] [PubMed]
10. Baron, V.O.; Chen, M.; Clark, S.O.; Williams, A.; Dholakia, K.; Gillespie, S.H. Detecting Phenotypically Resistant Mycobacterium tuberculosis Using Wavelength Modulated Raman Spectroscopy. *Methods Mol. Biol.* **2018**, *1736*, 41–50. [CrossRef] [PubMed]
11. Nickel, J.C.; Ruseska, I.; Wright, J.B.; Costerton, J.W. Tobramycin resistance of Pseudomonas aeruginosa cells growing as a biofilm on urinary catheter material. *Antimicrob. Agents Chemother.* **1985**, *27*, 619–624. [CrossRef]
12. Brunetti, G.; Conteduca, D.; Armenise, M.N.; Ciminelli, C. Novel Micro-Nano Optoelectronic Biosensor for Label-Free Real-Time Biofilm Monitoring. *Biosensors* **2021**, *11*, 361. [CrossRef]
13. Kim, S.; Yu, G.; Kim, T.; Shin, K.; Yoon, J. Rapid bacterial detection with an interdigitated array electrode by electrochemical impedance spectroscopy. *Electrochim. Acta* **2012**, *82*, 126–131. [CrossRef]
14. Wang, Y.; Reardon, C.P.; Read, N.; Thorpe, S.; Evans, A.; Todd, N.; Van Der Woude, M.; Krauss, T.F. Attachment and antibiotic response of early-stage biofilms studied using resonant hyperspectral imaging. *NPJ Biofilms Microbiomes* **2020**, *6*, 57. [CrossRef]
15. Ferrier, D.C.; Shaver, M.P.; Hands, P.J.W. Micro- and nano-structure based oligonucleotide sensors. *Biosens. Bioelectron.* **2015**, *68*, 798–810. [CrossRef] [PubMed]
16. Bedin, C.; Crotti, S.; Tasciotti, E.; Agostini, M. Diagnostic devices for circulating biomarkers detection and quantification. *Curr. Med. Chem.* **2018**, *25*, 4304–4327. [CrossRef] [PubMed]
17. Duncan, R.; Gaspar, R. Nanomedicine(s) under the microscope. *Mol. Pharm.* **2011**, *8*, 2101–2141. [CrossRef]
18. Wu, Y.; Bennett, D.; Tilley, R.D.; Gooding, J.J. How nanoparticles transform single molecule measurements into quantitative sensors. *Adv. Mater.* **2019**, *32*, e1904339. [CrossRef]
19. Cretich, M.; Daaboul, G.G.; Sola, L.; Unlu, M.S.; Chiari, M. Digital detection of biomarkers assisted by nanoparticles: Application to diagnostics. *Trends Biotechnol.* **2015**, *33*, 343–351. [CrossRef]
20. Kataoka, M.; Ishida, K.; Ogasawara, K.; Nozaki, T.; Satoh, Y.I.; Sata, T.; Sato, Y.; Hasegawa, H.; Nakajima, N. Serial section array scanning electron microscopy analysis of cells from lung autopsy specimens following fatal A/H1N1 2009 pandemic influenza virus infection. *J. Virol.* **2019**, *93*, e00644-19. [CrossRef]
21. Gopinath, S.C.B.; Awazu, K.; Fujimaki, M.; Shimizu, K.; Shima, T. Observations of immuno-gold conjugates on influenza viruses using waveguide-mode sensors. *PLoS ONE* **2013**, *8*, e69121. [CrossRef]
22. Jiang, C.; Lu, H.; Zhang, H.; Shen, Y.; Lu, Y. Recent advances on in situ SEM mechanical and electrical characterization of low-dimensional nanomaterials. *Scanning* **2017**, *2017*, 1985149. [CrossRef] [PubMed]
23. Gasparotto, G.; Costa, J.P.C.; Costa, P.I.; Zaghete, M.A.; Mazon, T. Electrochemical immunosensor based on ZnO nanorods-Au nanoparticles nanohybrids for ovarian cancer antigen CA-125 detection. *Mater. Sci. Eng. C Mater. Biol. Appl.* **2017**, *76*, 1240–1247. [CrossRef] [PubMed]
24. Moschou, D.; Trantidou, T.; Regoutz, A.; Carta, D.; Morgan, H.; Prodromakis, T. Surface and electrical characterization of Ag/AgCl pseudo-reference electrodes manufactured with commercially available PCB technologies. *Sensors* **2015**, *15*, 18102–18113. [CrossRef]
25. Yuan, P.; Ding, X.; Yang, Y.Y.; Xu, Q.-H. Metal nanoparticles for diagnosis and therapy of bacterial infection. *Adv. Healthc. Mater.* **2018**, *7*, e1701392. [CrossRef]
26. Mocan, T.; Matea, C.T.; Pop, T.; Mosteanu, O.; Buzoianu, A.D.; Puia, C.; Iancu, C.; Mocan, L. Development of nanoparticle based optical sensors for pathogenic bacteria detection. *J. Nanobiotechnol.* **2017**, *15*, 25. [CrossRef] [PubMed]
27. Lv, M.M.; Fan, S.F.; Wang, Q.L.; Lv, Q.Y.; Song, X.; Cui, H.F. An enzyme-free electrochemical sandwich DNA assay based on the use of hybridization chain reaction and gold nanoparticles: Application to the determination of the DNA of *Helicobacter pylori*. *Microchim. Acta* **2019**, *187*, 73. [CrossRef] [PubMed]
28. Wu, Y.; Tilley, R.D.; Gooding, J.J. The challenges and solutions in developing ultrasensitive biosensors. *J. Am. Chem. Soc.* **2019**, *141*, 1162–1170. [CrossRef] [PubMed]
29. Kim, H.; Takei, H.; Yasuda, K. Quantitative evaluation of a gold-nanoparticle labeling method for detecting target DNAs on DNA microarrays. *Sens. Actuators B Chem.* **2010**, *144*, 6–10. [CrossRef]
30. Hughes, D.; Andersson, D.I. Evolutionary trajectories to antibiotic resistance. *Annu. Rev. Microbiol.* **2017**, *71*, 579–596. [CrossRef] [PubMed]
31. Bush, K. Past and present perspectives of β-lactamases. *Antimicrob. Agents Chemother.* **2018**, *62*, e01076-18. [CrossRef] [PubMed]
32. Philippon, A.; Slama, P.; Dény, P.; Labia, R. A structure-based classification of class A β-lactamases, a broadly diverse family of enzymes. *Clin. Microbiol. Rev.* **2016**, *29*, 29–57. [CrossRef]
33. Avershina, E.; Shapovalova, V.; Shipulin, G. Fighting Antibiotic Resistance in Hospital-Acquired Infections: Current State and Emerging Technologies in Disease Prevention, Diagnostics and Therapy. *Front. Microbiol.* **2021**, *12*, 707330. [CrossRef]
34. Frens, G. Controlled nucleation for the regulation of the particle size in monodisperse gold suspensions. *Nat. Phys. Sci.* **1973**, *241*, 20–22. [CrossRef]
35. Presnova, G.V.; Rubtsova, M.Y.; Presnov, D.E.; Grigorenko, V.G.; Yaminsky, I.V.; Egorov, A.M. Streptavidin conjugates with gold nanoparticles for visualization of single DNA interactions on the silicon surface. *Biochem. Suppl. Ser. B Biomed. Chem.* **2014**, *8*, 164–167. [CrossRef]

36. Lamture, J.B.; Beattie, K.L.; Burke, B.E.; Eggers, M.D.; Ehrlich, D.J.; Fowler, R.; Hollis, M.A.; Kosicki, B.B.; Reich, R.K.; Smith, S.R. Direct detection of nucleic acid hybridization on the surface of a charge coupled device. *Nucleic Acids Res.* **1994**, *22*, 2121–2125. [CrossRef] [PubMed]
37. Rubtsova, M.Y.; Ulyashova, M.M.; Edelstein, M.V.; Egorov, A.M. Oligonucleotide microarrays with horseradish peroxidase-based detection for the identification of extended-spectrum β-lactamases. *Biosens. Bioelectron.* **2010**, *26*, 1252–1260. [CrossRef]
38. Dubrovin, E.V.; Presnova, G.V.; Rubtsova, M.Y.; Egorov, A.M.; Grigorenko, V.G.; Yaminsky, I.V. The use of atomic force microscopy for 3D analysis of nucleic acid hybridization on microarrays. *Acta Nat.* **2015**, *7*, 108–114. [CrossRef]
39. Park, S.-J.; Taton, T.A.; Mirkin, C.A. Array-Based Electrical Detection of DNA with Nanoparticle Probes. *Science* **2002**, *295*, 1503–1506. [CrossRef]
40. Kim, S.K.; Cho, H.; Jeong, J.; Kwon, J.N.; Jung, Y.; Chung, B.H. Label-free and naked eye detection of PNA/DNA hybridization using enhancement of gold nanoparticles. *Chem. Commun.* **2010**, *46*, 3315–3317. [CrossRef] [PubMed]
41. Cho, H.; Jung, J.; Chung, B.H. Scanometric analysis of DNA microarrays using DNA intercalator-conjugated gold nanoparticles. *Chem. Commun.* **2012**, *48*, 7601–7603. [CrossRef] [PubMed]
42. Ulyashova, M.M.; Rubtsova, M.Y.; Egorov, A.M. Colorimetric detection of immobilized horseradish peroxidase based on the co-oxidation of benzidine derivatives and 4-chloro-1-naphthol. *Russ. Chem. Bull.* **2011**, *60*, 656–661. [CrossRef]
43. Rizzi, G.; Lee, J.-R.; Guldberg, P.; Dufva, M.; Wang, S.X.; Hansen, M.F. Denaturation strategies for detection of double stranded PCR products on GMR magnetic biosensor array. *Biosens. Bioelectron.* **2017**, *93*, 155–160. [CrossRef] [PubMed]
44. Ravi, N.; Rizzi, G.; Chang, S.E.; Cheung, P.; Utz, P.J.; Wang, S.H. Quantification of cDNA on GMR biosensor array towards point-of-care gene expression analysis. *Biosens. Bioelectron.* **2019**, *130*, 338–343. [CrossRef] [PubMed]
45. Liu, H.H.; Cao, X.; Yang, Y.; Liu, M.G.; Wang, Y.F. Array-based nano-amplification technique was applied in detection of hepatitis E virus. *J. Biochem. Mol. Biol.* **2006**, *39*, 247–252. [CrossRef] [PubMed]
46. Kim, S.; Misra, A. SNP genotyping: Technologies and biomedical applications. *Annu. Rev. Biomed. Eng.* **2007**, *9*, 289–320. [CrossRef] [PubMed]

biosensors

MDPI

Article

Silent Antibodies Start Talking: Enhanced Lateral Flow Serodiagnosis with Two-Stage Incorporation of Labels into Immune Complexes

Dmitriy V. Sotnikov [1,*], Nadezhda A. Byzova [1], Anatoly V. Zherdev [1], Youchun Xu [2] and Boris B. Dzantiev [1]

[1] A.N. Bach Institute of Biochemistry, Research Center of Biotechnology of the Russian Academy of Sciences, Leninsky Prospect 33, 119071 Moscow, Russia; nbyzova@mail.ru (N.A.B.); zherdev@inbi.ras.ru (A.V.Z.); dzantiev@inbi.ras.ru (B.B.D.)

[2] State Key Laboratory of Membrane Biology, Department of Biomedical Engineering, School of Medicine, Tsinghua University, Beijing 100084, China; xyc2012@mail.tsinghua.edu.cn

* Correspondence: sotnikov-d-i@mail.ru; Tel.: +7-495-954-3142

Abstract: The presence of pathogen-specific antibodies in the blood is widely controlled by a serodiagnostic technique based on the lateral flow immunoassay (LFIA). However, its common one-stage format with an antigen immobilized in the binding zone of a test strip and a nanodispersed label conjugated with immunoglobulin-binding proteins is associated with risks of very low analytical signals. In this study, the first stage of the immunochromatographic serodiagnosis was carried out in its traditional format using a conjugate of gold nanoparticles with staphylococcal immunoglobulin-binding protein A and an antigen immobilized on a working membrane. At the second stage, a labeled immunoglobulin-binding protein was added, which enhanced the coloration of the bound immune complexes. The use of two separated steps, binding of specific antibodies, and further coloration of the formed complexes, allowed for a significant reduction of the influence of non-specific immunoglobulins on the assay results. The proposed approach was applied for the serodiagnosis using a recombinant RBD protein of SARS-CoV-2. As a result, an increase in the intensity of test zone coloration by more than two orders of magnitude was demonstrated, which enabled the significant reduction of false-negative results. The diagnostic sensitivity of the LFIA was 62.5% for the common format and 100% for the enhanced format. Moreover, the diagnostic specificity of both variants was 100%.

Keywords: immunochromatography; test strips; RBD protein; COVID-19; coronavirus

Citation: Sotnikov, D.V.; Byzova, N.A.; Zherdev, A.V.; Xu, Y.; Dzantiev, B.B. Silent Antibodies Start Talking: Enhanced Lateral Flow Serodiagnosis with Two-Stage Incorporation of Labels into Immune Complexes. *Biosensors* **2022**, *12*, 434. https://doi.org/10.3390/bios12070434

Received: 8 April 2022
Accepted: 15 June 2022
Published: 21 June 2022

1. Introduction

Detection of pathogen-specific antibodies in the blood (serodiagnosis) plays a key role in the diagnosis of many infectious diseases. The advantages of this method are the selection of a specific biomatrix for testing (in contrast with the determination of viruses or bacteria in different organs and tissues), and the similarity of the assay protocols for various infections [1]. In medical diagnostics, microplate-based enzyme-linked immunosorbent assays (ELISAs), agglutination tests, and other immunoassay formats are successfully used for this purpose [2–4]. The current trend in the development of diagnostic tools is simplification and acceleration of the testing procedure. The lateral flow immunoassay (LFIA) is one of the most relevant immunoassay formats, with the availability of industrial facilities for the wide manufacturing of diagnostic kits. Multimembrane composites (test strips) with preliminary applied immune reactants and colored nanoparticles as detected labels provide maximum reducing manipulations of an operator. Contact of a test strip with the tested sample initiates the lateral flow of the immune reactants along the membranes and the assay results can be visualized within 10–15 min by the coloration of the test strip zones [5]. Successful development of the LFIA-based serodiagnosis of various diseases is presented in several publications [6–10]. As an additional application, the assessment of the vaccine effectiveness and the organism's protection against infections can be mentioned [11,12].

The recent COVID-19 pandemic also caused the necessity of serodiagnostic LFIAs, which were successfully developed and commercialized by different groups and organizations [13,14]. Nevertheless, the development and practical application of a serodiagnostic LFIA are accomplished by significant limitations. The percentage of the infection cases revealed by the LFIA serodiagnosis is often inferior to the cases detected by instrumental immunoassays, such as the ELISA. The main factor causing this obstacle is the need to detect specific antibodies with low concentrations in the presence of great excess of total immunoglobulins [15].

Serodiagnosis is typically implemented using immunoglobulin-binding proteins, such as anti-species antibodies or bacterial protecting compounds (staphylococcal protein A, streptococcal protein G, etc.) [16]. In the common LFIA format, they are immobilized on colored nanoparticles, applied on a test strip, and form complexes with all immunoglobulins, including those specific to the target pathogen during lateral flow. The resulting complexes interact with the antigen immobilized in the test zone (TZ) of the strip, whereas unbound compounds move across this zone. Therefore, the coloration of the TZ indicates the presence of specific antibodies in the tested sample [8,17–19]. However, non-specific immunoglobulins block binding sites at the surface of nanoparticles and only a minor part of specific antibodies can form labeled complexes in the TZ. This differs from the ELISA-based serodiagnosis where non-bound antibodies can be removed (washed) prior to the inclusion of labels into immune complexes. This limited binding leads to a significant percentage of false-negative results, which are often indicated during the validation of these assays [20,21].

The possibility to change the order of the immune complexes formation was considered in alternative serodiagnostic LFIAs [20,22–24]. As a result, some limitations were noted in these approaches. For example, the inverted LFIA format, where the antigen is immobilized on label particles and the immunoglobulin-binding protein is adsorbed on the working membrane, allows for the binding of more immunoglobulins than the traditional LFIA format. Nevertheless, in this case, the interference of non-specific immunoglobulins is not eliminated [25–27]. One of the approaches to eliminate their influence is the "double antigen sandwich" LFIA format based on the antibody polyvalence [23,28–30]. However, due to the side formation of polyvalent complexes, the efficiency of this format strongly depends on the ratio of the reagents [15]. Moreover, the use of more sensitive labels or signal amplification by aggregation or catalysis was considered [31–34]. The improvement achieved by these methods is associated with significant changes in the production of test strips. Therefore, they can be implemented in mass practice only after solving additional tasks, such as stabilization of new reactants and prevention of their non-specific interactions.

Therefore, the improvement of the common serodiagnostic LFIA based on the conjugates of immunoglobulin-binding proteins with traditional nanodispersed labels that ensure an increase in labeled immune complexes in the TZ is in great demand. In the present study, a two-stage serodiagnostic LFIA is proposed. This method includes the common LFIA (the first stage) and additional labeling of bound specific immunoglobulins by a conjugate of gold nanoparticles with an immunoglobulin-binding protein (the second stage). This approach was performed for the serodiagnostic of COVID-19 infection and demonstrated the enhancement of the coloration intensity by two orders of magnitude, as compared with the common LFIA format, which significantly reduces the likelihood of false-negative results.

2. Materials and Methods

2.1. Chemicals, Materials, and Apparatuses

The recombinant receptor-binding domain (RBD) of SARS-CoV-2 spike protein [35] was kindly provided by Dr. I.I. Vorobiev. The monoclonal antibodies (MAb) to RBD, clone RBD5313, were from HyTest (Moscow, Russia). Goat anti-mouse immunoglobulins (GAMI) and a conjugate of recombinant staphylococcal protein A with gold nanoparticles (pA–GNP) were from Arista Biologicals (Allentown, PA, USA). The polyclonal anti-human antibodies labeled with horseradish peroxidase were from Imtek (Moscow, Russia). NIBSC Anti-SARS-CoV-2 Antibody Diagnostic Calibrant reagent was purchased from National Institute for Biological Standards and Control (Hertfordshire, UK). Human serum was kindly provided by Dr. S.F. Biketov (State Research Center of Applied Microbiology and Biotechnology, Obolensk, Russia) and collected during previous studies [33] from volunteers and patients after obtaining written and informed consent. The pooled negative serum was prepared by mixing 10 sera from donors without symptoms of respiratory diseases and antibodies against RBD of SARS-CoV-2, in accordance with enzyme immunoassay.

Bovine serum albumin (BSA), sucrose, poly(vinyl formal), Tris, Tween 20, Triton X-100, 3,3′,5,5′-tetramethylbenzidine (TMB), hydrogen peroxide (H_2O_2, 30%), were from Sigma-Aldrich (St. Louis, MO, USA). Salts and acids were all from Khimmed (Moscow, Russia). The solutions were all prepared using deionized water produced by Milli-Q (Billerica, MA, USA).

Nitrocellulose working membranes (CNPC-15), glass fiber conjugate membranes (PT-R7), sample membranes (GFB-R4), and adsorbent membranes (AP045) were purchased from Advanced Microdevices (Ambala Cantt, India). The 96-well transparent polystyrene microplates for ELISA were purchased from Corning Costar (Tewksbury, MA, USA).

2.2. Characterization of pA–GNP

For transmission electron microscopy (TEM), pA–GNP solution was applied to 300-mesh grids (Pelco International; Redding, CA, USA) coated with a poly(vinyl formal) film. Then, the film was placed on the glass and exposed to 0.15% *v*/*v* solution of formvar in chloroform. The images were obtained with a JEM CX-100 microscope (Jeol, Tokyo, Japan) at 80 kV and analyzed by the Image Tool software (University of Texas, Health Science Center, San Antonio, TX, USA).

The hydrodynamic size of pA–GNP was measured using a Zetasizer Nano (Malvern Pananlytical; Malvern, UK) at 25 °C for 10 s at a scattering angle of 173°.

2.3. Preparation of Test Strips

GAMI (0.5 mg/mL) and RBD (0.5–1.0 mg/mL) in 50 mM K-phosphate buffer with 100 mM NaCl, pH 7.4 (PBS), were applied to the control zone (CZ) and the TZ of the working membrane, respectively, by Image Technology IsoFlow dispenser (Lebanon, NH, USA) at a loading of 0.12 μL/mm. The pA–GNP conjugate (optical density at 520 nm (OD_{520}) = 2–20) containing 1.0% *v*/*v* Tween 20 was applied to the conjugate membrane (0.8 μL/mm). Both membranes were dried at room temperature for 12 h and composed together with sample and absorbent membranes to multimembrane sheets. Finally, test strips (of 3.5 mm width) were obtained by cutting sheets with an automatic Index Cutter-1 guillotine (A-Point Technologies; Gibbstown, NJ, USA) and stored at room temperature in zipper bags.

2.4. Lateral Flow Immunoassay

The common LFIA was performed at room temperature as follows:

- The test strip was placed on a horizontal surface;
- 60 μL of a sample was applied to the sample membrane;
- the strip was incubated for 5 min;
- 20 μL of TTBSA buffer (10 mM Tris, 0.25% *w*/*v* BSA, 0.25% *w*/*v* sucrose, 1.0% *v*/*v* Tween 20, 0.05% *w*/*v* NaN_3, pH 8.5) was applied to the sample membrane;
- the strip was incubated for 5 min.

The enhanced LFIA was started as described above for the common LFIA. Then, 2 μL of the pA–GNP conjugate was applied to the bottom edge of the working membrane. Thereafter, 60 μL of TTBSA was applied and the test strip was incubated for 5 min.

Following the LFIA, test strips were scanned using a CanoScan 9000F scanner (Canon, Tochigi, Japan). The obtained digital images were processed by TotalLab TL120 software (Nonlinear Dynamics, Newcastle, UK) to measure the intensity of TZ coloration. The processing of test strips by a TotalLab application included the finding of TZ in the scanned images, generation of color intensity profiles, subtraction of background coloration using the "Background Subtraction" tool, and final registration of the "Volume" value as the analytical signal using the "Band Detection" tool.

Each sample was tested in duplicate. The limit of detection (LOD) was determined as the concentration providing the coloration of the TZ, which is higher than the sum of the average coloration intensity and three standard deviations for a blank probe.

2.5. ELISA of Human Serum

RBD (1 µg/mL, in PBS) was incubated in the microplate wells at 4 °C overnight. The wells were washed 4 times to remove unbound molecules using PBS with 0.05% *v*/*v* Triton X-100 (PBST). Thereafter, serum diluted with PBST (1:25–1:50,000) was added and the microplate was incubated for 1 h at 37 °C. Then, the microplate was washed again and anti-human antibodies labeled with horseradish peroxidase (dilution of 1:3000 in PBST) were added to each well. The microplate was incubated for 1 h at 37 °C. After washing the microplate as described above, the enzyme activity of the bound peroxidase label was determined. For this purpose, the substrate mixture containing 0.4 mM TMB and 3 mM H_2O_2 in 40 mM sodium citrate buffer, pH 4.0, was added. Following incubation at room temperature for 15 min, the reaction was stopped by adding 1 M H_2SO_4 to the substrate mixture (*v*/*v* = 1:2). Finally, A_{450} was measured using a Zenyth 3100 microplate photometer (Anthos Labtec Instruments, Wals, Austria).

3. Results and Discussion

3.1. Consideration of the Proposed Assay Format

The serodiagnostic LFIA can be implemented in several formats, which differ in the order of arranging the reagents on the membranes and the composition of the resulting complexes [20,22–24]. The most common format is based on the use of labeled immunoglobulin-binding proteins as reagents to visualize immune complexes. Pathogen antigens are immobilized in the TZ and interact with specific antibodies, which leads to label binding.

The assay proposed in this study was first implemented in the same manner as the common LFIA (Figure 1, stage 1). In the case of the presence of specific antibodies in a tested blood sample, colored complexes of the immobilized antigen, specific antibodies, and a labeled immunoglobulin-binding protein are formed in the TZ. To clarify points for improvement, let us consider this process in more detail. The concentration of IgG as the main class of immunoglobulins in human blood is typically about 6–20 mg/mL [36], while the sorption capacity of gold nanoparticles with a diameter of 30 nm (as the optimal LFIA label) does not exceed 5 µg/mL per unit of the OD [37]. Therefore, even if the OD of the conjugate is 10, it binds less than 1% of the IgG fraction from the sample. The same proportion is typical for specific anti-pathogen antibodies. Theoretically, the problem can be solved by concentrating the conjugate. However, an increase in the concentration of gold nanoparticles promotes their aggregation, which leads to overlapping of binding centers and a stability decrease. Therefore, a radical increase in the conjugate concentration is restricted. As a result, the majority of specific antibodies that bind in the TZ are not associated with the label. To involve them in the generation of the detected signal, an additional stage was proposed.

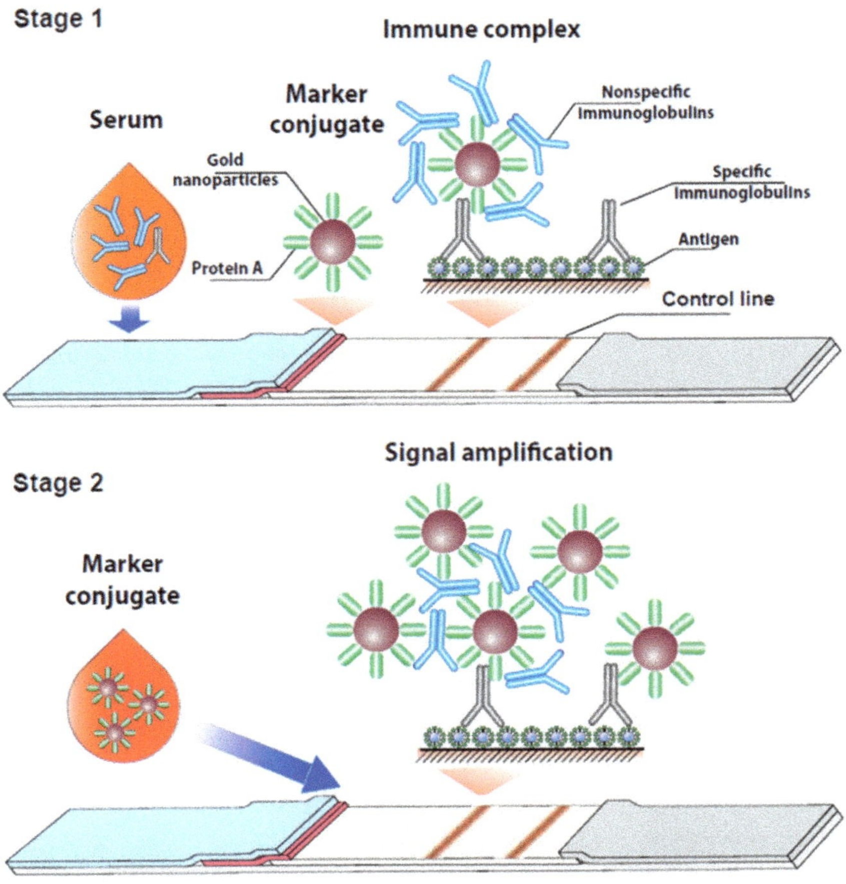

Figure 1. Scheme of the proposed enhanced serodiagnostic LFIA.

At the second stage of the assay, a labeled immunoglobulin-binding protein is added to the working membrane of the test strip. This complex flows along the membranes after the sample, thus washing out the unbound immunoglobulins from the working membrane and interacting with the immunoglobulins bound in the TZ. At this stage, the release of the analytical signal occurs in two ways (Figure 1, stage 2). On the one hand, the labeled immunoglobulin-binding protein interacts with the antibody-antigen complexes in the TZ. On the other hand, the label conjugate bound in the TZ at the first stage provides available sites to bind the added reagent. Since the IgG molecules have two symmetric binding sites for immunoglobulin-binding proteins, they could serve in the formation of various polycompound complexes based on the labeled immunoglobulin-binding protein, which is initially bound at the TZ and contains a large number of IgG molecules at its surface. In this way, non-specific immunoglobulins interfere with signal generation at the first stage, but provide signal enhancement at the second stage.

Aggregation of functionalized nanoparticles was previously used in LFIA to enhance the analytical signal [38,39]. In these studies, bifunctional conjugates of nanoparticles were used. In these conjugates, a part of the nanoparticle surface was occupied by molecules with affinity to the analyte. The other part was occupied by molecules that provided aggregation (for example, streptavidin and biotinylated protein). This approach requires strict control of the conjugate composition to fulfill the efficiency of both processes. In our approach, aggregation-based enhancement of analytical signal in LFIA serodiagnosis is provided by a monofunctional conjugate of a nanoparticle and an immunoglobulin-binding protein. This simplification becomes possible due to the fact that the antibodies are both analytes and triggers of the aggregation process.

In the present study, the approach proposed for signal amplification and improvement of diagnostics was tested to develop a new LFIA format for the detection of antibodies against the RBD antigen of the SARS-CoV-2.

3.2. Characterization of the Conjugate of Gold Nanoparticles with Protein A

Protein A of *Staphylococcus aureus* labeled with gold nanoparticles was used as a reagent for the binding of immunoglobulins. To characterize the size and homogeneity of nanoparticles, microphotographs were obtained by TEM. Images of nanoparticles are presented in Figure 2a and the particle size distribution is shown in Figure 2b. The average size of nanoparticles (171 particles were processed) was 31.51 ± 9.27 nm (minimum value—16.78 nm, maximum value—65.16 nm) with an ellipticity of 1.28 ± 0.27. Conjugated nanoparticles were not aggregated in solution, but rather, they were stable during storage and drying on a membrane.

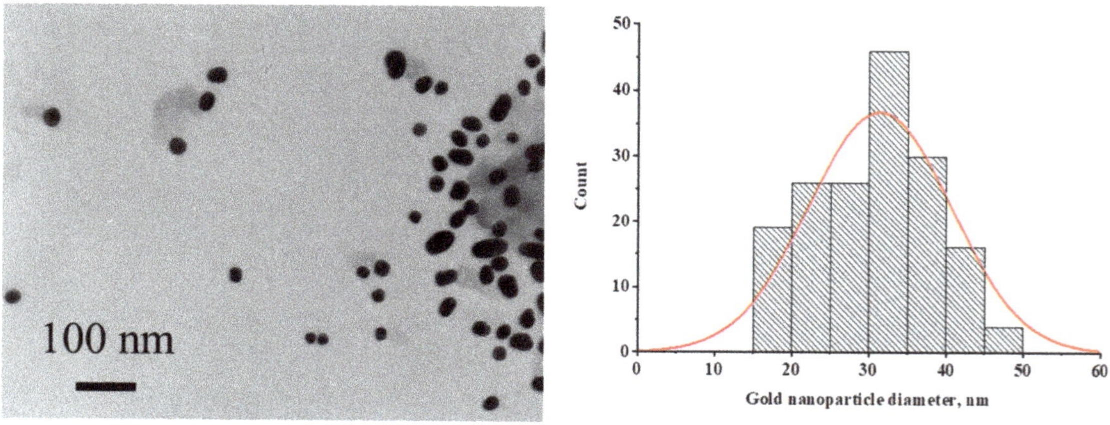

(a) (b)

Figure 2. (**a**) Micrograph of the conjugate of gold nanoparticles with staphylococcal protein A. (**b**) Histogram of particle size distribution.

3.3. Detection of Specific IgG in Model Solutions

To estimate the process which occurs in the proposed two-stage serodiagnostic LFIA and the possibility of detecting the lower concentration of specific antibodies in blood samples, the knowledge concerning the minimal concentration of detectable antibodies is very important. For this purpose, we detected specific antibodies against the SARS-CoV-2 RBD in model solutions. Humanized rat monoclonal antibodies against the RBD were used as analytes. They contained human Fc fragments to most accurately reproduce the behavior of real immunoglobulins in human biosamples. Figure 3 shows the results of antibody testing by the common LFIA. It was demonstrated that up to 10 ng/mL of specific antibodies can be detected by this method in the absence of non-specific immunoglobulins. The concentration of IgG specific to individual antigens in the blood is typically in the range of 3–50 µg/mL [40–42]. Therefore, the test system has a large margin in sensitivity for antibody detection. Additionally, the most likely reason for a possible false-negative test result is signal suppression by non-specific immunoglobulins.

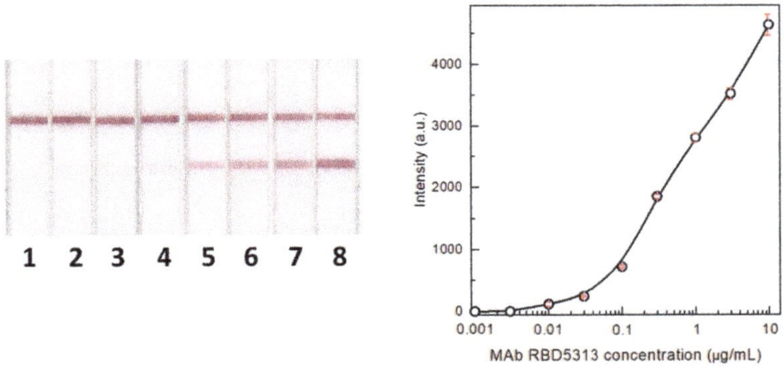

(a) (b)

Figure 3. Images of test strips (**a**) after the common LFIA of samples containing 0.003 (1), 0.01 (2), 0.03 (3), 0.1 (4), 0.3 (5), 1 (6), 3 (7), and 10 (8) μg/mL of MAb RBD5313 in PBS and the concentration dependence (**b**) of the assay (n = 3). Hereafter, top and bottom colored lines on the images of test strips correspond to the CZ and TZ, respectively.

3.4. The Influence of Serum Components on the LFIA

To evaluate the effect of biosamples' components, primarily, non-specific immunoglobulins, on the LFIA results, we used the pooled serum. The pooled serum was diluted with PBST (1:3–1:100) and MAb against RBD (in the same way as the previous section), and then the mixture was added to the model solutions to a final concentration of 10 μg/mL. The mixture was tested using the common and enhanced LFIAs. The obtained results demonstrated that in the case of the common LFIA, the use of undiluted serum or one serum diluted less than 6 times led to the decrease in the signal to an undetectable value (Figure 4a,c (curve I)). Even when the serum was diluted 100 times, the signal was reduced by 8 times relative to the signal registered after the LFIA in the buffer. Following the second stage of the assay, the LFIA signal increased by more than 3 times (Figure 4b,c (curve II)). The comparison of different serum dilutions demonstrates that a 25-fold dilution of serum is sufficient to achieve the maximum signal.

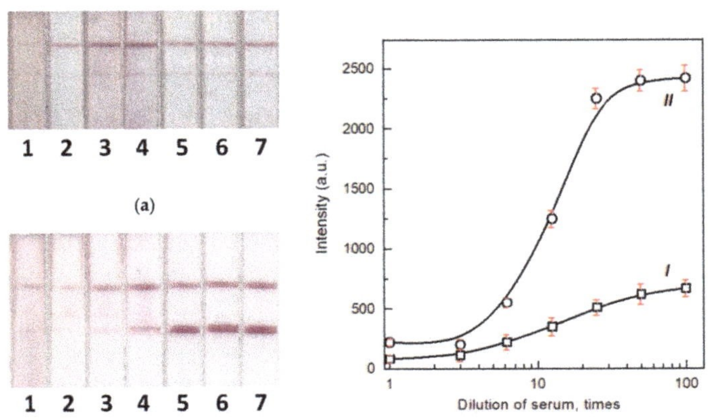

(b) (c)

Figure 4. Images of test strips after the common (**a**) and enhanced (**b**) LFIA of samples containing 1- (1), 3- (2), 6.2- (3), 12.5- (4), 25- (5), 50- (6), and 100- (7) -fold dilutions of pooled negative human serum and 10 μg/mL of monoclonal antibodies against RBD. (**c**) Concentration dependence of the common (I) and enhanced (II) LFIA (n = 3).

The obtained data are correlated with the regularities described in the previous study [21], which recommends the dilution of serum from 10 to 100 times to minimize the effect of non-specific antibodies on the results of common serodiagnostic LFIA. For further studies, sera diluted 25 times in PBST were used in the analyzed samples.

3.5. Comparison of LODs for Two Formats of the LFIA

Using the selected dilution of serum samples, the LODs for specific MAb RBD5313 were estimated for the common and enhanced LFIA. The final concentration of the added antibodies varied from 3 ng/mL to 10 μg/mL. To be specific, when performing the common LFIA, the LOD of specific antibodies in serum was 300 ng/mL, which was 30-fold higher than when detecting antibodies in a buffer solution with the same test system (Figure 5a,c (curve I)). Following the second stage, the LOD decreased by 30 times (Figure 5b,c (curve II)). Therefore, in our case, the proposed enhancement strongly eliminates the effect of non-specific antibodies on the LODs: A 30-fold loss was compensated by a 30-fold improvement. Moreover, the results demonstrated that at lower antibody concentrations in the sample, the enhancement at the second stage of LFIA was increasingly manifested, as compared with high antibody concentrations (values above 1 μg/mL). This fact is particularly important for the detection of the anti-pathogen antibodies in weakly positive samples.

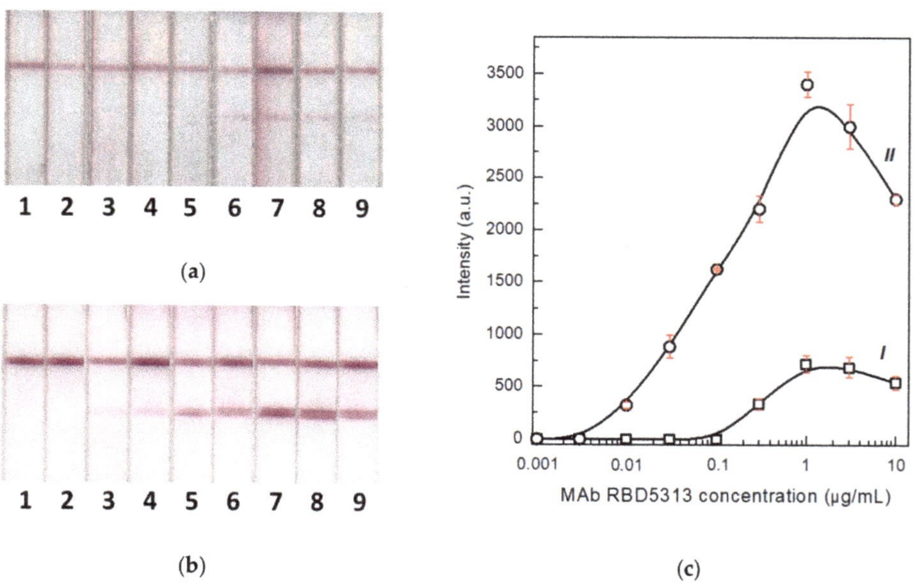

Figure 5. Images of test strips after the common (**a**) and enhanced (**b**) LFIA of samples containing 0 (1), 0.003 (2), 0.01 (3), 0.03 (4), 0.1 (5), 0.3 (6), 1 (7), 3 (8), and 10 (9) μg/mL of MAb RBD5313. (**c**) Concentration dependence of the common (I) and enhanced (II) LFIA ($n = 3$).

3.6. Determination of the Optimal Concentration of the Label Conjugate

As previously noted, blood immunoglobulins are in large quantitative excess relative to the binding capacity of the label conjugate (considering real ratios of these reactants in the test system). Therefore, the higher the concentration of the conjugate, the higher the sensitivity of the test system. However, an increase in the concentration of colloidal particles increases the likelihood of their aggregation, which reduces the availability of immunoglobulin-binding centers. Therefore, the optimal concentration of the conjugate will provide the maximum binding capacity. To obtain this concentration, we performed a series of test strips that differed in the concentration of the pA–GNP conjugate applied prior to the first stage of the assay. Figure 6 shows the testing results of serum samples with a known concentration of specific antibodies by LFIAs differing in concentrations of pA–GNP applied on the test strips (OD_{520} of GNP solutions were 2, 5, 10, and 20). The optimal concentration of the conjugate, which provided the maximum analytical signal and the minimum LOD, corresponded to OD_{520} = 10. This conjugate concentration was used in all our experiments in the future.

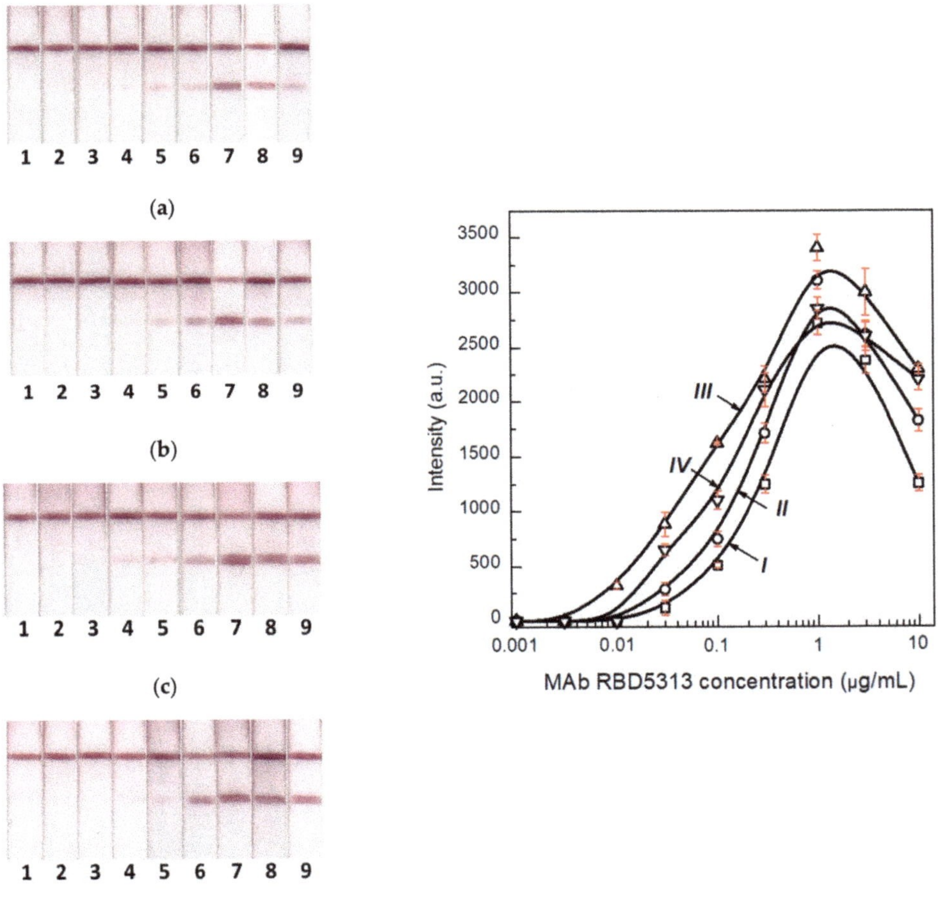

Figure 6. Images of test strips manufactured with the pA–GNP conjugate with OD_{520} = 2 (**a**), 5 (**b**), 10 (**c**), and 20 (**d**) after LFIA of samples containing 0 (1), 0.003 (2), 0.01 (3), 0.03 (4), 0.1 (5), 0.3 (6), 1 (7), 3 (8), and 10 (9) ng/mL of MAb RBD5313. (**e**) Concentration dependence for test systems containing the pA–GNP conjugate with OD_{520} = 2 (I), 5 (II), 10 (III), and 20 (IV) (n = 3).

3.7. Testing of the Developed LFIAs on Blood Samples of Patients with a Confirmed COVID-19 Diagnosis

To validate the developed analytical systems, the positive standard of the National Institute for Biological Standards and Control (Hertfordshire, UK) was first used. As shown in Figure 7, the proposed enhancing approach increased the coloration in the TZ from 6 times to 3 orders of magnitude. The maximum signal was observed in the enhanced LFIA when the serum was diluted from 20 to 40 times.

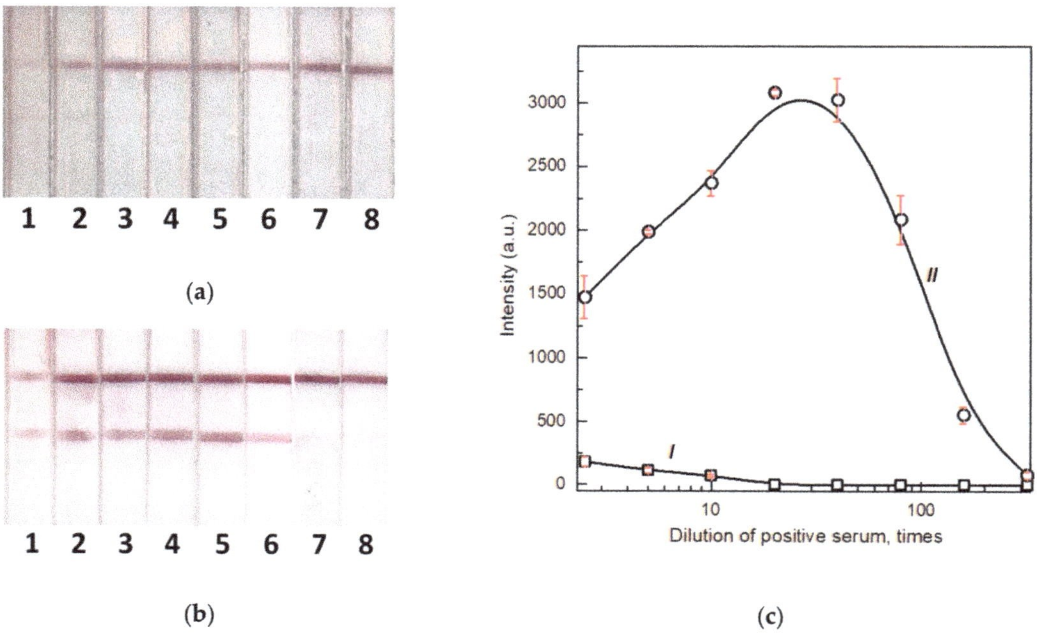

Figure 7. Images of test strips after the common (**a**) and enhanced (**b**) LFIAs of samples of pooled human serum containing antibodies to RBD SARS-CoV-2 diluted by 2.5 (1), 5 (2), 10 (3), 20 (4), 40 (5), 80 (6), 160 (7), and 320 (8) times. (**c**) Concentration dependence of the common (I) and enhanced (II) LFIA ($n = 3$).

Following this step, the developed LFIAs were compared with the ELISA in terms of the efficiency of detecting specific antibodies from patients in the samples. For this purpose, a panel of 24 sera was collected, which included 16 sera from patients with confirmed COVID-19 and eight negative sera. This panel has been tested by ELISA.

The resulting panel of positive and negative samples was tested by the LFIA in two formats (see Table 1). Samples providing the color intensities of the TZ above 100 arbitrary units (a.u.), which roughly corresponded to the visual detection threshold, were considered as positive. To be specific, when using the common LFIA, six out of 16 positive samples provided negative results; namely, the diagnostic sensitivity was 62.5%. This value is in accordance with the middle of the sensitivity interval of SARS-CoV-2 LFIAs. Moreover, in accordance with the published reviews, the sensitivity reached levels that vary from 49 to 85% [14,43–45]. To be specific, when using the enhanced LFIA, all of the 16 ELISA-positive sera provided positive results (with a diagnostic sensitivity of 100%). Moreover, when testing a panel of negative sera, no positive results were observed (i.e., the diagnostic specificity was 100%). Notably, the presented analytical parameters characterize the assay with an instrumental registration of the results. Visual assessment may lead to inaccurate conclusions in the case of low coloration.

Table 1. Comparison of ELISA and LFIA data for 24 human serum samples.

Positive Sera																
	1	2	3	4	5	6	7	8	9	10	11	12	13	14	15	16
Common LFIA																
I (a.u.)	208	135	140	135	155	130	142	180	90	85	0	0	0	0	0	35
	+	+	+	+	+	+	+	+	+	-	-	-	-	-	-	-
Enhanced LFIA	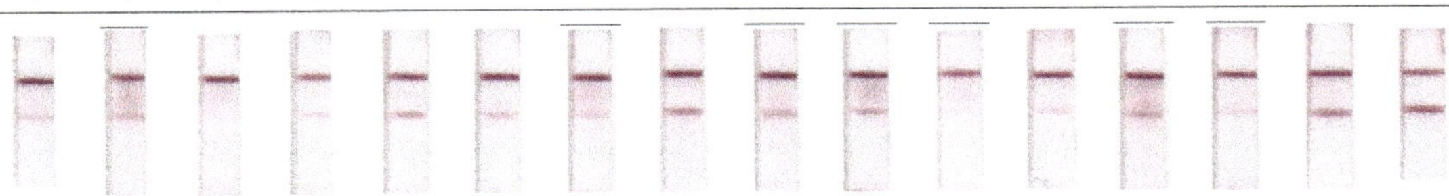															
I (a.u.)	585	635	175	525	1650	1285	880	2520	775	1780	125	956	1430	785	2350	3575
	+	+	+	+	+	+	+	+	+	+	+	+	+	+	+	+
ELISA (A 450)	0.807	1.034	0.299	1.004	1.114	0.957	0.747	1.038	0.848	1.124	0.256	0.730	0.932	0.659	0.899	1.072

Negative Sera								
	1	2	3	4	5	6	7	8
Common LFIA	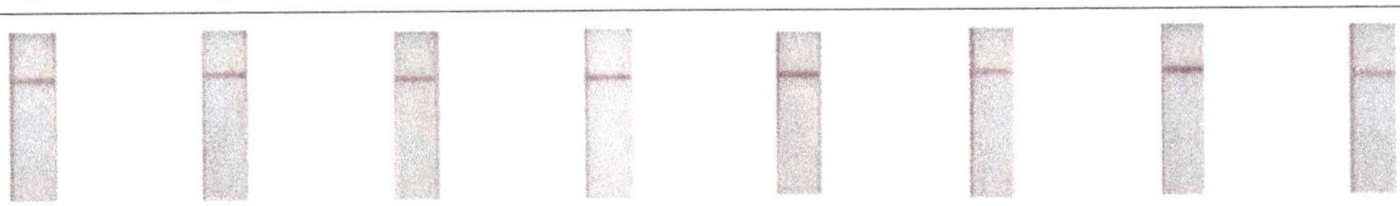							
I (a.u.)	0	0	0	0	0	0	0	0
	-	-	-	-	-	-	-	-
Enhanced LFIA	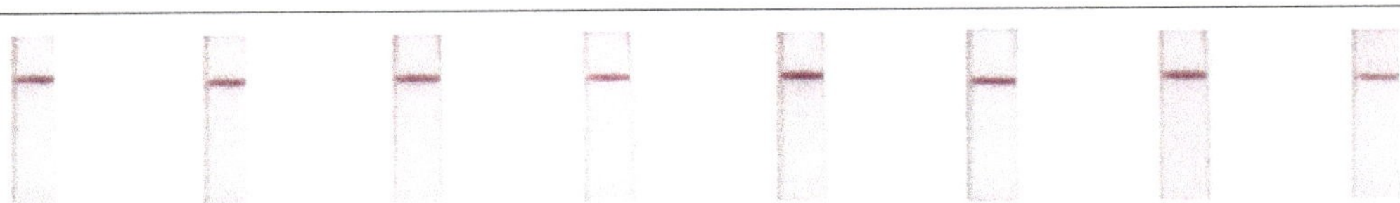							
I (a.u.)	0	65	75	0	55	0	0	0
	-	-	-	-	-	-	-	-
ELISA (A 450)	0.125	0.184	0.198	0.093	0.128	0.120	0.191	0.228

4. Conclusions

The present study demonstrates the critical influence of non-specific immunoglobulins on the diagnostic sensitivity of immunochromatographic serodiagnosis and suggests a viable method to eliminate this influence. The proposed format of the serodiagnostic LFIA with two stages of signal generation was tested for the COVID-19 serodiagnosis and showed a signal increase up to three orders of magnitude. At the same time, the 16 positive sera that were tested showed a decrease in the percentage of false-negative results from 37.5% to 0, as compared with the common serodiagnostic LFIA. The implications for the applied two-stage strategy are evident, operator actions are somewhat more complicated and the assay time was extended by 5 min. However, the modified assay still meets the criteria of rapid testing. The applicability of the described approach is not limited to a specific pathogen. Additionally, the proposed solution can be employed to increase the serodiagnostic sensitivity of other diseases.

Author Contributions: Conceptualization, D.V.S., N.A.B., A.V.Z. and B.B.D.; investigation, D.V.S. and N.A.B.; methodology, D.V.S., N.A.B., A.V.Z. and B.B.D.; resources, D.V.S. and N.A.B.; validation, D.V.S. and N.A.B.; visualization, D.V.S., N.A.B. and A.V.Z.; writing—original draft, D.V.S., N.A.B., A.V.Z. and B.B.D.; writing—review and editing, D.V.S., N.A.B., A.V.Z., Y.X. and B.B.D. All authors have read and agreed to the published version of the manuscript.

Funding: The reported study was funded by Russian Foundation for Basic Research and National Natural Science Foundation of China in accordance with the research project 20-58-55001.

Institutional Review Board Statement: Biosamples from patients were obtained on the basis of their informed consent that was confirmed by filling and signing the templates approved by Russian authorized agencies (statement letter from State Research Center for Applied Microbiology & Biotechnology No. 150-50/9-13-1053-2021 dated 11 May 2021).

Informed Consent Statement: Informed consent was obtained from all subjects involved in the study.

Data Availability Statement: Data are contained within the article. Initial data of instrumental measurements are available on request from the corresponding author.

Conflicts of Interest: The authors declare no conflict of interest.

References

1. Champoux, J.J.; Lawrence, D.W.; Neidhardt, F.C.; Plorde, J.J. *Sherris Medical Microbiology: An Introduction to Infectious Diseases*, 4th ed.; McGraw-Hill: New York, NY, USA, 2004; pp. 247–249.
2. Lau, S.K.P.; He, Z.; Tsang, C.C.; Chan, T.T.Y.; Luk, H.K.H.; Chan, E.; Li, K.S.M.; Fung, J.; Chow, F.W.N.; Tam, A.R.; et al. A sensitive and specific competitive enzyme-linked immunosorbent assay for serodiagnosis of COVID-19 in animals. *Microorganisms* **2021**, *9*, 1019. [CrossRef] [PubMed]
3. Elschner, M.C.; Laroucau, K.; Singha, H.; Tripathi, B.N.; Saqib, M.; Gardner, I.; Saini, S.; Kumar, S.; El-Adawy, H.; Melzer, F.; et al. Evaluation of the comparative accuracy of the complement fixation test, Western blot and five enzyme-linked immunosorbent assays for serodiagnosis of glanders. *PLoS ONE* **2019**, *14*, e0214963. [CrossRef] [PubMed]
4. Fortes-Gabriel, E.; Guedes, M.S.; Shetty, A.; Gomes, C.K.; Carreira, T.; Vieira, M.L.; Esteves, L.; Mota-Vieira, L.; Gomes-Solecki, M. Enzyme immunoassays (EIA) for serodiagnosis of human leptospirosis: Specific IgG3/IgG1 isotyping may further inform diagnosis of acute disease. *PLoS Negl. Trop. Dis.* **2022**, *16*, e0010241. [CrossRef] [PubMed]
5. Goryacheva, I.Y.; Lenain, P.; de Saeger, S. Nanosized labels for rapid immunotests. *TrAC Trends Anal. Chem.* **2013**, *46*, 30–43. [CrossRef]
6. Kwizera, R.; Bongomin, F.; Olum, R.; Worodria, W.; Bwanga, F.; Meya, D.B.; Kirenga, B.J.; Gore, R.; Denning, D.W.; Fowler, S.J. Evaluation of an *Aspergillus* IgG/IgM lateral flow assay for serodiagnosis of fungal asthma in Uganda. *PLoS ONE* **2021**, *16*, e0252553. [CrossRef]
7. Lu, J.; Wu, Z.; Liu, B.; Wang, C.; Wang, Q.; Zhang, L.; Wang, Z.; Chen, C.; Fu, Y.; Li, C.; et al. A time-resolved fluorescence lateral flow immunoassay for rapid and quantitative serodiagnosis of *Brucella* infection in humans. *J. Pharm. Biomed. Anal.* **2021**, *200*, 114071. [CrossRef]
8. Jain, M.; Singh, A.K.; Kumar, A.; Gupta, S.; Polavarapu, R.; Sohal, J.S. Comparative performance of different antigens on the lateral flow assay (LFA) platform for the rapid serodiagnosis of paratuberculosis. *J. Microbiol. Methods* **2021**, *192*, 106367. [CrossRef]
9. Lempp, F.A.; Roggenbach, I.; Nkongolo, S.; Sakin, V.; Schlund, F.; Schnitzler, P.; Wedemeyer, H.; le Gal, F.; Gordien, E.; Yurdaydin, C.; et al. A rapid point-of-care test for the serodiagnosis of hepatitis delta virus infection. *Viruses* **2021**, *13*, 2371. [CrossRef]

10. Anil, M.; Navaneetha, P.; Reddy, P.V.J.; Chand, M.P.; Kumar, A.V.; Kishor, P.B.K.; Rao, K.R.S.S.; Rathnagiri, P. Development of point-of-care lateral flow immunochromatographic assay for foot and mouth disease diagnosis. *Curr. Trends Biotechnol. Pharm.* **2021**, *15*, 15–21. [CrossRef]
11. Higgins, R.L.; Rawlings, S.A.; Case, J.; Lee, F.Y.; Chan, C.W.; Barrick, B.; Burger, Z.C.; Yeo, K.J.; Marrinucci, D. Longitudinal SARS-CoV-2 antibody study using the Easy Check COVID-19 IgM/IgG™ lateral flow assay. *PLoS ONE* **2021**, *16*, e0247797. [CrossRef]
12. Srivastav, S.; Dankov, A.; Adanalic, M.; Grzeschik, R.; Tran, V.; Pagel-Wieder, S.; Gessler, F.; Spreitzer, I.; Scholz, T.; Schnierle, B.; et al. Rapid and sensitive SERS-based lateral flow test for SARS-CoV2-specific IgM/IgG antibodies. *Anal. Chem.* **2021**, *93*, 12391–12399. [CrossRef]
13. Fu, Y.; Pan, Y.; Li, Z.; Li, Y. The utility of specific antibodies against SARS-CoV-2 in laboratory diagnosis. *Front. Microbiol.* **2021**, *11*, 603058. [CrossRef]
14. Vengesai, A.; Midzi, H.; Kasambala, M.; Mutandadzi, H.; Mduluza-Jokonya, T.L.; Rusakaniko, S.; Mutapi, F.; Naicker, T.; Mduluza, T. A systematic and meta-analysis review on the diagnostic accuracy of antibodies in the serological diagnosis of COVID-19. *Syst. Rev.* **2021**, *10*, 155. [CrossRef]
15. Sotnikov, D.V.; Zherdev, A.V.; Dzantiev, B.B. Lateral flow serodiagnosis in the double-antigen sandwich format: Theoretical consideration and confirmation of advantages. *Sensors* **2021**, *21*, 39. [CrossRef]
16. Cavalera, S.; di Nardo, F.; Chiarello, M.; Serra, T.; Colitti, B.; Guiotto, C.; Fagioli, F.; Cagnazzo, C.; Denina, M.; Palazzo, A.; et al. Bacterial ligands as flexible and sensitive detectors in rapid tests for antibodies to SARS-CoV-2. *Anal. Bioanal. Chem.* **2022**, 1–10. [CrossRef]
17. Yunus, M.H.; Arifin, N.; Balachandra, D.; Anuar, N.S.; Noordin, R. Lateral flow dipstick test for serodiagnosis of strongyloidiasis. *Am. J. Trop. Med. Hyg.* **2019**, *101*, 432–435. [CrossRef]
18. Prakash, C.; Kumar, B.; Singh, R.P.; Singh, P.G.; Shrinet, G.; Das, A.R.; Ashmi, M.; Abhishek, S.K.P.; Singh, M.; Gupta, V.K. Development and evaluation of a gold nanoparticle based lateral flow assay (LFA) strip test for detection of *Brucella* spp. *J. Microbiol. Methods* **2021**, *184*, 106185. [CrossRef]
19. Manasa, M.; Revathi, P.; Chand, M.P.; Maroudam, V.; Navaneetha, P.; Raj, G.D.; Kishor, P.B.K.; De, B.; Rathnagiri, P. Protein-G-based lateral flow assay for rapid serodiagnosis of brucellosis in domesticated animals. *J. Immunoass. Immunochem.* **2019**, *40*, 149–158. [CrossRef]
20. Di Nardo, F.; Chiarello, M.; Cavalera, S.; Baggiani, C.; Anfossi, L. Ten years of lateral flow immunoassay technique applications: Trends, challenges and future perspectives. *Sensors* **2021**, *21*, 5185. [CrossRef]
21. Sotnikov, D.V.; Zherdev, A.V.; Dzantiev, B.B. Mathematical model of serodiagnostic immunochromatographic assay. *Anal. Chem.* **2017**, *89*, 4419–4427. [CrossRef]
22. Sotnikov, D.V.; Berlina, A.N.; Zherdev, A.V.; Eskendirova, S.Z.; Mukanov, K.K.; Ramankulov, Y.R.; Mukantayev, K.N.; Dzantiev, B.B. Comparison of three schemes of quantum dots-based immunochromatography for serodiagnosis of brucellosis in cattle. *J. Eng. Appl. Sci.* **2019**, *14*, 3711–3718. [CrossRef]
23. Karakus, C.; Salih, B.A. Comparison of the lateral flow immunoassays (LFIA) for the diagnosis of *Helicobacter pylori* infection. *J. Immunol. Methods* **2013**, *396*, 8–14. [CrossRef]
24. Cavalera, S.; Di Nardo, F.; Forte, L.; Marinoni, F.; Chiarello, M.; Baggiani, C.; Anfossi, L. Switching from multiplex to multimodal colorimetric lateral flow immunosensor. *Sensors* **2020**, *20*, 6609. [CrossRef]
25. Sotnikov, D.V.; Zherdev, A.V.; Dzantiev, B.B. Theoretical and experimental comparison of different formats of immunochromatographic serodiagnostics. *Sensors* **2018**, *18*, 36. [CrossRef] [PubMed]
26. Martínez-Sernández, V.; Muiño, L.; Perteguer, M.J.; Gárate, T.; Mezo, M.; González-Warleta, M.; Muro, A.; da Costa, J.M.C.; Romarís, F.; Ubeira, F.M. Development and evaluation of a new lateral flow immunoassay for serodiagnosis of human fasciolosis. *PLoS Negl. Trop. Dis.* **2011**, *5*, e1376. [CrossRef]
27. Ben-Selma, W.; Harizi, H.; Boukadida, J. Immunochromatographic IgG/IgM test for rapid diagnosis of active tuberculosis. *Clin. Vaccine Immunol.* **2011**, *18*, 2090–2094. [CrossRef]
28. Wu, H.S.; Chiu, S.C.; Tseng, T.C.; Lin, S.F.; Lin, J.H.; Hsu, Y.H.; Wang, M.C.; Lin, T.L.; Yang, W.Z.; Ferng, T.L.; et al. Serologic and molecular biologic methods for SARS-associated coronavirus infection, Taiwan. *Emerg. Infect. Dis.* **2004**, *10*, 304–310. [CrossRef]
29. Sato, N.S.; de Melo, C.S.; Zerbini, L.C.; Silveira, E.P.; Fagundes, L.J.; Ueda, M. Assessment of the rapid test based on an immunochromatography technique for detecting anti-*Treponema pallidum* antibodies. *Rev. Inst. Med. Trop. Sao Paulo* **2003**, *45*, 319–322. [CrossRef]
30. Vrublevskaya, V.V.; Afanasyev, V.N.; Grinevich, A.A.; Skarga, Y.Y.; Gladyshev, P.P.; Ibragimova, S.A.; Krylsky, D.V.; Dezhurov, S.V.; Morenkov, O.S. A sensitive and specific lateral flow assay for rapid detection of antibodies against glycoprotein B of Aujeszky's disease virus. *J. Virol. Methods* **2017**, *249*, 175–180. [CrossRef]
31. Jia, J.; Ao, L.; Luo, Y.; Liao, T.; Huang, L.; Zhuo, D.; Jiang, C.; Wang, J.; Hu, J. Quantum dots assembly enhanced and dual-antigen sandwich structured lateral flow immunoassay of SARS-CoV-2 antibody with simultaneously high sensitivity and specificity. *Biosens. Bioelectron.* **2022**, *198*, 113810. [CrossRef]
32. Liu, H.; Dai, E.; Xiao, R.; Zhou, Z.; Zhang, M.; Bai, Z.; Shao, Y.; Qi, K.; Tu, J.; Wang, C.; et al. Development of a SERS-based lateral flow immunoassay for rapid and ultra-sensitive detection of anti-SARS-CoV-2 IgM/IgG in clinical samples. *Sens. Actuators B Chem.* **2021**, *329*, 129196. [CrossRef] [PubMed]

33. Panferov, V.G.; Byzova, N.A.; Biketov, S.F.; Zherdev, A.V.; Dzantiev, B.B. Comparative study of in situ techniques to enlarge gold nanoparticles for highly sensitive lateral flow immunoassay of SARS-CoV-2. *Biosensors* **2021**, *11*, 229. [CrossRef] [PubMed]
34. Sotnikov, D.V.; Barshevskaya, L.V.; Zherdev, A.V.; Eskendirova, S.Z.; Mukanov, K.K.; Mukantayev, K.K.; Ramanculov, Y.M.; Dzantiev, B.B. Immunochromatographic system for serodiagnostics of cattle brucellosis using gold nanoparticles and signal amplification with quantum dots. *Appl. Sci.* **2020**, *10*, 738. [CrossRef]
35. Sinegubova, M.V.; Orlova, N.A.; Kovnir, S.V.; Dayanova, L.K.; Vorobiev, I.I. High-level expression of the monomeric SARS-CoV-2 S protein RBD 320-537 in stably transfected CHO cells by the EEF1A1-based plasmid vector. *PLoS ONE* **2021**, *16*, e0242890. [CrossRef]
36. Meulenbroek, A.J. *Human IgG Subclasses Useful Diagnostic Markers for Immunocompetence*, 3rd ed.; Sanquin: Amsterdam, The Netherlands, 2008; 55p.
37. Sotnikov, D.V.; Byzova, N.A.; Zherdev, A.V.; Dzantiev, B.B. Retention of activity by antibodies immobilized on gold nanoparticles of different sizes: Fluorometric method of determination and comparative evaluation. *Nanomaterials* **2021**, *11*, 3117. [CrossRef]
38. Choi, D.H.; Lee, S.K.; Oh, Y.K.; Bae, B.W.; Lee, S.D.; Kim, S.; Shin, Y.-B.; Kim, M.G. A dual gold nanoparticle conjugate-based lateral flow assay (LFA) method for the analysis of troponin I. *Biosens. Bioelectron.* **2010**, *25*, 1999–2002. [CrossRef]
39. Taranova, N.A.; Urusov, A.E.; Sadykhov, E.G.; Zherdev, A.V.; Dzantiev, B.B. Bifunctional gold nanoparticles as an agglomeration-enhancing tool for highly sensitive lateral flow tests: A case study with procalcitonin. *Microchim. Acta* **2017**, *184*, 4189–4195. [CrossRef]
40. Kemper, M.J.; Altrogge, H.; Ganschow, R.; Müller-Wiefel, D.E. Serum levels of immunoglobulins and IgG subclasses in steroid sensitive nephrotic syndrome. *Pediatr. Nephrol.* **2002**, *17*, 413–417. [CrossRef]
41. Tuerlinckx, D.; Florkin, B.; Ferster, A.; De Schutter, I.; Chantrain, C.; Haerynck, F.; Philippet, P.; Strengers, P.; Laub, R. Pneumococcal antibody levels in children with PID receiving immunoglobulin. *Pediatrics* **2014**, *133*, e154–e162. [CrossRef]
42. Turner, P.; Turner, C.; Green, N.; Ashton, L.; Lwe, E.; Jankhot, A.; Day, N.P.; White, N.J.; Nosten, F.; Goldblatt, D. Serum antibody responses to pneumococcal colonization in the first 2 years of life: Results from an SE Asian longitudinal cohort study. *Clin. Microbiol. Infect.* **2013**, *19*, E551–E558. [CrossRef]
43. Gong, F.; Wei, H.X.; Li, Q.; Liu, L.; Li, B. Evaluation and comparison of serological methods for COVID-19 diagnosis. *Front. Mol. Biosci.* **2021**, *8*, 682405. [CrossRef]
44. Bastos, M.L.; Tavaziva, G.; Abidi, S.K.; Campbell, J.R.; Haraoui, L.P.; Johnston, J.C.; Lan, Z.; Law, S.; MacLean, E.; Trajman, A.; et al. Diagnostic accuracy of serological tests for COVID-19: Systematic review and meta-analysis. *BMJ* **2020**, *370*, m2516. [CrossRef] [PubMed]
45. Mohit, E.; Rostami, Z.; Vahidi, H.A. A comparative review of immunoassays for COVID-19 detection. *Expert Rev. Clin. Immunol.* **2021**, *17*, 573–599. [CrossRef]

biosensors

Article

Quantitation of MicroRNA-155 in Human Cells by Heterogeneous Enzyme-Linked Oligonucleotide Assay Coupled with Mismatched Catalytic Hairpin Assembly Reaction

Oleg L. Bodulev [1], Ivan I. Galkin [2], Shulin Zhao [3], Olga Y. Pletyushkina [2] and Ivan Y. Sakharov [1,*]

[1] Department of Chemistry, Lomonosov Moscow State University, Leninskie Gory, Bldg. 1, 119991 Moscow, Russia; bodulevoleg@mail.ru
[2] A.N. Belozersky Research Institute of Physico-Chemical Biology, Lomonosov Moscow State University, Leninskie Gory, Bldg. 1, 119992 Moscow, Russia; ddeathseller@gmail.com (I.I.G.); pletjush@genebee.msu.ru (O.Y.P.)
[3] State Key Laboratory for the Chemistry and Molecular Engineering of Medicinal Resources, Guangxi Normal University, Guilin 541004, China; zhaoshulin001@163.com
* Correspondence: sakharovivan@gmail.com; Tel.: +7-495-939-3407

Abstract: In the present work, we describe the development of a chemiluminescent enzyme-linked oligonucleotide assay coupled with mismatched catalytic hairpin assembly (mCHA) amplification for the quantitative determination of microRNA-155. To improve its sensitivity, a polymerase-free mCHA reaction was applied as an isothermal amplification method. The detection limit of the proposed assay was 400 fM. In addition, the high specificity of the assay was demonstrated. The proposed assay allowed assessment of the content of microRNA-155 in human cancer lines such as HepG2, Caco2, MCF7, and HeLa. The quantitation of microRNA-155 was performed after purification of short RNAs (less than 200 nt) from cell lysates since a high matrix effect was observed without this pre-treatment. The results of the quantitative determination of the microRNA content in cells were normalized using nematode microRNA-39, the concentration of which was determined using a heterogeneous assay developed by us using a strategy identical to that of the microRNA-155 assay.

Keywords: microRNA detection; polymerase-free isothermal amplification of nucleic acids; mismatched catalytic hairpin assembly

Citation: Bodulev, O.L.; Galkin, I.I.; Zhao, S.; Pletyushkina, O.Y.; Sakharov, I.Y. Quantitation of MicroRNA-155 in Human Cells by Heterogeneous Enzyme-Linked Oligonucleotide Assay Coupled with Mismatched Catalytic Hairpin Assembly Reaction. *Biosensors* **2022**, *12*, 570. https://doi.org/10.3390/bios12080570

Received: 31 May 2022
Accepted: 23 July 2022
Published: 26 July 2022

Publisher's Note: MDPI stays neutral with regard to jurisdictional claims in published maps and institutional affiliations.

1. Introduction

MicroRNAs (miRNAs) are short (17–25 nucleotides in length) non-coding RNA oligonucleotides that regulate gene expression by interacting with specific sites of messenger RNA causing translational repression or its degradation [1,2]. MiRNAs were established to be involved in the regulation of important biological processes such as the development, differentiation, metabolism, and the formation of an immunological response [3]. Currently, about 2600 miRNAs have been detected in the human organism [4]. In addition, previous studies have shown that the miRNA content in tissues, cells, and biological fluids undergo changes in a wide range of diseases [5]. In particular, a relationship was observed between fluctuations in the miRNA expression levels and the development of human oncological diseases. A comparative analysis of samples of healthy individuals and cancer patients revealed upregulation or downregulation of the expression of some miRNAs [6]. This effect depended on the type and stage of the disease. Moreover, a change in the concentration of miRNA was also noted during treatment of patients. Considering the above, numerous efforts are being carried out to develop analytical methods for the determination of miRNA biomarkers.

It should be noted that the concentration of miRNAs in biological tissues and liquids is extremely low (in the fM–pM range [7,8]) and their determination in clinical samples is a challenge for analytical chemists. This means that practically significant miRNA methods should have a low detection limit and high sensitivity coefficient. Often, this is achieved by introducing an amplification stage into the analysis scheme [9,10]. Presently, the most used method of amplification is the reverse transcription polymerase chain reaction (RT-PCR), which is widely applied to copy miRNA sequences. However, RT-PCR has some disadvantages associated with the need to vary the temperature during the analysis, which requires the use of expensive equipment. In addition, miRNAs are short oligonucleotides, and they must be extended in RT-PCR.

These shortcomings have forced researchers to actively look for alternative methods for the amplification of miRNAs. Currently, the prospects of using isothermal methods of miRNA amplification have been demonstrated [11–13]. Among them, a special niche is occupied by enzyme-free methods since they are cheaper. Catalytic hairpin assembly (CHA) and its improved variant, called mismatched CHA (mCHA), are most commonly used.

Another important parameter of miRNA assays is their specificity. This is due to the fact that the number of known miRNAs is high, and their sequences are short, i.e., miRNAs show high homology. Therefore, only highly sensitive and highly specific miRNA assays can be useful for their practical application.

One of the miRNAs considered as a promising biomarker in early diagnosis of cancer is miRNA-155 [5,14]. Its concentration was found to increase in the serum of breast cancer patients [15]. Overexpression of miRNA-155 significantly promotes the proliferation of breast cancer cells, which was explained by the silencing of the Socs1 and RhoA tumor-suppressing genes [16]. Interestingly, the microRNA-155 concentration was reduced after surgery or chemotherapy in patients with breast cancer [17], i.e., the concentration of miRNA-155 can be used to assess the condition of patients during treatment.

In this work, a chemiluminescent heterogeneous method for the determination of miRNA-155 was developed. To enhance the assay's sensitivity, mCHA reaction was used as a polymerase-free amplification method. Furthermore, the use of streptavidin-polyperoxidase conjugate and an enhanced chemiluminescence reaction additionally amplified the chemiluminescent signal. The proposed assay made it possible to quantify miRNA-155 in cancer cell lines such as HeLa, HepG2, Caco2, and MCF-7.

2. Experimental

2.1. Reagents, Materials, and Equipment

Milk casein, Tween-20, 4-morpholinopyridine, Tris, and luminol were obtained from Sigma-Aldrich (St. Louis, MO, USA). Monoclonal anti-fluorescein antibody (anti-FluAb) belonging to the IgG class was produced by Bialexa (Moscow, Russia). Further, 30% H_2O_2 was purchased from ChimMed (Moscow, Russia). Chemiluminescent enhancer for HRP, 3-(10′-phenothiazinyl)-propionic acid was produced as described in [18]. Conjugate of streptavidin and horseradish polyperoxidase (Str–PolyHRP80) was obtained from SDT GmbH (Baesweiler, Germany). Black multiwell microplates (High Binding) were obtained from Thermo Fisher Scientific (Waltham, MA, USA). All oligonucleotides used in this work (Table S1) were obtained from Syntol (Moscow, Russia).

The human cells (MCF-7, HeLa, Caco2, and HepG2) were obtained from the Russian Collection of Cell Cultures (Institute of Cytology, St. Petersburg, Russia). All cells were cultured in Dulbecco's Modified Eagle's Medium with 10% fetal bovine serum at 37 °C in an atmosphere containing 5% CO_2 as described in our previous work [19].

2.2. Measurement of MicroRNA Concentration

First, 96-well black microplates with adsorbed anti-FluAb were prepared as described in [20]. For this, 50 μL of 6 μg/mL monoclonal anti-FluAb in 50 mM carbonate buffer, pH 9.5 was added to the plate wells and incubated for 20 h at 4 °C. Then, the plates were rinsed three times with 10 mM Tris-HCl, pH 7.2 (TB) supplemented with 300 mM NaCl (TBS) and 50 μL of 1 mg/mL casein in TBS was added to the wells. After incubation at 37 °C for 60 min, the plates were rinsed three times with TBS supplemented with 0.05% Tween X100 (TBST). The prepared plates were stored at 4 °C.

Before being used in the assay, Flu-HP1 and B-HP2 were annealed at 88 °C for 15 min and cooled to room temperature for 1 h. The Flu-HP1 probe was annealed at a concentration of 30 nM in 10 mM Tris-HCl, pH 7.2 (TB) supplemented with 20 mM $MgCl_2$. The B-HP2 probe was annealed at a concentration of 100 nM in TB supplemented with 10 mM $MgCl_2$ [21]. The determination of the miRNA concentration in the samples was carried out as follows: In total, 50 µL of Flu-HP1 (30 nM in TB supplemented with 20 mM $MgCl_2$) was added to the plate wells to bind with the adsorbed anti-FITC antibodies. After a 60-min incubation at 37 °C, the plates were rinsed three times with TBST. Afterward, 25 µL of aqueous solution of target miRNA (0–200 pM) or purified small RNA solution and 25 µL 160–1280 nM biotin-H2 in TB supplemented with 20–160 mM $MgCl_2$ was added to the wells. The plates were incubated for 60 min at 25 °C and washed three times with TBST. Thereafter, the plates were filled with 50 µL of Str–PolyHRP80 conjugate solution (diluted 1:100,000 in TBS, supplemented with 1 mg/mL milk casein) and incubated for 60 min at 37 °C. The plates were rinsed thrice again with TBST and 100 µL of freshly prepared solution containing 100 mM Tris, pH 8.3, with 5.2 mM 3-(10′-phenothiazinyl) propionic acid, 9.3 mM 4-morpholinopyridine, and 1 mM luminol, and 3 mM H_2O_2 [22] was introduced to the wells. Chemiluminescence was analyzed using a SpectraMax L luminometer (Molecular Devices, Sunnyvale, CA, USA) at room temperature.

2.3. Purification and Quantitation of Cellular MicroRNAs

The purification of short RNAs from cell lysates was carried out using the LRU-100-50 kit from "BioLabMix" (Russia, Novosibirsk) as described previously [19]. Purified miRNA samples were stored at −20 °C. The quantitation of the microRNA-155 and microRNA-39 in the miRNA samples purified from the cell lysates was performed as described in Section 2.2.

3. Results and Discussion

3.1. Design of the MicroRNA Assay

The principle of the chemiluminescent enzyme-linked oligonucleotide assay for miRNA-155 determination is schematically presented in Figure 1. The mCHA reaction was employed for isothermal amplification of the analytical signal. The sequences of the hairpins used in this study were modeled according to the mCHA theory [23]. Their secondary structures are presented in Figure 2.

Immobilization of the HP1 hairpin probe labeled with fluorescein (Flu-HP1) occurred due to its interaction with the commercial anti-fluorescein antibody (anti-FluAb) preadsorbed on the surface of multiwell plates. Preliminarily, Flu-HP1 was annealed to form a hairpin structure [24]. During the immobilization of Flu-HP1, its concentration was 30 nM, since it was shown that this concentration allows the maximum chemiluminescent signal to be obtained [25].

In the presence of the analyte, the captured Flu-HP1 sequence containing an overhang located at the 3′ end and adjacent stem strand reacts with the miRNA-155 sequence, forming a primary duplex structure (Figure 3A). As a result, the HP1 region at the 5′ end becomes accessible due to its release from the stem. The released sequence then hybridizes with the protruding fragment of the biotin-labeled HP2 (B-HP2), resulting in the formation of the Flu-HP1/B-HP2 complex (Figure 3B). During the formation of the secondary complex, the miRNA-155 molecule is displaced by B-HP2 from the primary complex into the solution. Consequently, miRNA-155 is in a free state and is able to initiate the next cycle of mCHA amplification. The higher the number of amplification cycles, the more molecules of the Flu-HP1/B-HP2 complex are formed.

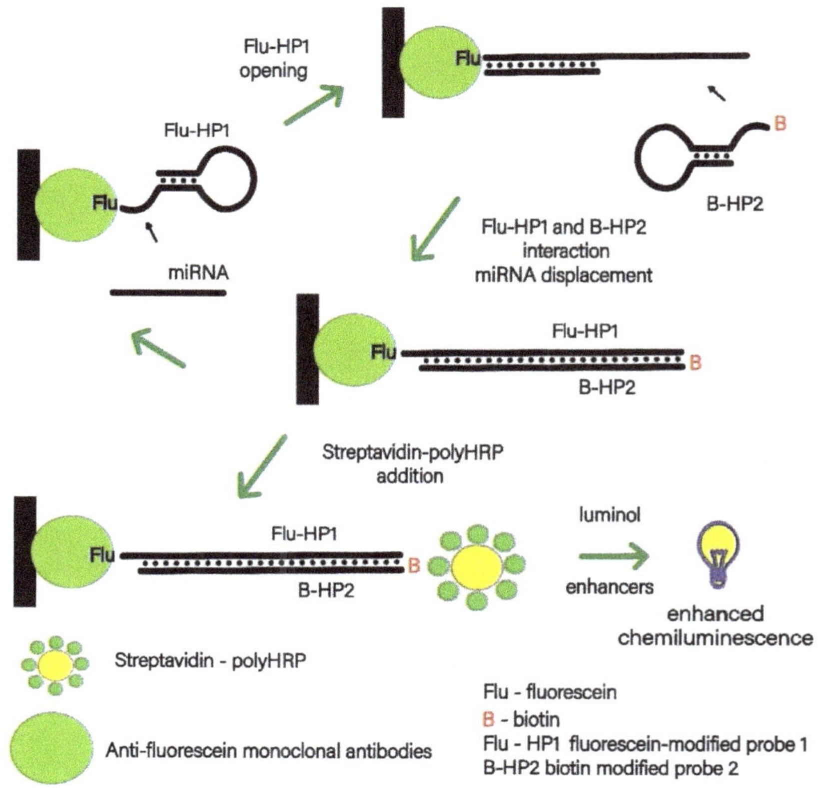

Figure 1. Scheme of the chemiluminescent heterogenous microRNA-155 assay coupled with mismatched catalytic hairpin assembly amplification.

The amount of Flu-HP1/B-HP2 duplex produced was quantified by a reaction with the streptavidin-polyHRP conjugate, the enzyme activity of which was estimated by an enhanced chemiluminescence reaction [22]. The application of the triple amplification strategy led to a multiple increase in the intensity of the chemiluminescent signal.

In the absence of the target, the hairpins used that are in closed forms should not react with each other. Unfortunately, this reaction, depending on the assay conditions, can be observed to one degree or another. This results in the appearance of a background signal. To minimize the background, the experimental conditions for the mCHA-assisted assays should be optimized.

3.2. Optimization of the Assay Conditions for MicroRNA-155 Detection

As mCHA represents a catalytic interaction of hairpins, the main factor affecting the effectiveness of the amplification step in the mCHA-based assays and, therefore, its sensitivity is the concentration of hairpins used. The use of H1/H2 reactive hairpins at low concentrations slows down the reaction, which reduces the assay sensitivity. On the other hand, a sharp increase in the hairpin concentration also reduces the sensitivity due to high background. Thus, it is necessary to find the optimal concentration of hairpins in the reaction medium. It should be noted that the optimal concentrations of hairpins may also depend on the time of the amplification reaction.

In order for the proposed analysis to not take much time, the amplification was carried out for 1 h. The concentration of the B-HP2 probe in the reaction solution varied within 80–640 nM. Under these conditions, a set of calibration curves were constructed for the determination of miRNA-155 (Figure 4A). For each curve, the detection limit was calculated by the 3σ rule. As seen in Figure 4B, a minimum detection limit was obtained using the B-HP2 probe at a concentration of 160 nM. At other concentrations of B-HP2, the sensitivity of the assay was lower.

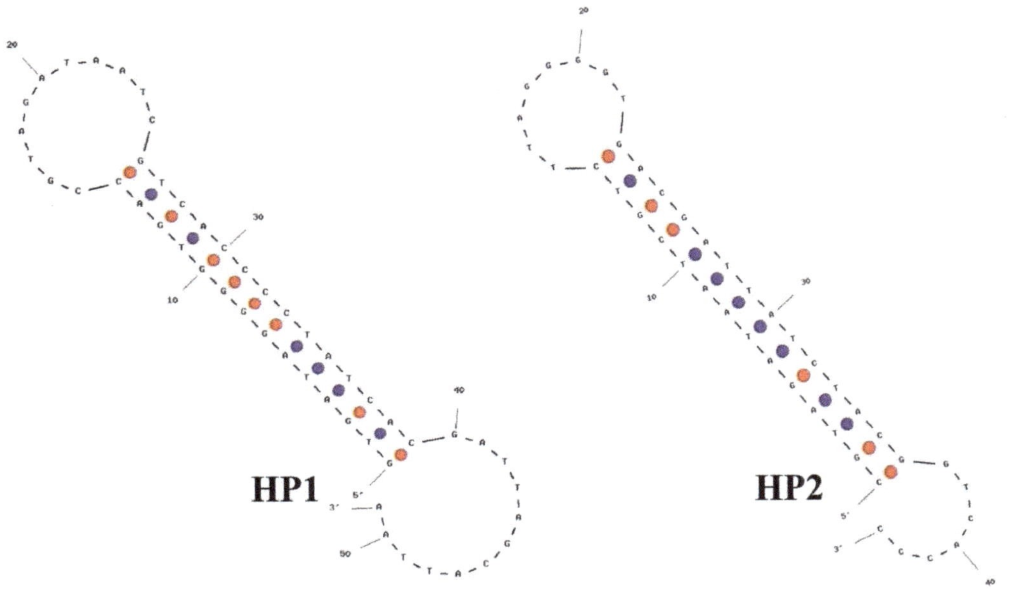

ΔG (HP1 formation) = −13.4 kCal/mole **ΔG (HP2 formation) = −13 kCal/mole**

Figure 2. Secondary structure of the hairpins used in the chemiluminescent heterogenous assay based on the mismatched catalytic hairpin assembly reaction for microRNA-155 detection. Modeling of the hairpin structures and ΔG assessment were performed using OligoAnalyzer 3.1 software.

(A)

```
5' GTGATAGGGGTGACCGTAGATAATCGTCACCCCTATCACGATTAGCATTAA-Flu
                                                   ||||||||||||||||||||||||
3'                                                 UGGGGAUAGUGCUAAUCGUAAUU
```

[ΔG (primary duplex formation) = −42 kCal/mole]

(B)

```
5' GTGATAGGGGTGACCGTAGATAATCGTCACCCCTATCACGATTAGCATTAA-Flu
              |||||||||||||||||||||||||||||   |||||||
3' biotin-CCCACTGGCATCTATTAGCAGTGGGGATTCTGCTAATAGATGC
```

[ΔG (secondary duplex formation) = −53.5 kCal/mole]

Figure 3. Schematic illustration of the formation of (**A**) primary and (**B**) secondary duplexes produced in the chemiluminescent assay based on the mismatched catalytic hairpin assembly reaction for microRNA-155 detection. The underlined nucleotide fragments are involved in the stems of the hairpin structures. Modeling of the hairpin structures and ΔG assessment were performed using OligoAnalyzer 3.1 software.

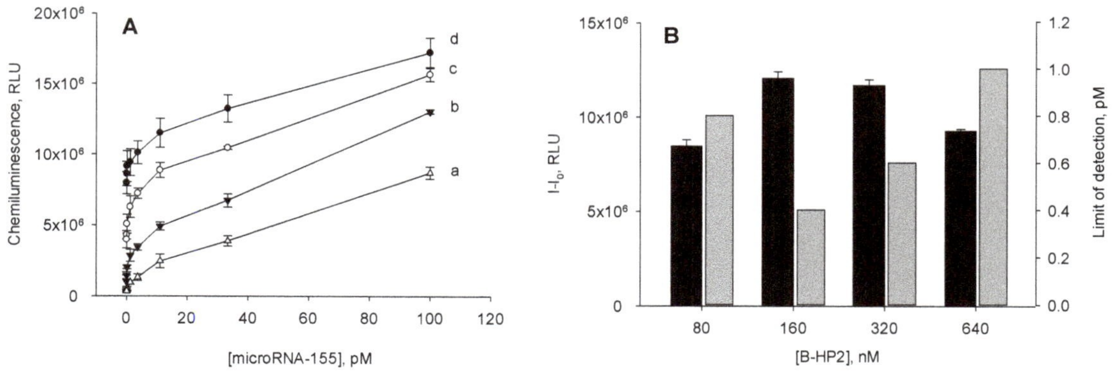

Figure 4. Effect of the B-HP2 probe concentration in the reaction solution on (**A**) the behavior of the calibration curves, (**B**) chemiluminescence signal ($I–I_0$) (black columns), and limit of detection (gray columns) of the amplified microRNA-155 assay. The mismatched catalytic hairpin assembly reaction was carried out in 10 mM Tris-HCl with pH 7.2 containing 20 mM $MgCl_2$ at 25 °C for 1 h. The concentration of the B-HP2 probe was (a) 80, (b) 160, (c) 320, and (d) 640 mM. The value of the chemiluminescence signal ($I–I_0$) was calculated as the difference between the chemiluminescence intensities recorded in the presence (100 pM) and in the absence of microRNA-155. The detection limit was calculated using the 3σ rule.

We also analyzed the dependence of the chemiluminescent signal formed in the presence of 100 pM of miRNA-155 on the concentration of the B-HP2 probe (Figure 4). When the concentration of the B-HP2 probe was in the range of 160–320 nM, a maximum signal was observed. The use of the B-HP2 probe outside this concentration range resulted in a decrease in the chemiluminescence. Considering the factors mentioned above, in the further work, we used the B-HP2 probe at a concentration of 160 nM.

It is well known that the salt composition of the reaction solution strongly affects the affinity of the interaction of complementary nucleic acids. NaCl and $MgCl_2$ are salts that are usually used in the reactions of nucleic acids. The salt effect is especially significant when carrying out heterogeneous reactions [26]. In our work, we showed that the mCHA reaction of the immobilized Flu-H1 and soluble B-HP2 did not proceed in 10 mM Tris-HCl, pH 7.2. The introduction of $MgCl_2$ to the reaction buffer intensified the mCHA and this effect was concentration dependent (Figure 5). The favorable concentration of $MgCl_2$ was 20 mM. It should be noted that in 10 mM Tris-HCl, pH 7.2 containing 20 mM $MgCl_2$, we observed not only the maximum chemiluminescent signal but also the maximum sensitivity (a minimum value of detection limit) for miRNA-155 detection (Figure 5). The addition of NaCl to the $MgCl_2$-containing buffer had no positive effect on the assay sensitivity (data not shown).

3.3. Analytical Parameters of the Heterogenous mCHA-Based Assay of MicroRNA-155

Under favorable conditions (10 mM Tris-HCl of pH 7.2 with 20 mM $MgCl_2$ and 160 nM B-HP2 probe), the behavior of the calibration curve for miRNA-155 detection (Figure 6) obeyed the following equation:

$$\gamma = \frac{ax}{b+x} + \frac{cx}{d+x}\ (\mathrm{R}^2\ 0.9980)$$

where *a*, *b*, *c* and *d* are 2.5×10^6, 0.22, 1.1×10^8, and 1.06×10^3, respectively (curve fitting was performed using SigmaPlot 12.5 software). The detection limit of miRNA-155 was 400 fM. The coefficient of variation of the chemiluminescent signal within the working range was less than 12%.

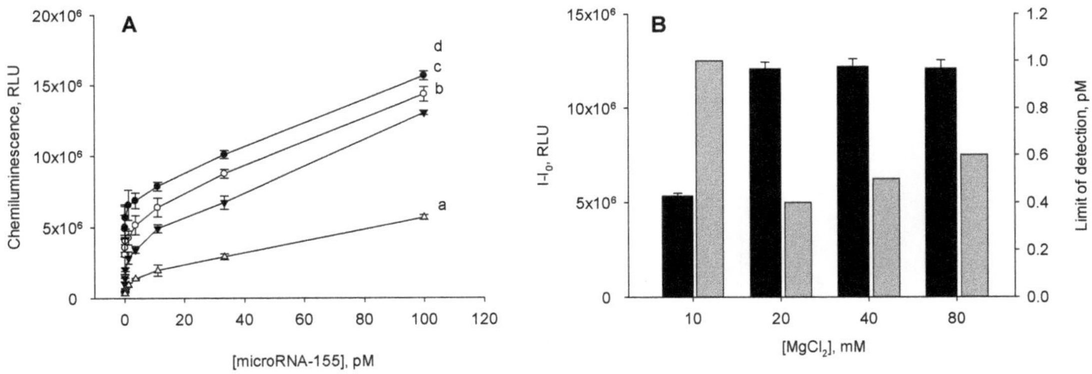

Figure 5. Effect of the $MgCl_2$ concentration in the reaction solution at the stage of the mismatched catalytic hairpin assembly reaction (mCHA) amplification on (**A**) the behavior of the calibration curves constructed, (**B**) chemiluminescence signal ($I–I_0$) (black columns), and limit of detection (gray columns) of the amplified microRNA-155 assay. The mCHA reaction was carried out at 25 °C for 1 h using the 160 nM B-HP2 probe in 10 mM Tris-HCl with pH 7.2 containing $MgCl_2$ concentrations of (a) 10, (b) 20, (c) 40, and (d) 80 mM. The value of the chemiluminescence signal ($I–I_0$) was calculated as the difference between the chemiluminescence intensities recorded in the presence (100 pM) and in the absence of the microRNA-155. The detection limit was calculated based on the 3σ rule.

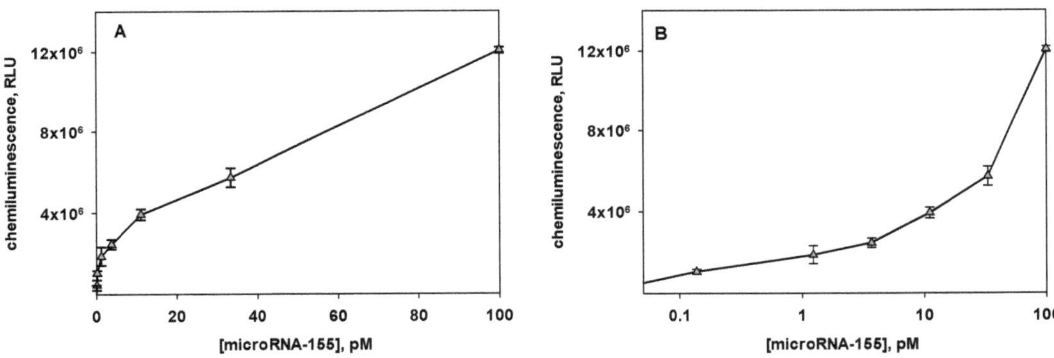

Figure 6. Calibration curve for the determination of microRNA-155 by the chemiluminescent heterogenous assay based on mismatched catalytic hairpin assembly amplification (n = 6) presented in (**A**) linear and (**B**) semi-logarithmic coordinates.

3.4. Specificity

To assess the specificity of the proposed assay, the cross-reactivity of target microRNA-155 and five widely studied miRNAs (miRNA-141, miRNA-319a, miRNA-21, miRNA-205, and miRNA-39) was investigated. As seen in Figure 7, only in the case of miRNA-155 a high chemiluminescent signal was observed.

For the other miRNAs studied, the signals were negligible. This result allowed us to characterize the developed method as highly specific.

3.5. Detection of MicroRNAs in Human Cells

The high matrix effect observed when analyzing samples with a complex composition, such as biological real samples, often does not allow a high sensitivity and the required accuracy to be obtained. In order to evaluate the matrix effect observed during the detection of miRNAs in cell lysates by the developed assay, we prepared spiked samples of miRNA-39 added to MCF-7 lysates. MiRNA-39 was chosen in this experiment because this nematode miRNA is absent in human tissues and fluids [27]. The determination of the concentration of miRNA-39 in the assay buffer and the spiked samples was carried out by the chemiluminescent heterogeneous assay, which was constructed using a strategy similar to that of the assay of miRNA-155 (Figure S1).

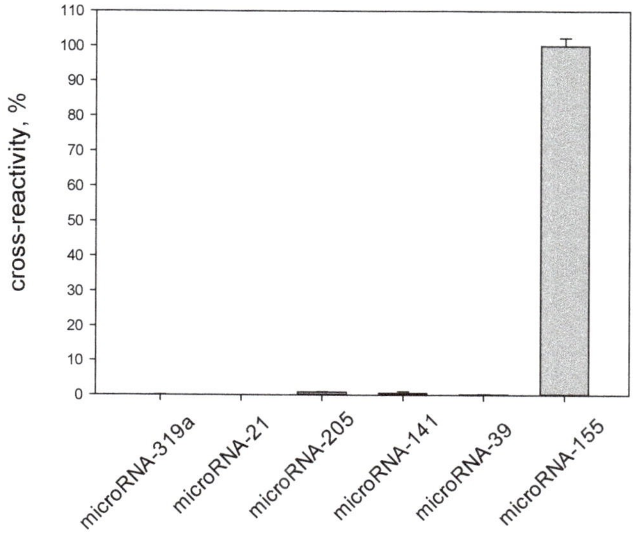

Figure 7. Specificity of the chemiluminescence heterogeneous method for the determination of microRNA-155 based on mismatched catalytic hairpin assembly amplification (n = 3). The concentration of the studied microRNAs was 100 pM.

The results of the optimization of the experimental conditions of the miRNA-39 assay are presented in the Supplementary Materials (Figures S2 and S3). The calibration curve of the heterogenous mCHA-based assay of miRNA-39 obeyed the following equation:

$$\gamma = \frac{ax}{b+x} + \frac{cx}{d+x} \ (R^2\ 0.9976)$$

where a, b, c and d are 5.03×10^6, 6.5, 6.3×10^{14}, and 5.8×10^9, respectively (Figure S4). The detection limit and coefficient of variation of the chemiluminescent signal in the working range were 300 fM and less than 11%, respectively. Similar to the miRNA-155 assay, the assay of miRNA-39 was also highly specific (Figure S5).

When we tried to determine miRNA-39 in the crude MCF-7 lysate using the mCHA-based assay, we observed a very high matrix effect. Therefore, using the proposed assay, the direct quantitation of miRNA-39 cannot be carried out. To prevent this effect, a commercial kit LRU-100-50 (BioLabMix) for the purification of short RNAs (less than 200 nts in length), including miRNAs (see Section 2.3), was used. Actually, the matrix effect was not observed when the spiked samples were prepared by introducing miRNA-39 to a solution of short RNAs isolated from MCF-7 lysates (data not shown). The results obtained permitted us to conclude that the calibration curve generated when the analysis was performed in the assay buffer can be used to calculate the miRNA concentrations in pre-treated samples.

Although the BioLabMix kit prevents the matrix effect, its use results in a partial loss of miRNAs. To determine the purification yield, we prepared two spiked samples. One of the samples was prepared by adding miRNA-39 to the crude MCF-7 lysate followed by purification using the BioLabMix kit. Another sample was prepared by adding miRNA-39 to the solution of purified small RNAs prepared from the crude MCF-7 lysate with the same kit. The purification yield determined by comparing the chemiluminescent signals generated by the analysis of both spiked samples was in the range of 50–60%. Since miRNA molecules are highly homologous, we believe that the obtained values of the purification yield can be used to normalize the results of the detection of all miRNAs, including miRNA-155.

Using the BioLabMix kit, the content of miRNA-155 in some cultured human cells was evaluated by the proposed assay. The average content of miRNA-155 in HepG2 cells (cell line isolated from a hepatocellular carcinoma) was 180 ± 40 copies per cell. In Caco2 cells (epithelial cells isolated from colon cancer tissue), the content of the target was lower and equal to 68 ± 26 copies per cell. In contrast, in MCF7 cells (breast cancer cells) and HeLa cells (cell line isolated from a cervical carcinoma), miRNA-155 was not detected.

4. Conclusions

In the present work, we described the development of a chemiluminescent heterogeneous assay for the quantitative determination of miRNA-155. To improve its sensitivity, a polymerase-free mCHA was applied as an isothermal amplification method. This allowed the construction of an assay with a detection limit of 400 fM. The proposed assay also showed high specificity. It should be noted that the use of commercially available microplates as a carrier for the heterogeneous assay (in our work, we used 96-well plates, though 384- or 1536-well plates can also be used) makes it possible to easily standardize and automate the determination of miRNA-155.

The proposed assay allowed successful determination of the content of miRNA-155 in human cancer cell lines such as HepG2, Caco2, MCF7, and HeLa. The quantitation of miRNA-155 was performed after the purification of short RNAs (less than 200 nts) from cell lysates since without such pre-treatment, a high matrix effect was observed. The results of the evaluation of the miRNA content in cells were normalized using nematode miRNA-39, the concentration of which was determined using a heterogeneous assay developed by us using a strategy identical to that of the miRNA-155 assay.

Supplementary Materials: The following supporting information can be downloaded at: https://www.mdpi.com/article/10.3390/bios12080570/s1, Structure of oligonucleotides used in this work; the experimental results of the optimization of the chemiluminescent assay of microRNA-39. Figure S1. Secondary structures of (A) the hairpins used in chemiluminescent heterogenous mCHA-based microRNA-39 assay, (B) Flu-HP1- microRNA-39, and (C) Flu-HP1- B-HP2 duplexes, and ΔG values of their formation. Underlined nucleotides are involved in the stem of hairpin structures. Modeling of the hairpin structures and ΔG assessment were performed using OligoAnalyzer 3.1 software. Figure S2. Effect of the B-HP2 probe concentration in the reaction solution on (A) the behavior of the calibration curves, (B) chemiluminescence signal ($I–I_0$) (black columns), and limit of detection (gray columns) of the amplified microRNA-39 assay. The mismatched catalytic hairpin assembly reaction was carried out in 10 mM Tris-HCl with pH 7.2 containing 20 mM MgCl2 at 25 °C for 1 h. The concentration of the B-HP2 probe was (a) 80, (b) 160, (c) 320, and (d) 640 mM. The value of the chemiluminescence signal ($I–I_0$) was calculated as the difference between the chemiluminescence intensities recorded in the presence (100 pM) and in the absence of microRNA-39. The detection limit was calculated using the 3σ rule. Figure S3. Effect of the MgCl2 concentration in the reaction solution at the stage of mCHA amplification on (A) the behavior of the calibration curves constructed, (B) chemiluminescence signal ($I–I_0$) (black columns), and limit of detection (gray columns) of the amplified microRNA-39 assay. The mismatched catalytic hairpin assembly reaction was carried out at 25 °C for 1 h using the 160 nM B-HP2 probe in 10 mM Tris-HCl with pH 7.2 containing MgCl2 concentrations of (a) 10, (b) 20, (c) 40, and (d) 80 mM. The value of the chemiluminescence signal ($I–I_0$) was calculated as the difference between the chemiluminescence intensities recorded in the presence (100 pM) and in the absence of the microRNA-39. The detection limit was calculated based on the 3σ rule. Figure S4. Calibration curve for the determination of microRNA-39 by the chemiluminescent heterogenous assay based on the mismatched catalytic hairpin assembly reaction (n = 6) presented in (A) linear and (B) semi-logarithmic coordinates. The mCHA reaction was carried out using 160 nM B-HP2 in 10 mM Tris-HCl with pH 7.2 containing 20 mM MgCl2 at 25 °C for 1 h. Figure S5. Specificity of the chemiluminescence heterogeneous method for the determination of microRNA-39 based on the mismatched catalytic hairpin assembly reaction (n = 3). The concentration of the studied miRNAs was 100 pM. Table S1. The list of oligonucleotides used in this work (5′-3′).

Author Contributions: Conceptualization, I.Y.S. and S.Z.; methodology, O.L.B., I.I.G. and O.Y.P., writing—original draft preparation, O.L.B. and I.Y.S.; writing—review, O.L.B. and I.Y.S. All authors have read and agreed to the published version of the manuscript.

Funding: This research was funded by the Russian Foundation for Basic Research (Grant No. 21-54-53007) and the National Natural Science Foundation of China (Grant No. 2201101378).

Institutional Review Board Statement: Not applicable.

Informed Consent Statement: Not applicable.

Data Availability Statement: Not applicable.

Conflicts of Interest: The authors declare no conflict of interest.

References

1. Pritchard, C.C.; Cheng, H.H.; Tewari, M. MicroRNA profiling: Approaches and considerations. *Nat. Rev. Gen.* **2012**, *13*, 358–369. [CrossRef] [PubMed]
2. Kozomara, A.; Birgaoanu, M.; Griffiths-Jones, S. MiRBase: From miRNA sequences to function. *Nucl. Acids Res.* **2019**, *47*, D155–D162. [CrossRef] [PubMed]
3. Alvarez-Garcia, I.; Miska, E.A. MicroRNA functions in animal development and human disease. *Development* **2005**, *132*, 4653–4662. [CrossRef] [PubMed]
4. Plotnikova, O.; Baranova, A.; Skoblov, M. Comprehensive analysis of human microRNA–mRNA interactome. *Front. Gen.* **2019**, *10*, 933. [CrossRef] [PubMed]
5. Due, H.; Svendsen, P.; Bødker, J.S.; Schmitz, A.; Bøgsted, M.; Johnsen, H.E.; El-Galaly, T.C.; Stidsholt Roug, A.; Dybkær, K. MiR-155 as a Biomarker in B-Cell Malignancies. *Biomed. Res. Intern.* **2016**, *2016*, 9513037. [CrossRef] [PubMed]
6. Lu, J.; Getz, G.; Miska, E.A.; Alvarez-Saavedra, E.; Lamb, J.; Peck, D.; Sweet-Cordero, A.; Ebert, B.L.; Mak, R.H.; Ferrando, A.A.; et al. MicroRNA expression profiles classify human cancers. *Nature* **2005**, *435*, 834–838. [CrossRef] [PubMed]
7. Williams, Z.; Ben-Dov, I.Z.; Elias, R.; Mihailovic, A.; Brown, M.; Rosenwaks, Z.; Tuschl, T. Comprehensive profiling of circulating miRNA via small RNA sequencing of cDNA libraries reveals biomarker potential and limitations. *Proc. Natl. Acad. Sci. USA* **2013**, *110*, 4255–4260. [CrossRef]
8. Labib, M.; Berezovski, M.V. Electrochemical sensing of miRNAs: Avenues and paradigms. *Biosens. Bioelectron.* **2015**, *68*, 83–94. [CrossRef]
9. Tang, Y.; Liu, M.; Zhao, Z.; Li, Q.; Liang, X.; Tian, J.; Zhao, S. Fluorometric determination of microRNA-122 by using ExoIII-aided recycling amplification and polythymine induced formation of copper nanoparticles. *Microchim. Acta* **2019**, *186*, 133. [CrossRef]
10. Chen, S.; Zhao, J.; Xu, C.; Sakharov, I.Y.; Zhao, S. Absolute Quantification of MicroRNAs in a Single Cell with CL Detection Based on Rolling Circle Amplification on a Microchip Platform. *Anal. Chem.* **2021**, *93*, 9218–9225. [CrossRef]
11. Bodulev, O.L.; Sakharov, I.Y. Modern Methods for Assessment of microRNAs. *Biochemistry* **2022**, *87*, 425–442. [CrossRef] [PubMed]
12. Zhao, Y.; Chen, F.; Li, Q.; Wang, L.; Fan, C. Isothermal Amplification of Nucleic Acids. *Chem. Rev.* **2015**, *115*, 12491–12545. [CrossRef] [PubMed]
13. Ye, J.; Xu, M.; Tian, X.; Cai, S.; Zeng, S. Research advances in the detection of miRNA. *J. Pharm. Anal.* **2019**, *9*, 217–226. [CrossRef] [PubMed]
14. Wang, F.; Hou, J.; Jin, W.; Li, J.; Yue, Y.; Jin, H.; Wang, X. Increased circulating microRNA-155 as a potential biomarker for breast cancer screening: A meta-analysis. *Molecules* **2014**, *19*, 6282–6293. [CrossRef] [PubMed]
15. Zeng, H.; Fang, C.; Nam, S.; Cai, Q.; Long, X. The clinicopathological significance of microRNA-155 in breast cancer: A meta-analysis. *BioMed Res. Int.* **2014**, *2014*, 724209. [CrossRef]
16. Jiang, S.; Zhang, H.W.; Lu, M.H.; He, X.H.; Li, Y.; Gu, H.; Liu, M.F.; Wang, E.D. MicroRNA-155 Functions as an OncomiR in Breast Cancer by Targeting the Suppressor of Cytokine Signaling 1 Gene. *Cancer Res.* **2010**, *70*, 31193127. [CrossRef] [PubMed]
17. Sun, Y.; Wang, M.; Lin, G.; Sun, S.; Li, X.; Qi, J.; Li, J. Serum MicroRNA-155 as a Potential Biomarker to Track Disease in Breast Cancer. *PLoS ONE* **2012**, *7*, e47003. [CrossRef]
18. Han, F.; Zhao, J. Colorimetric Fluoride-Anion Sensor Based on Intramolecular Hydrogen Bonding and Enol-Keto Tautomerization of a Phenothiazine Derivative. *Helv. Chim. Acta* **2008**, *91*, 635–645. [CrossRef]
19. Solovjev, A.M.; Galkin, I.I.; Pletjushkina, O.Y.; Medvedko, A.V.; Zhao, S.; Sakharov, I.Y. Isothermal chemiluminescent assay based on circular stand-displacement polymerization reaction amplification for cel-miRNA-39-3p determination in cell extracts. *Int. J. Biol. Macromol.* **2021**, *182*, 987–992. [CrossRef]
20. Solovjev, A.M.; Kurzeev, S.A.; Sakharov, I.Y. Chemiluminescent microplate-based assay of DNA based on isothermal circular strand-displacement polymerization reaction (ICSDPR). *Talanta* **2020**, *215*, 120895. [CrossRef]
21. Bodulev, O.L.; Zhao, S.; Sakharov, I.Y. Improving the sensitivity of the miRNA assay coupled with the mismatched catalytic hairpin assembly reaction by optimization of hairpin annealing conditions. *Anal. Chem.* **2021**, *93*, 6824–6830. [CrossRef]

22. Sakharov, I.Y.; Demiyanova, A.S.; Gribas, A.V.; Uskova, N.A.; Efremov, E.E.; Vdovenko, M.M. 3-(10′-Phenothiazinyl)propionic acid is a potent primary enhancer of peroxidase-induced chemiluminescence and its application in sensitive ELISA of methylglyoxal-modified low density lipoprotein. *Talanta* **2013**, *115*, 414–417. [CrossRef]
23. Jiang, Y.S.; Bhadra, S.; Li, B.; Ellington, A.D. Mismatches improve the performance of strand-displacement nucleic acid circuits. *Angew. Chem. Int. Ed. Engl.* **2014**, *126*, 1876–1879. [CrossRef]
24. Braunlin, W.H.; Giri, I.; Beadling, L.; Breslauer, K.J. Conformational screening of oligonucleotides by variable-temperature high performance liquid chromatography: Dissecting the duplex–hairpin–coil equilibria of d(CGCGAATTCGCG). *Biopolymers* **2004**, *74*, 221–231. [CrossRef]
25. Bodulev, O.L.; Burkin, K.M.; Efremov, E.E.; Sakharov, I.Y. One-pot microplate-based chemiluminescent assay coupled with catalytic hairpin assembly amplification for DNA detection. *Anal. Bioanal. Chem.* **2020**, *412*, 5105–5111. [CrossRef]
26. Špringer, T.; Šípová, H.; Vaisocherová, H.; Štěpánek, J.; Homola, J. Shielding effect of monovalent and divalent cations on solid-phase DNA hybridization: Surface plasmon resonance biosensor study. *Nucl. Acids Res.* **2010**, *38*, 7343–7351. [CrossRef]
27. Wang, Y.N.; Yu, L.; Zhao, X.S.; Zhang, W.; Wan, J.; Yu, B. Establishment of plasma microRNA detection method by using Taqman probe based quantitative reverse transcription PCR. *Cell. Mol. Biol.* **2015**, *61*, 51–56. [CrossRef]

biosensors

MDPI

Article

Complementary Strategy Enhancing Broad-Specificity for Multiplexed Immunoassay of Adulterant Sulfonylureas in Functional Food

Zhaodong Li [1,†], Haihuan Xie [2,†], Tingdan Fu [3], Yingying Li [4], Xing Shen [1], Xiangmei Li [1], Yi Lei [5], Xiaojun Yao [6], Anastasios Koidis [7], Yingju Liu [1], Xinan Huang [8] and Hongtao Lei [1,*]

1 Guangdong Provincial Key Laboratory of Food Quality and Safety/National-Local Joint Engineering Research Center for Machining and Safety of Livestock and Poultry Products, South China Agricultural University, Guangzhou 510642, China; scaulizhaodong@scau.edu.cn (Z.L.); shenxing325@163.com (X.S.); lixiangmei12@163.com (X.L.); yingjuliu@scau.edu.cn (Y.L.)
2 Wuzhou Institute for Food and Drug Control, Wuzhou 543099, China; 704596242@foxmail.com
3 Institute of Tropical Medicine, Guangzhou University of Chinese Medicine, Guangzhou 510400, China; tingsane@163.com
4 Wuzhou Product Quality Inspection Institue, Wuzhou 543001, China; liyingyingwz@163.com
5 Guangdong Institute of Food Inspection, Zengcha Road, Guangzhou 510435, China; leiy04@foxmail.com
6 Dr. Neher's Biophysics Laboratory for Innovative Drug Discovery, State Key Laboratory of Quality Research in Chinese Medicine, Macau Institute for Applied Research in Medicine and Health, Macau University of Science and Technology, Macau 999078, China; xjyao@must.edu.mo
7 Institute for Global Food Security, Queen's University Belfast, 19 Chlorine Gardens, Belfast BT9 5DJ, UK; t.koidis@qub.ac.uk
8 Tropical Medicine Institute & South China Chinese Medicine Collaborative Innovation Center, Guangzhou University of Chinese Medicine, Guangzhou 510405, China; xinanhuang@gzucm.edu.cn
* Correspondence: hongtao@scau.edu.cn; Tel.: +86-20-8528-3925
† These authors contributed equally to this work.

Citation: Li, Z.; Xie, H.; Fu, T.; Li, Y.; Shen, X.; Li, X.; Lei, Y.; Yao, X.; Koidis, A.; Liu, Y.; et al. Complementary Strategy Enhancing Broad-Specificity for Multiplexed Immunoassay of Adulterant Sulfonylureas in Functional Food. *Biosensors* **2022**, *12*, 591. https://doi.org/10.3390/bios12080591

Received: 13 June 2022
Accepted: 28 July 2022
Published: 2 August 2022

Abstract: Sulfonylureas, a family of anti-diabetic drugs widely used in the clinical treatment of type 2 diabetes, have recently emerged as an illegal adulterant in functional foods, to enhance the claimed anti-diabetic activity. To establish a screening assay method against their adulteration, with the aid of molecular simulation of hapten, two antibodies were raised and complementarily used to enhance the broad-specificity of an enzyme-linked immunosorbent assay (ELISA), which demonstrated simultaneous detection capability to 6 sulfonylureas; the detection limits ranged from 0.02 to 1.0 ng/mL, and recoveries were between 78.3% to 104.5%. Liquid chromatography with tandem mass spectrometry (LC-MS/MS) confirmed the reliability of the proposed ELISA, based on real samples. These results suggest that the proposed ELISA could be an ideal method for screening to monitor for illicit adulteration of sulfonylureas in functional pill products.

Keywords: sulfonylureas; broad-specificity; antibody; immunoassay; adulteration; functional foods

1. Introduction

More than 463 million people worldwide are afflicted by a chronic disease characterized the persistent hyperglycemia known as type 2 diabetes [1]. These patients need to stay on medication throughout their lives to help their bodies metabolize glucose. The common hypoglycemic drugs include insulin, metformin, and sulfonylureas. However, the body may develop resistance if these medications are given in the long term [2]. As a result, to reduce dependence on glucose-lowering drugs, many diabetics prefer to consume functional foods or natural medicines that are supposedly "natural, side-effect-free, and efficacious" [3]. However, some so-called strongly active products have recently been found to be illegally added synthetic drugs, such as sulfonylureas, added to functional foods or natural medicines [4] in order to "boost" their effect.

Sulfonylureas are a typical class of drugs used orally for the treatment of type 2 diabetes, including first-generation drugs that are no longer used (acetohexamide, tolbutamide, chlorpropamide, carbutamide, and tolazamide), and second-generation drugs (gliclazide, glyburide, glipizide, gliquidone, glimepiride, and glibornuride) [5]. The addition of these therapeutic drugs to dietary supplements illegally violates the law and the original intent of consumers. However, on a more important note, it will lead to a profound and serious public health problem, exposing consumers to the risk of serious adverse reactions in the form of acute hypoglycemia, diarrhea, and erythrocytic anemia [6]. Currently, analytical methods for sulfonylureas are generally instrumental methods, such as ultra-performance liquid chromatography (UPLC) [7], liquid chromatography-tandem mass spectrometry (LC-MS/MS) [8], real-time mass spectrometry (DART-MS) [6,9], and on-line two-dimensional liquid chromatography (2D-MS) [10], etc. Instrumental methods are very sensitive and reliable. However, they require complex sample pretreatment, expensive equipment support, and large amounts of organic solvents that cannot be used for the rapid detection of sulfonylureas in functional foods. In contrast, immunoassay, such as the enzyme-linked immunosorbent assay (ELISA), has been widely used in the rapid screening of illegal additives in functional foods, due to its simplicity, rapidity, and ease of interpretation [11,12]. However, no ELISA has been reported for sulfonylurea detection in functional foods. Besides, it is known that the sulfonylurea family contains a class compound with similar structures and activity; therefore, there is an urgent need to enhance assay efficiency to establish a immunoassay that possesses the simultaneous detection capability for sulfonylureas in one, rather than repeatedly.

In this study, four novel haptens were rationally designed by aligning the common structure of molecular modeling of 11 sulfonylureas, and then immunizing rabbits to generate two polyclonal antibodies with broad-specificity that can recognize many sulfonylureas. An ELISA (Figure 1) was established based on a complementary strategy, the combination of the obtained two antibodies for the simultaneous detection of 6 sulfonylureas, which has recently been in functional pills with claimed anti-diabetic properties.

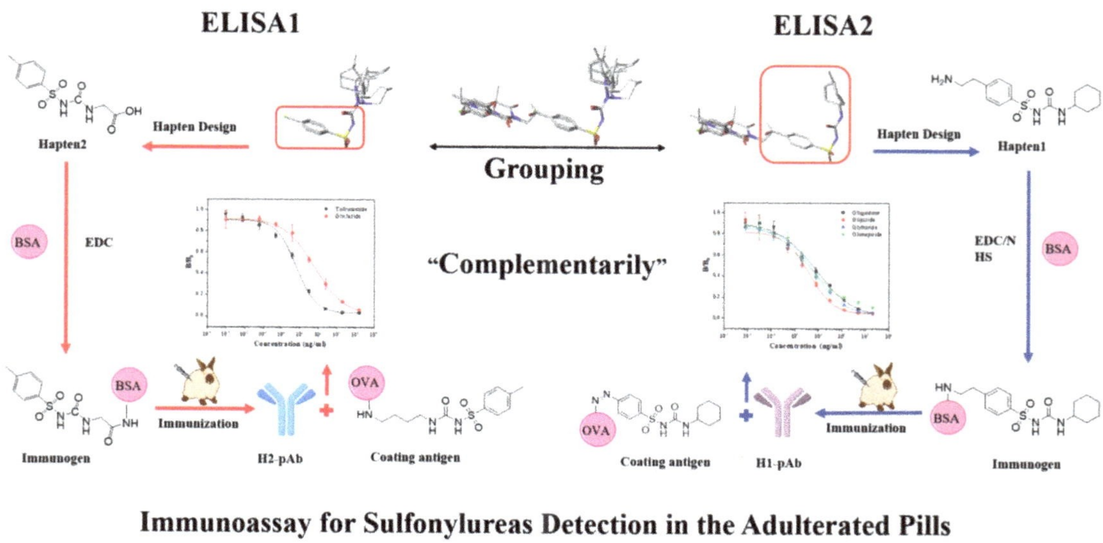

Figure 1. Schematic illustration of complementary ic-ELISA for Sulfonylureas. H1-pAb, antibody to Hapten 1. H2-pAb, antibody to Hapten 2. BSA, bovine serum albumin. EDC/NHS, two coupling regents EDC and NHS for active ester method.

2. Materials and Methods

2.1. Chemicals and Reagents

The chemicals 4-(2-aminoethyl) benzenesulfonamide (CAS 35303-76-5), ethyl tosylcarbamate (CAS 5577-13-9), di-tert-butyl decarbonate (CAS 24424-99-5), 4-methylbenzenesulfonyl isocyanate (CAS 4083-64-1), cyclohexyl isocyanate (CAS 3173-53-3), tert-butyl (4-aminobutyl) carbamate (CAS 1088779-66-1), glycine methyl ester hydrochloride (CAS 5680-79-5), N-(4-sulfamoylphenyl)acetamide (CAS 121-61-9), acetohexamide, tolbutamide, chlorpropamide, carbutamide, tolazamide, glibornuride, gliclazide, glimepiride, glyburide, gliquidone, glipizide, rosiglitazone, phenformin, metformin hydrochloride, and repaglinide, were purchased from Shanghai Aladdin Biochemical Technology Co., Ltd. (Shanghai, China). Bovine serum albumin (BSA), ovalbumin (OVA), 1-(3-Dimethylaminopropyl)-3-ethyl carbodiimide hydrochloride (EDC), hydrolyzed protein, N-hydroxylsuccinamide (NHS), Freund's complete adjuvant, Freund's incomplete adjuvant, 3,3′,5,5′-tetramethylbenzidine (TMB), and goat anti-rabbit IgG-HRP were purchased from Sigma-Aldrich (St. Louis, MO, USA). Dichloromethane, trifluoroacetic acid, N, N-Dimethylformamide (DMF), Tween-20 and methanol, were bought from Damao Chemical Reagent Co., Ltd. (Tianjin, China). All reagents are in analytical reagent grade or higher purity.

New Zealand White rabbits (2–3 months old about 2 kg) were purchased from Guangdong Medical Laboratory Animal Center and kept in the Animal Experiment Centre of South China Agriculture University (Animal Experiment Ethical Approval Number: 2020e009, Supplementary Materials Figure S9). All necessary animal work licenses were obtained prior to the start of the work.

2.2. Instruments

Ultraviolet−visible spectra were documented on a Nanodrop 2000C spectrophotometer (Thermo, MA, USA). Absorbance was measured on a Multiskan Spectrum (Thermo, Waltham, MA, USA). Polystyrene 96-well plates (KE-96−8) were acquired from Yijiamei (Xiamen, China). The LC-MS/MS experiment was performed on a AB QTRAP4500 triple quadrupole mass spectrometer (AB SCIEX, Framingham, MA, USA), and POROSHELL HPH-C18 (2.1 × 150 mm, 4 μm, Agilent, Santa Clara, CA, USA) was used to separate compounds.

2.3. Buffers and Solutions

The following buffers and solutions were used: (A) 0.1 mol/L 2-(N-morpholine) ethanesulfonic acid solution, pH 4.5–5.0, as binding buffer for the EDC method. (B) 0.1 mol/L carbonate buffer (CB), pH 9.6, as coating buffer. (C) Washing solution consisting of PBST solution in phosphate-buffered saline (PBS, 0.01 mol/L, pH 7.4) containing 0.5% Tween-20. (D) Blocking buffer to be prepared with 0.1% hydrolyzed protein (w/v) and 0.01 mol/L PBST. (E) 2 mol/L sulfuric acid as the termination solution. (F) Standard stock solution (1 mg/mL) produced by dissolving an appropriate amount of standards in methanol and stored at 4 °C until use.

2.4. Conjugation

The hapten synthesis is described in detail in the Supplementary Materials.

2.4.1. Hapten 1-OVA/BSA

Hapten 1 was coupled to BSA and OVA by a direct EDC method [13] with a slight change. Briefly, 8 mg Hapten 1, 6 mg EDC, and 12 mg BSA or OVA were solubilized in 1000 μL of conjugation buffer (2-(N-morpholine) ethanesulfonic acid solution, 0.1 mol/L, pH 4.5–5.0). The solution was stirred at 4 °C for 12 h. Finally, the prepared conjugates were purified by dialysis with PBS (0.01 M, pH 7.4) at 4 °C for 3 days to remove unreacted reactants and non-coupled free hapten.

2.4.2. Hapten 2-OVA/BSA

The Hapten 2-OVA/BSA was prepared using the active ester method [13]. Briefly, 3 mg Hapten 2, 5.7 mg EDC, and 1.7 mg NHS were solubilized in 200 μL of DMF. After reacting at 4 °C overnight, the activated hapten solution was added dropwise to 2 mL solution of 12 mg BSA or OVA in PBS (0.01 M, pH 7.4) under stirring and reacted for another 4 h at room temperature. The prepared conjugates were purified by dialyzing with 0.01 M PBS (pH 7.4) for 3 days at 4 °C.

2.4.3. Hapten 3-OVA

The coupling method is the same as that of hapten 1-OVA/BSA, with hapten 1 changed to an equal mole of hapten 3.

2.4.4. Hapten 4-OVA/BSA

The preparation of SUs4-OVA/BSA was carried out by the diazotization method. All procedures described below were performed at 0~4 °C. Six miligrams of hapten 4 were dissolved in 1 mL of methanol and adjusted to a pH of 1 with pre-cooled 1 mol/L HCl. To the solution of hapten 4 was added dropwise 0.5 M sodium nitrite, with constant stirring until the starch potassium iodide test paper turned blue and black, then continuing to react for 40 min in the dark, and obtaining a dark yellow turbid liquid. To remove the excess nitrous acid, the pH of the solution was adjusted to 8 with 3 M NaOH. The diazonium salt solution of hapten 4 was added dropwise to 12 mg of OVA or BSA in 3 mL of PBS (0.01 M, pH 7.4), and the pH of the solution was maintained at around 9.5 with 3 M NaOH. The solution was stirred for 12 h after the addition of the final drop. The prepared conjugates were purified by dialyzing with 0.01 M PBS (pH 7.4) for 3 days at 4 °C.

A full wavelength UV-Vis spectroscopy scan was applied to verify the conjugation (Supplementary Materials Figure S4) and final preservation at −20 °C until use. Hapten 1-BSA and Hapten 2-BSA Hapten 4-BSA were used as the immunogen, while Hapten 1-OVA, Hapten 2-OVA, Hapten 3-OVA, and Hapnten 4-OVA were used as the coating antigens.

2.5. *Antibody Generation*

The immunization process in animals was described in our former publication [14]. The collected antisera were assessed by ELISA, with titer and inhibition rate [15] and antisera purified by ammonium sulfate precipitation for use in further process development [16]. The purified antibodies were characterised by sodium dodecyl sulfate-polyacrylamide gel electrophoresis (SDS-PAGE) and named as H1-Ab (immunized by Hapten 1-BSA) H2-Ab (immunized by Hapten 2-BSA), and H4-Ab (immunized by Hapten 4-BSA).

2.6. *ELISA Procedure*

The detection method for sulfonylureas was made up of two independent ELISAs established by H1-Ab and H2-Ab and the corresponding coating antigen. The individual ELISA methods were the same as the general ELISA method, which is described in detail in the Supporting Information.

An inhibition rate ($B/B_0 \times 100$) was utilized to evaluate the binding ability of the antibody to the coating agent and sulfonylureas, where B0 and B represent the absorbance values of the negative solution and sulfonylureas standard solution (1 μg/mL), respectively. Then, a four-parameter logistic function plotted by Origin 8.5 (Origin Lab Corp., Northampton, MA, USA) was fitted to establish calibration curves of 11 sulfonylureas under optimized antibody/coating antigen concentration. The detection limit (LOD) was defined as the inhibitory concentration at 10% (IC_{10}), while the dynamic detectable range was defined as the values ranging from IC_{20}–IC_{80}.

2.7. *Specificity*

The specificity of the ELISA was determined by the half inhibitory concentration (IC_{50}) of the rosiglitazone, phenformin, metformin hydrochloride, repaglinide, acetohexamide, tolbutamide, chlorpropamide, carbutamide, tolazamide, gliclazide, glyburide, glipizide, gliquidone, glimepiride, and glibornuride under optimized conditions. The cross-reactivity (*CR*) was calculated according to the following equation:

$$CR(\%) = \frac{IC_{50}(most\ sensitivity\ SUs,\ \text{nmol/mL})}{IC_{50}(other\ SUs,\ \text{nmol/mL})} \times 100\% \quad (1)$$

2.8. Sample Preparation

Healthcare capsules were acquired from the locally available market and determined to be "true negative" by LC-MS/MS (see Section 2.10, Recovery). The shell of the capsule samples was removed to collect the powder. For each sample, 1.000 g of the sample powder was accurately weighed into a 15 mL centrifuge plastic tube, 4 mL of methanol was added and vortex-mixed for 2 min, centrifuged at 4000× *g* for 5 min, and the supernatant was aspirated to obtain the sample's working solution. For the tablet sample, the powder was obtained by grinding in a mortar, and the remaining treatment was the same as the capsule.

2.9. Matrix Effect

Blank capsule samples were prepared as described above. The working solutions of the samples were diluted to 40, 60 and 80 times with PBST, respectively. Glipizide (for H1-Ab ELISA) or tolbutamide (for H2-Ab ELISA) was then diluted with PBST and three different dilution extracts to establish calibration curves for the ELISA. To ensure that matrix effects were negligible, *t*-tests were used to compare calibration curves and confirm the most appropriate dilution.

2.10. Recovery and Confirmation

Recovery is the addition of a standard of known content, i.e., the component being measured, to a blank sample or to a background of known content, and the ratio of the measured value to the added value calculated using an established method. Six sulfonylureas, often employed as illicit additions, were spiked in capsule samples and compared with the LC-MS/MS method to evaluate the detection accuracy of the developed ELISA. The capsule samples were added to three concentration levels of glyburide (0.32, 1.6, and 6.4 mg/kg), glipizide (0.32, 1.6, and 4.8 mg/kg), glimepiride (0.32, 1.6, and 6.4 mg/kg), gliquidone (0.64, 3.2, and 12.8 mg/kg), tolbutamide (0.32, 1.6, and 6.4 mg/kg), and gliclazide (3.2, 16, and 64 mg/kg), respectively. Three replicates were tested. The samples were then subjected to the extraction method described above. The accuracy and precision were estimated using the recovery and coefficient of variation (CV), respectively. The determination coefficient (R^2) between ELISA and LC-MS/MS method was used to evaluate the reliability of the ELISA.

The standard tolbutamide, glyburide, glipizide, glimepiride, gliquidone, and gliclazide were dissolved in MeOH with a concentration from 1.0~1000.0 ng/mL, respectively. As mentioned in a previous article [17], LC-MS/MS analysis was performed on AB QTRAP4500 triple quadrupole mass spectrometer.

$$\text{recovery}(\%) = \frac{Detected\ concentration}{Actual\ concentration\ added} \times 100\% \quad (2)$$

2.11. Analysis of Blind Samples

Eight samples (Figure 2) were purchased from the local market, after being prepared by the above extraction method, and the extract was diluted at appropriate concentration; the extraction was then analyzed by both the established ELISA and LC-MS/MS simultaneously.

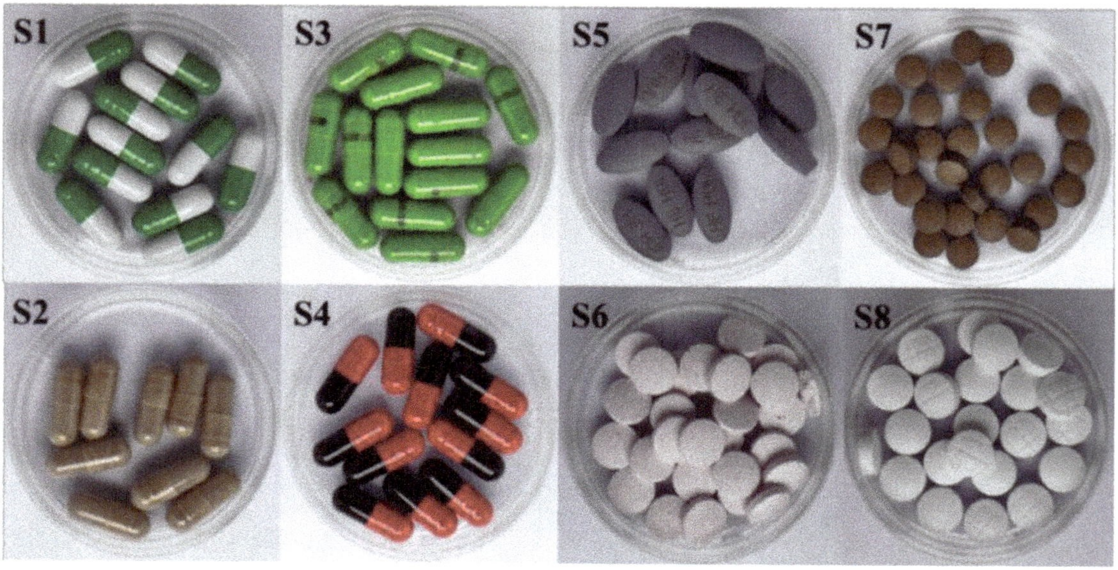

Figure 2. Eight real samples claiming anti-diabetic properties. S1~S4 were capsule samples with different matrices. S5~S8 were tablet samples with different matrices.

3. Results

3.1. Antibody Production

3.1.1. Hapten Design

Hapten design plays a key role in the whole process of producing a broad-specificity antibody and developing a broad-specificity immunoassay. To obtain the broad-specific antibody, that can widely recognize a group of serial compounds with different but usually related structures [18], structural similarity is often considered the basis of immune hapten selection that induces broad-specific antibody production. For the structural analysis of 11 sulfonylureas, it was found that the molecular structures can be divided into three parts, R1, R2, and general structural (S-arylsulfonylurea) (Table 1). Further analysis showed that the 11 sulfonylureas can be further grouped into two parts. One part is R2 with cyclohexane (named Group 1), while the other is R1 with the single-atom substituent (named Group 2). In addition, from the alignment result of 11 sulfonylureas (Figure 3), grouping can significantly reduce group structure differences, indicating that designing a broad-specificity hapten for each group will help improve the total affinity of broad-specificity antibodies. Thus, Hapten 1–4 was designed (Table 1). Hapten 1 and Hapten 3 retain the S-arylsulfonylurea and cyclohexane features of Group 1. Besides, the p-substituent on the phenyl ring of Hapten 1 was ethylamine, which could be coupled to BSA by forming an amide bond with the one-step EDC method. The amide bond-forming was consistent with most of the structures in Group 1. To improve the affinity of the antibody prepared with the Hapten 1 to Group 1. Hapten 3 is designed as a heterogeneous coating hapten to reduce the influence of "linking arm antibodies" [19], thereby increasing the detection sensitivity to the ELISA process. The design of Hapten 2 and Hapten 4 was similar to that of Hapten 1 and Hapten 3. However, since it is hard to determine which structure is more suitable for antibody preparation, both Hapten 2 and Hapten 4 were used as immunogens synthesis and as heterogeneous coating haptens to each other.

Table 1. Structures and classification of 11 sulfonylureas and the structures of 4 designed hapten.

Name	General Structural	R_1 [a]	R_2 [b]	Sulfonylureas	R_1	R_2
Glyburide		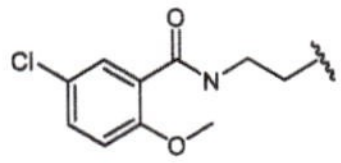	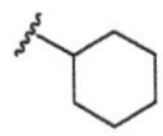	Tolbutamide		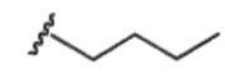
Glipizide	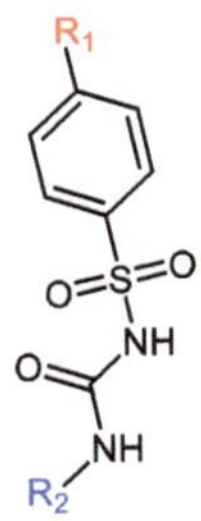	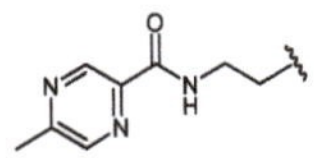	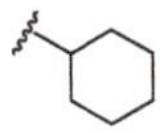	Carbutamide	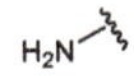	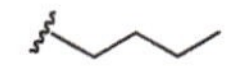
Glimepiride		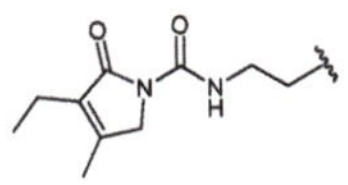	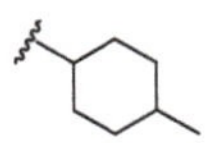	Gliclazide	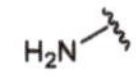	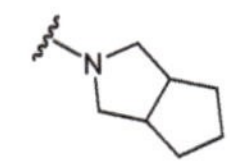
Gliquidone		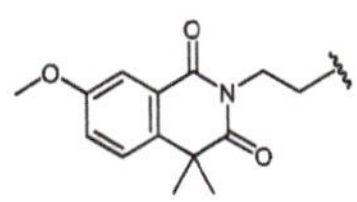	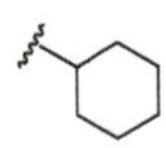	Tolazamide		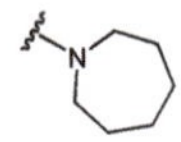
Acetohexamide		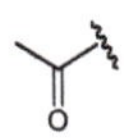	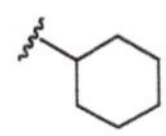	Glibornuride		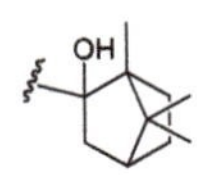
				Chlorpropamide	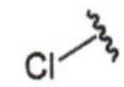	
Hapten	Hapten 1	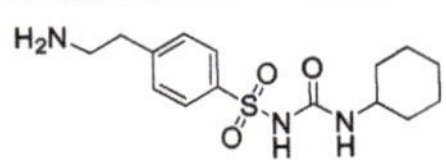		Hapten 2	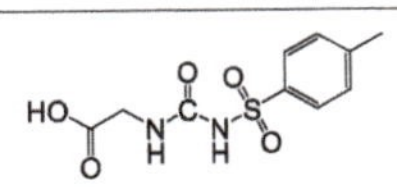	
	Hapten 3	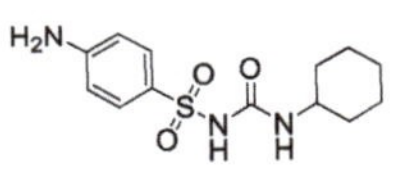		Hapten 4	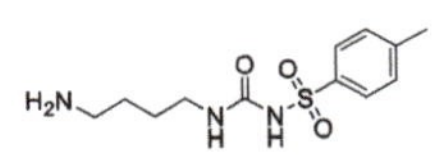	

[a] R1, The upper part of a structure common to sulfonylureas and connected by wavy lines; [b] R2, The lower part of the structure common to sulfonylureas and connected by wavy lines.

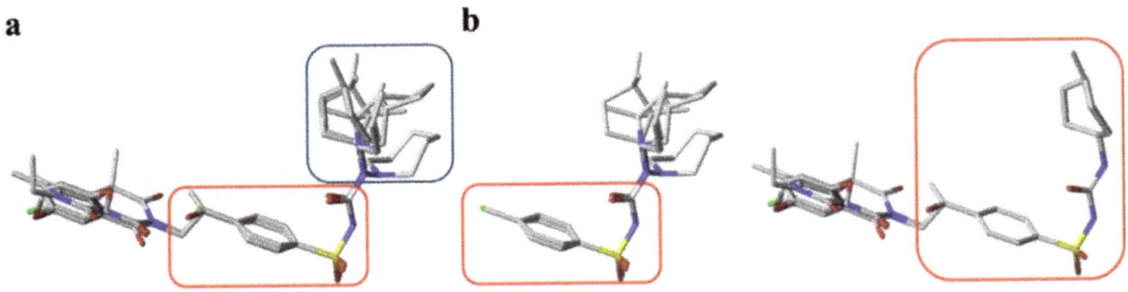

Figure 3. The alignment result of 11 sulfonylureas with the lowest energy conformations. (**a**) total alignment result of gliquidone, glipizide, acetohexamide, glyburide, glimepiride, tolazamide, tolbutamide, chlorpropamide, gliclazide, carbutamide, and glibornuride. (**b**) grouping alignment result. The left side included tolbutamide, carbutamide, gliclazide, tolazamide, glibornuride, and chlorpropamide; The right side included glyburide, glipizide, glimepiride, gliquidone, and acetohexamide. Blue and red rectangular boxes represent the variant and the common skeleton structure of sulfonylureas.

3.1.2. Conjugate Preparation

A UV–vis spectrophotometer was utilized to determine the effectiveness of the conjugation reaction of Hapten 1-OVA/BSA, Hapten 2-OVA/BSA, Hapten 3-OVA, and Hapten 4-OVA/BSA. As shown in Figure S10a,b,e,f, the proteins modified with Hapten 1 and

Hapten 4 demonstrated a slight shift in the valley at 250 nm, compared with the natural protein. Furthermore, Hapten 1 and 4 also demonstrated a pronounced absorption peak at 240 nm. Therefore, the shift of the UV absorption peak of the protein can be ascribed to the successful coupling of the hapten and carrier protein [20]. Similarly, Hapten 2-OVA/BSA (Figure S10c,d), due to the active ester method, is more effective than the direct EDC method; the conjugation showed a significant shift compared with the natural protein. While the Hapten 4 coupled to the BSA/OVA by the diazotization method (Figure S10g,h), the formation of diazonium bonds caused Hapten 3-BSA/OVA to display a new absorption peak at 350 nm. This indicated that Hapten 4 was successfully coupled to the carrier protein [20].

3.1.3. ELISA Optimization

To obtain a high sensitivity, 12 combinations with three antibodies and four coating antigens were evaluated by titer and inhibition rate. The result (Figure S11) shows that the combinations of H1-Ab/Hapten 3-OVA and H2-Ab/Hapten 4-OVA had a superior inhibition rate and higher titer, which were 83.3%/128,000 and 88.8%/256,000, respectively. Therefore, the heterologous assay combinations of H1-Ab/Hapten 3-OVA and H2-Ab/Hapten 4-OVA were selected for further optimization.

It is well known that the concentration of the coated antigen and antibody could affect the sensitivity of an immunoassay [13]. Thus, a chessboard titration was used to optimize the dilute multiple of coating antigen and purified antibody (Figure S13), by comparing the values of the IC_{50} and Amax/IC_{50} ratio (Figure S13), calculated from the calibration curve [14], where Amax represents the maximum signal at zero calibrator concentration. The optimal dilution multiple of coating antigen and antibody for H1-Ab/Hapten 3-OVA and H2-Ab/Hapten 4-OVA were found to be 1:32,000/1:512,000 and 1:64,000/1:4000, respectively.

3.1.4. Broad-Specificity

The specificity of the two antibodies was evaluated by cross-reactivity of 11 sulfonylureas and four other oral hypoglycemic drugs usually adulterated in the capsule (Table 2); the H1-Ab and H2-Ab showed a desirable broad-specificity with sulfonylureas and no obvious cross-reactivity (CR) for other drugs. Since hapten 1 and 3 bind to the protein at R1 and hapten 2 and 4 bind to the protein at R2, the R1 terminus of glimepiride, glipizide, glyburide, gliquidone, and acetohexamide are similar and complex, while the R2 terminus is essentially the same. Therefore, the attachment of the protein at the R1 end exposes the common R2 end, so that antibodies immunized with hapten 1 recognize all five drugs simultaneously. Their IC_{50} were 7.1, 8.5, 12.1, 14.8 and 29.4 nmol/μL, and their LOD (limit of detection) were 0.1, 0.7, 0.4, 0.04, and 2.1 nmol/μL, respectively. Conversely, tolbutamide, chlorpropamide, and gliclazide have similar and complex R2 ends and essentially identical R1 ends, so that the attachment of the protein at the R2 end exposes the common R1 end and therefore the antibody from semi-antigen 2 recognizes all three drugs simultaneously. The IC_{50} of tolbutamide, chlorpropamide, and gliclazide were 23.3, 33.6 and 179.6 nmol/μL, respectively, and 4, 23, and 4 times more sensitivity than that of H1-Ab. Hence, these data indicated the effectiveness of the hapten design and proved that this "complementary" format (combination of H1-Ab and H2-Ab) could be used to establish a more broad-specific ELISA.

Table 2. Specificity and sensitivity of the ic-ELISA.

Name	H1-Ab				H2-Ab			
	LOD [a] nmol/μL	Dynamic Working Range (nmol/μL)	IC_{50} [b] nmol/μL	CR [c] %	LOD nmol/μL	Dynamic Working Range nmol/μL	IC_{50} nmol/μL	CR %
Glimepiride	0.1	0.6–102.5	7.1	100.0	1672.4	-	17,228.5	0.1 [d] (0.04) [e]
Glipizide	0.7	1.6–45.8	8.5	83.6	1823.3	-	>10,000	<0.01
Glyburide	0.4	1.6–96.4	12.1	58.7	3788.8	-	>10,000	<0.01
Gliquidone	0.04	1.1–190.7	14.8	48.3	2455.7	-	>10,000	<0.01
Acetohexamide	2.1	5.3–161.9	29.4	24.2	23.5	-	4173.0	0.6 (0.2)
Tolazamide	7.7	19.9–530.5	103.1	6.9	2.2	11.2–3221.6	189.5	12.3 (3.8)
Tolbutamide	32.2	81.7–1947.8	398.7	1.8	1.5	4.4–127.2	23.3	100 (30.6)
Carbutamide	66.3	161.1–3331.6	732.3	1.0	15.8	62.7–6911.3	658.6	3.5 (1.1)
Gliclazide	37.7	115.3–5224.3	775.8	0.9	3.1	13.9–2347.8	179.6	13.0 (4.0)
Chlorpropamide	80.2	267.8–16,444.3	2098.7	0.3	0.7	6.9–163.7	33.6	69.3 (21.2)
Glibornuride	228.1	-	>10,000	<0.01	9.8	52.4–15,498.0	901.0	2.6 (0.8)
Repaglinide	-	-	>10,000	<0.01	-	-	>10,000	<0.01
Rosiglitazone	-	-	>10,000	<0.01	-	-	>10,000	<0.01
Phenformin	-	-	>10,000	<0.01	-	-	>10,000	<0.01
Metformin hydrochloride	-	-	>10,000	<0.01	-	-	>10,000	<0.01

[a] LOD, detection limit. [b] IC_{50}, half inhibitory concentration. [c] CR, cross-reactivity of H1-Ab was calculated according to IC_{50} (other SUs) divided by IC_{50} (glimepiride). [d] CR of H1-Ab calculated with the IC_{50} (tolbutamide). [e] CR of H1-Ab calculated with the IC_{50} (glimepiride).

3.2. Matrix Effect

The two ELISA methods were evaluated for their matrix effect by comparing the calibration curve of gliquidone or tolbutamide in PBST, which was used as a control with those calibration curves of gliquidone or tolbutamide in capsule extracts (Figure 4). The *t*-test results showed that the significance values of 80 times dilution were more than 0.05, within the dynamic working range, indicating that no significant differences were encountered between the two matrixes at the 0.05 level. This indicated that the matrix effect could be negligible with 80 times dilution of the capsule extracts. Considering sample preparation and the matrix effect, the LODs for glimepiride, glipizide, glyburide, gliquidone, tolbutamide, and gliclazide were respectively 16.0, 86.4, 70.4, 4.48, 137.6, and 316.8 μg/kg, lower than the LODs of the statutory instrument detection method established by China [21]. Hence, the proposed ELISA can be used to detect both spiked and authentic samples.

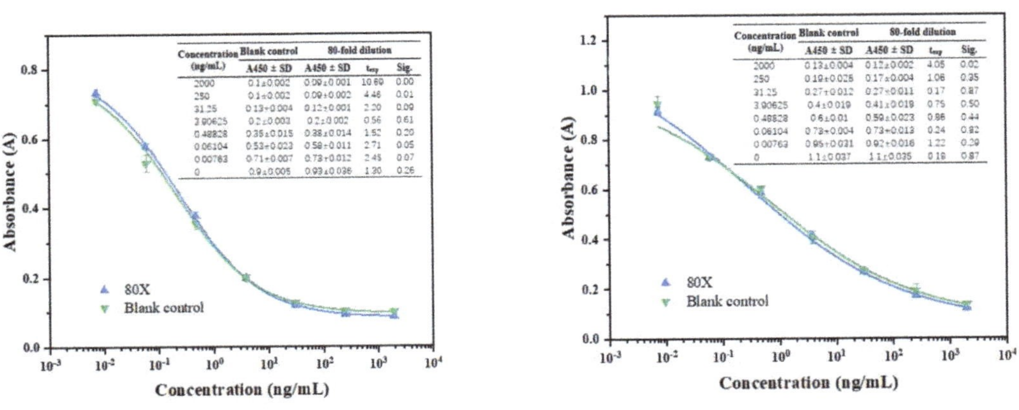

(a) (b)

Figure 4. Calibration curves and *t*-test results of ic-ELISA in PBST and dilution 80 times (**a**) Glipizide as the drug for H1-Ab/Hapten 3-OVA ic-ELISA; (**b**) Tolbutamide as the drug for H2-Ab/Hapten 4-OVA ic-ELISA; The texp and Sig. determined by the *t* test (p = 0.05).

3.3. Recovery and Confirmation

The ELISA recoveries for the six sulfonylureas ranged from 79.2% to 110.0%, with the CVs ranged from 3.7% to 18.5% (Table 3), indicating that the developed ELISA method for six sulfonylureas possessed an acceptable accuracy and excellent reproducibility. The recoveries of LC-MS/MS ranged from 76.4% to 112.1%, with CVs of 0.4% to 10.4%. In addition, the R2 ranged from 0.978 to 0.999, which indicated a good consistency between ELISA and LC-MS/MS.

Table 3. Recovery of sulfonylureas in capsule by ELISA and LC-MS/MS (n = 3).

Name	Spiked Level (μg/kg)	ELISA			LC-MS/MS			$R^{2\,b}$
		Observed Level (μg/kg)	Average Recovery %	CV [a] %	Observed Level (μg/kg)	Average Recovery %	CV %	
Glipizide	320	259.5	81.1	10.9	244.3	76.4	1.0	0.995
	1600	1272.0	79.5	7.9	1543.5	96.5	0.8	
	4800	4434.7	92.4	4.1	4701.4	98.0	2.4	
Glimepiride	320	318.8	99.6	10.8	253.5	79.2	9.4	0.994
	1600	1548.8	96.8	18.5	1340.3	83.8	3.1	
	6400	5526.3	86.3	14.2	5745.4	89.8	1.6	
Gliquidone	640	703.8	110.0	12.4	647.8	101.2	0.4	0.995
	3200	3092.0	96.6	5.7	3366.3	105.2	2.1	
	12,800	11,472.8	89.6	7.4	12,304.1	96.1	1.8	
Glyburide	320	301.9	94.3	6.8	267.4	83.6	10.4	0.978
	1600	1554.4	97.1	6.2	1409.5	88.1	1.3	
	6400	5621.5	87.8	3.7	6363.2	99.4	2.8	
Tolbutamide	320	319.2	99.7	10.4	315.9	98.7	5.8	0.999
	1600	1550.3	96.9	9.0	1722.5	107.7	0.9	
	6400	6494.9	101.5	6.5	6539.4	102.2	0.7	
Gliclazide	3200	2534.1	79.2	18.1	3013.1	94.2	2.7	0.994
	16,000	16,034.9	100.2	6.8	14,888.0	93.1	1.1	
	64,000	67,399.1	105.3	4.1	71,709.9	112.1	3.8	

[a] CV, coefficient of variation. [b] R^2, determination coefficient.

3.4. Analysis of Blind Samples

Analysis of blind samples can verify the adaptability of the established method in different pill matrices. As displayed in Table S2, all samples were sulfonylureas free, and this may be due to fewer or no cases of sulfonylurea adulteration in functional foods recently, or a limitation of the small number of samples collected. However, this does not mean that sulphonylureas are not added to functional foods. Sulphonylureas are still a potential risk for illegal addition to functional foods [22]. Sampling of functional foods for sulphonylureas will continue in follow-up work. It was also found that there were no false positive results for the detected eight samples, which indicates that the developed ELISA methods may be reliably used for different kinds of pills claiming anti-diabetic properties.

4. Conclusions

To summarize, a high sensitivity ELISA was developed using two anti-sulfonylureas antibodies with broad-specificity, based on a complementary strategy. The developed method could be used for the screening assay of illegally added sulfonylurea drugs in pills with a simple methanol extraction and dilution, demonstrating consistency with results obtained using the LC-MS/MS method. The proposed method could also potentially be used for multi-analysis of sulfonylureas in other similar products.

Supplementary Materials: The following supporting information can be downloaded at: https://www.mdpi.com/article/10.3390/bios12080591/s1, Figure S1: The synthetic routes of Hapten 1; Figure S2: The result of ESI-MS analysis (negative) for Hapten1; Figure S3: The synthetic routes of Hapten 2; Figure S4: The result of ESI-MS analysis (negative) for Hapten 2; Figure S5: The synthetic routes of Hapten 3; Figure S6: The result of ESI-MS analysis (negative) for Hapten 3; Figure S7: The synthetic routes of Hapten 4; Figure S8: The result of ESI-MS analysis (negative) for Hapten 4; Figure S9: Ethical review of animal experiments; Figure S10: The UV-VIS spectroscopy of Hapten1-BSA (a), Hapten1-OVA (b), Hapten2-BSA (c), Hapten2-OVA (d), Hapten3-BSA (e), Hapten3-OVA (f), Hapten4-BSA (g), Hapten4-OVA (h), and corresponding carrier protein and hapten; Figure S11: Titer and inhibition test for H1-Ab, H2-Ab, and H3-Ab with homologous and heterologous assay formats; Figure S12: The SDS-PAGE of purified antibodies. Lane M, standard protein markers; Lane 1, the reduction result for H1-Ab; Lane2, the reduction result for H2-Ab; Figure S13: Optimization of ELISA concentration of the coated antigen and antibody. H1-Ab with Hapten3-OVA (a, b) and H2-Ab with Hapten4-OVA (c, d); Table S1: MS/MS conditions; Table S2: Analysis of Sulfonylureas in Real Samples by ELISA and LC-MS/MS.

Author Contributions: Z.L.: Methodology, Investigation, Writing—original draft. H.X.: Investigation, Resources. T.F.: Investigation, Resources. Y.L. (Yingying Li): Resources. X.S.: Validation. X.L., X.Y., A.K., Y.L. (Yi Lei), Y.L. (Yingju Liu) and X.H.: Investigation. H.L.: Project administration, Writing—original review & editing, Funding acquisition. All authors have read and agreed to the published version of the manuscript.

Funding: This work was financially supported by the National Scientific Foundation of China (31871883, 21874048), the Heyuan Planned Program in Science and Technology (2019041), the Generic Technique Innovation Team Construction of Modern Agriculture of Guangdong Province (2021KJ130), and the National Key Research and Development Program of the Thirteenth Five-Year Plan (2017YFC1601700).

Institutional Review Board Statement: All patients involved in this study gave their informed consent. Institutional review board approval of our hospital was obtained for this study.

Informed Consent Statement: Informed consent was obtained from all subjects involved in the study.

Data Availability Statement: Not applicable.

Conflicts of Interest: We declare that we have no financial and personal relationships with other people or organizations that can inappropriately influence our work.

Abbreviations

UPLC: ultra-performance liquid chromatography; LC-MS/MS: liquid chromatography-tandem mass spectrometry; DART-MS: real-time mass spectrometry; 2D-MS: on-line two-dimensional liquid chromatography; ELISA: enzyme-linked immunosorbent assay; Ic-ELISA: Indirect competitive enzyme-linked immunosorbent assay; BSA: bovine serum albumin; OVA: ovalbumin; EDC: 1-(3-Dimethylaminopropyl)-3-ethyl carbodiimide hydrochloride; NHS: hydrolyzed protein, N-hydroxyl- succinamide; TMB: freund's incomplete adjuvant, 3,3′,5,5′-tetramethylbenzidine; DMF: N, N-Dimethy- lformamide; CB: coating buffer (0.1 mol/L carbonate buffer, pH 9.6); PBS: phosphate buffered solution (0.01 mol/L, pH 7.4); PBST: PBS solution containing 0.5% Tween-20; SDS-PAGE: sodium dodecyl sulfate-polyacrylamide gel electrophoresis; LOD: the detection limit; CR: the cross-reactivity; CV: coefficient of variation; IC50: half of inhibitory concentration.

References

1. International Diabetes Federation. DIABETES ATLAS 9th Edition. Available online: https://diabetesatlas.org/en/resources (accessed on 18 March 2021).
2. Emoto, M.; Nishizawa, Y.; Maekawa, K.; Hiura, Y.; Kanda, H.; Kawagishi, T.; Shoji, T.; Okuno, Y.; Morii, H. Homeostasis model assessment as a clinical index of insulin resistance in type 2 diabetic patients treated with sulfonylureas. *Diabetes Care* **1999**, *22*, 818–822. [CrossRef] [PubMed]
3. Mahwish Saeed, F.; Arshad, M.S.; Nisa, M.U.; Nadeem, M.T.; Arshad, M.U. Hypoglycemic and hypolipidemic effects of different parts and formulations of bitter gourd (*Momordica charantia*). *Lipids Health Dis.* **2017**, *16*, 211. [CrossRef] [PubMed]

4. Calahan, J.; Howard, D.; Almalki, A.J.; Gupta, M.P.; Calderón, A.I. Chemical adulterants in herbal medicinal products: A review. *Planta Med.* **2016**, *82*, 505–515. [CrossRef] [PubMed]
5. Sola, D.; Rossi, L.; Schianca, G.P.C.; Maffioli, P.; Bigliocca, M.; Mella, R.; Corlianò, F.; Fra, G.P.; Bartoli, E.; Derosa, G. Sulfonylureas and their use in clinical practice. *Arch. Med. Sci.* **2015**, *11*, 840–848. [CrossRef] [PubMed]
6. Cao, J.; Jiang, Q.; Li, R.; Xu, Q.; Li, H. Nanofibers mat as sampling module of direct analysis in real time mass spectrometry for sensitive and high-throughput screening of illegally adulterated sulfonylureas in antidiabetic health-care teas. *Talanta* **2019**, *204*, 753–761. [CrossRef] [PubMed]
7. Kim, N.S.; Kim, K.Y.; Yoo, G.J.; Lee, J.H.; Park, H.N.; Park, S.-K.; Baek, S.Y. Determination of 26 anti-diabetic compounds in dietary supplements using a validated UPLC method. *Food Addit. Contam. Part A-Chem.* **2018**, *35*, 387–394. [CrossRef]
8. Ma, J.; Pawar, R.S.; Grundel, E. Validation of an LC-MS/MS method for analysis of anti-diabetic drugs in botanical dietary supplements labeled for blood sugar management. *Drug Test. Anal.* **2018**, *10*, 609–617. [CrossRef] [PubMed]
9. Zhou, Z.; Zhang, J.; Zhang, W.; Bai, Y.; Liu, H. Rapid screening for synthetic antidiabetic drug adulteration in herbal dietary supplements using direct analysis in real time mass spectrometry. *Analyst* **2011**, *136*, 2613–2618. [CrossRef] [PubMed]
10. Guo, P.; Xu, X.; Chen, G.; Bashir, K.; Shu, H.; Ge, Y.; Jing, W.; Luo, Z.; Chang, C.; Fu, Q. On-Line two dimensional liquid chromatography based on skeleton type molecularly imprinted column for selective determination of sulfonylurea additive in Chinese patent medicines or functional foods. *J. Pharm. Biomed. Anal.* **2017**, *146*, 292–301. [CrossRef] [PubMed]
11. He, F.; Yang, J.; Zou, T.; Xu, Z.; Tian, Y.; Sun, W.; Wang, H.; Sun, Y.; Lei, H.; Chen, Z.; et al. A gold nanoparticle-based immunochromatographic assay for simultaneous detection of multiplex sildenafil adulterants in health food by only one antibody. *Anal. Chim. Acta.* **2021**, *1141*, 1–12. [CrossRef] [PubMed]
12. Suryoprabowo, S.; Liu, L.; Kuang, H.; Cui, G.; Xu, C. Fluorescence based immunochromatographic sensor for rapid and sensitive detection of tadalafil and comparison with a gold lateral flow immunoassay. *Food Chem.* **2021**, *342*, 128255. [CrossRef] [PubMed]
13. Zeng, H.; Chen, J.; Zhang, C.; Huang, X.-A.; Sun, Y.; Xu, Z.; Lei, H. Broad-Specificity Chemiluminescence Enzyme Immunoassay for (Fluoro)quinolones: Hapten Design and Molecular Modeling Study of Antibody Recognition. *Anal. Chem.* **2016**, *88*, 3909–3916. [CrossRef] [PubMed]
14. Zhang, Y.; Wang, L.; Shen, X.; Wei, X.; Huang, X.; Liu, Y.; Sun, X.; Wang, Z.; Sun, Y.; Xu, Z. Broad-specificity immunoassay for simultaneous detection of ochratoxins A, B, and C in millet and maize. *J. Agric. Food Chem.* **2017**, *65*, 4830–4838. [CrossRef] [PubMed]
15. Luo, L.; Xu, Z.-L.; Yang, J.-Y.; Xiao, Z.-L.; Li, Y.-J.; Beier, R.C.; Sun, Y.-M.; Lei, H.-T.; Wang, H.; Shen, Y.-D. Synthesis of novel haptens and development of an enzyme-linked immunosorbent assay for quantification of histamine in foods. *J. Agric. Food Chem.* **2014**, *62*, 12299–12308. [CrossRef] [PubMed]
16. Page, M.; Thorpe, R. Purification of IgG by precipitation with sodium sulfate or ammonium sulfate. In *The Protein Protocols Handbook*; Springer: Berlin/Heidelberg, Germany, 1996; pp. 721–722.
17. Xie, H.; Li, Y.; Wang, J.; Lei, Y.; Koidis, A.; Li, X.; Shen, X.; Xu, Z.; Lei, H. Broad-specific immunochromatography for simultaneous detection of various sulfonylureas in adulterated multi-herbal tea. *Food Chem.* **2022**, *370*, 131055. [CrossRef] [PubMed]
18. Xu, Z.-L.; Zeng, D.-P.; Yang, J.-Y.; Shen, Y.-D.; Beier, R.C.; Lei, H.-T.; Wang, H.; Sun, Y.-M. Monoclonal antibody-based broad-specificity immunoassay for monitoring organophosphorus pesticides in environmental water samples. *J. Environ. Monit.* **2011**, *13*, 3040–3048. [CrossRef] [PubMed]
19. Chen, Z.-J.; Wu, H.-L.; Xiao, Z.-L.; Fu, H.-J.; Shen, Y.-D.; Luo, L.; Wang, H.; Lei, H.-T.; Hongsibsong, S.; Xu, Z.-L. Rational hapten design to produce high-quality antibodies against carbamate pesticides and development of immunochromatographic assays for simultaneous pesticide screening. *J. Hazard. Mater.* **2021**, *412*, 125241. [CrossRef] [PubMed]
20. Li, Y.; Zhao, F.; Zhao, L.; Yang, Z. Development of a broad-specificity immunoassay for determination of organophosphorus pesticides using dual-generic hapten antigens. *Food Anal. Method.* **2015**, *8*, 420–427. [CrossRef]
21. Supervision, G.A.O.Q. Detection of 75 Illegally Added Chemical Drugs in Health Food. 2017; BJS 201710. Available online: https://www.samr.gov.cn/spcjs/bcjyff/201905/P020190516332773491737.doc (accessed on 26 July 2022).
22. Ross, J.; Downs, J.; Bazydlo, L.; Bordwine, P.H.; Gineste, C.E.; Kopatic, M.C.; Rege, A.V.; Saady, D.M.; Utah, O.F.; Wyatt, S.A. Outbreak of Severe Hypoglycemia After Ingestion of a Male Enhancement Supplement-Virginia, August November 2019. *Morb. Mortal. Wkly. Rep.* **2020**, *69*, 740–743. [CrossRef] [PubMed]

biosensors

MDPI

Article

In Situ SERS Sensing by a Laser-Induced Aggregation of Silver Nanoparticles Templated on a Thermoresponsive Polymer

Larisa V. Sigolaeva [1,*], Natalia L. Nechaeva [2], Anton I. Ignatov [3,4], Lyubov Y. Filatova [1], Timur Z. Sharifullin [1], Jonas Eichhorn [5], Felix H. Schacher [5,6,7], Dmitry V. Pergushov [1,8], Alexander M. Merzlikin [9] and Ilya N. Kurochkin [1,2,*]

1 Department of Chemistry, M.V. Lomonosov Moscow State University, 119991 Moscow, Russia
2 N.M. Emanuel Institute of Biochemical Physics of Russian Academy of Sciences, 119334 Moscow, Russia
3 All-Russia Research Institute of Automatics, 127055 Moscow, Russia
4 National Research Moscow State University of Civil Engineering, 129337 Moscow, Russia
5 Institute of Organic Chemistry and Macromolecular Chemistry (IOMC), Friedrich-Schiller-University Jena, D-07743 Jena, Germany
6 Jena Center for Soft Matter (JCSM), Friedrich-Schiller-University Jena, D-07743 Jena, Germany
7 Center for Energy and Environmental Chemistry (CEEC), Friedrich-Schiller-University Jena, D-07743 Jena, Germany
8 N.N. Semenov Federal Research Center of Chemical Physics of Russian Academy of Sciences, 119991 Moscow, Russia
9 Institute of Theoretical and Applied Electromagnetics of Russian Academy of Sciences, 125412 Moscow, Russia
* Correspondence: lsigolaeva@belozersky.msu.ru (L.V.S.); ikur@sky.chph.ras.ru (I.N.K.); Tel.: +7-495-939-40-42 (L.V.S.); +7-495-939-43-91 (I.N.K.)

Abstract: A stimuli-responsive (pH- and thermoresponsive) micelle-forming diblock copolymer, poly(1,2-butadiene)$_{290}$-*block*-poly(*N*,*N*-dimethylaminoethyl methacrylate)$_{240}$ (PB-*b*-PDMAEMA), was used as a polymer template for the in situ synthesis of silver nanoparticles (AgNPs) through Ag^+ complexation with PDMAEMA blocks, followed by the reduction of the bound Ag^+ with sodium borohydride. A successful synthesis of the AgNPs on a PB-*b*-PDMAEMA micellar template was confirmed by means of UV–Vis spectroscopy and transmission electron microscopy, wherein the shape and size of the AgNPs were determined. A phase transition of the polymer matrix in the AgNPs/PB-*b*-PDMAEMA metallopolymer hybrids, which results from a collapse and aggregation of PDMAEMA blocks, was manifested by changes in the transmittance of their aqueous solutions as a function of temperature. A SERS reporting probe, 4-mercaptophenylboronic acid (4-MPBA), was used to demonstrate a laser-induced enhancement of the SERS signal observed under constant laser irradiation. The local heating of the AgNPs/PB-*b*-PDMAEMA sample in the laser spot is thought to be responsible for the triggered SERS effect, which is caused by the approaching of AgNPs and the generation of "hot spots" under a thermo-induced collapse and the aggregation of the PDMAEMA blocks of the polymer matrix. The triggered SERS effect depends on the time of a laser exposure and on the concentration of 4-MPBA. Possible mechanisms of the laser-induced heating for the AgNPs/PB-*b*-PDMAEMA metallopolymer hybrids are discussed.

Keywords: thermoresponsive; silver nanoparticles; SERS; 4-mercaptophenylboronic acid; amphiphilic diblock copolymer; poly(*N*,*N*-dimethylaminoethyl methacrylate); laser-induced aggregation; plasmonic heating; local laser exposure

Citation: Sigolaeva, L.V.; Nechaeva, N.L.; Ignatov, A.I.; Filatova, L.Y.; Sharifullin, T.Z.; Eichhorn, J.; Schacher, F.H.; Pergushov, D.V.; Merzlikin, A.M.; Kurochkin, I.N. In Situ SERS Sensing by a Laser-Induced Aggregation of Silver Nanoparticles Templated on a Thermoresponsive Polymer. *Biosensors* **2022**, *12*, 628. https://doi.org/10.3390/bios12080628

Received: 16 June 2022
Accepted: 5 August 2022
Published: 11 August 2022

1. Introduction

Surface-enhanced Raman scattering (SERS) is an ultrasensitive vibrational spectroscopic technique, which aims to detect analytes in close proximity to a surface of plasmonic nanostructures. It is now well understood that the plasmonic coupling effect among particles induces enormous electromagnetic enhancement that allows SERS signals to be detected with very high or even single-molecule sensitivity [1,2].

Tremendous progress in fabricating nanostructured surfaces as SERS substrates, such as ordered arrays, fractal films, metal clusters, or colloidal metal nanoparticles (NPs), has taken place due to the intensive development of nanotechnology [3,4]. In addition to the ordering of micro- or nanostructures, the distance dependence of the enhancement effect has also been reported, including the effects of the distance between adjacent nanostructures and of the distance between analyte molecules and SERS substrate. The higher surface enhancement factors show a near exponential decay with the gap distance between NPs [5].

SERS-responsive assemblies of gold or silver nanoparticles (AuNPs or AgNPs) can be formed in situ, wherein the overall dimensions and interparticle spatial distances can respond and adapt to external stimuli. Importantly, the significant enhancement of SERS spectra arises from the compulsory aggregation of AuNPs or AgNPs, leading to the formation of "hot spots" in the gap between NPs. The aggregation of NPs can be achieved by increasing the ionic strength [6], ultrasound treatment [7], or application of gradient electric field [8].

Several research groups have reported on AuNPs or AgNPs associated with a thermosensitive polymer, that is, poly(*N*-isopropylacrylamide) (PNIPAAm). These new hybrid metallopolymer materials provide a controllable spatial distribution of metal NPs in the polymer matrix via temperature variation [9–11] as PNIPAAm undergoes a volume phase transition from a swollen hydrated state to a shrunken collapsed state when the temperature rises above 32 °C—a lower critical solution temperature (LCST) [12]. The temperature-controlled changes in the interparticle distances and the spatial distribution of the metal NPs in thermoresponsive PNIPAAm-based metallopolymer hybrid systems considerably enhanced the SERS sensing of 4-mercaptobenzoic acid [13], 1-naphthol [14,15], dyes [16–18], or biologically relevant molecules such as L-tyrosine or DNA [18].

An alternative to the external heating of temperature-sensitive metallopolymer hybrids would be laser-induced local heating, which would result in the spatial approaching of metal NPs, thereby enhancing the SERS signal. A phenomenon of microscopically observed phase transition was already previously described [19,20], wherein the microparticles of collapsed PNIPAAm were formed under direct heating with a focused infrared laser within 10–700 s [19] in aqueous solutions, depending on the polymer concentration, molecular weight of the polymer, temperature, or laser power. Furthermore, the localized surface plasmon resonance of single AuNPs attached to a transparent substrate can monitor the phase transition of PNIPAAm in aqueous solution upon a laser exposure [21]. Additionally, the local laser-induced plasmonic heating of AgNPs in the thermoresponsive dextran-grafted PNIPAAm copolymer/AgNPs hybrid system has been observed [22]. Worth noting, however, is that such local SERS triggering effects in temperature-sensitive metallopolymer hybrids have not been exploited for SERS sensing applications to date.

Poly(*N*,*N*-dimethylaminoethyl methacrylate) (PDMAEMA) is also classified as a thermoresponsive polymer. Its LCST is dependent on the pH of the surrounding aqueous medium [23,24], thereby rendering this polymer dual-stimuli-sensitive. Thus, the hydrophilic–hydrophobic balance of PDMAEMA can be changed by varying both the pH and temperature. Linear PDMAEMAs and PDMAEMA-based diblock copolymers have been used as efficient surface modifiers for the design of various types of electrochemical sensors, e.g., for choline [25–27], phenol [28], myoglobin [29], dsDNA [30,31], and drugs [32], as well as to reveal drug–DNA interactions [33]. PDMAEMA-based (co/ter)polymers were applied for the non-viral delivery of genetic material into cells [34,35]. PDMAEMA-based polymers were also reported as proper polymeric templates for the synthesis of metal NPs [36,37], wherein a strong stabilizing effect of PDMAEMA segments was reported [38].

In this work, we prepared metallopolymer hybrids consisting of AgNPs, which were templated on a poly(1,2-butadiene)$_{290}$-*block*-poly(*N*,*N*-dimethylaminoethyl methacrylate)$_{240}$ (PB-*b*-PDMAEMA) diblock copolymer (where the subscripts denote the number-average degrees of polymerization of the corresponding blocks). Apart from that, for the first time, we demonstrated local laser-induced enhancement of the SERS signal of 4-mercaptophenylboronic acid (4-MPBA) contacting with the obtained metallopolymer hybrid. Indeed, the possible local plasmonic heating of the AgNPs on PDMAEMA segments of the diblock copolymer under a laser irradiation might result in the heating of the polymer matrix in the laser spot and consequently initiate its phase transition (coil-to-globule transition of PDMAEMA segments followed by their subsequent aggregation). As assumed, the latter induces the approaching of AgNPs and thereby leads to the generation of "hot spots". We also show herein that such a SERS triggering effect depends on the time of the laser exposure and on the concentration of 4-MPBA. Thus, we highlight a new approach to a SERS analysis that could imply potential applications in areas such as nanosensors for the detection and quantification of biologically relevant molecules.

2. Materials and Methods

2.1. *Materials*

4-MPBA, $AgNO_3$, and $NaBH_4$ were obtained from Sigma-Aldrich. The PB-*b*-PDMAEMA diblock copolymer was synthesized as described earlier [29]. All other chemicals were of analytical grade and used without further purification. All aqueous solutions were prepared using Milli-Q water (18.2 MΩ·cm) purified with a Milli-Q water purification system by Millipore.

2.2. *Polymer-Templated Synthesis of AgNPs*

For the synthesis of an AgNPs/PB-*b*-PDMAEMA sample, an aqueous solution of $AgNO_3$ (24 μL, 0.1447 M) was first added to an alkaline solution of the PB-*b*-PDMAEMA diblock copolymer (3 mL, 0.023 M with respect to the monomer units of the polymer) at a Ag:N molar ratio of 1:20. The pH value of the mixture was carefully adjusted to pH 9 and the mixture was allowed to stay under stirring at room temperature for 30 min. Then, an aqueous solution of $NaBH_4$ (130 μL, 2 g/L) was added into the mixture to provide a $NaBH_4$:$AgNO_3$ molar ratio of 2:1. Afterwards, the reaction mixture was stirred for 30 min at room temperature.

2.3. *UV–Vis Spectroscopy*

UV–Vis spectra were recorded using a Shimadzu UV-1800 double-beam spectrophotometer (Shimadzu, Kyoto, Japan) in quartz cuvettes with an optical path length of 10 mm. The spectra were acquired from 200 to 800 nm with a 1 nm resolution. Before acquisition, an AgNPs/PB-*b*-PDMAEMA sample was diluted 10 times with deionized water. The spectral data were treated with the ORIGIN software.

2.4. *Transmission Electron Microscopy (TEM)*

For TEM measurements, copper grids were rendered hydrophilic by Ar plasma cleaning for 2 min (Diener Electronics, Ebhausen, Germany). Ten microliters (10 μL) of either AgNPs/PB-*b*-PDMAEMA sample or control sample (AgNPs prepared in the absence of the PB-*b*-PDMAEMA diblock copolymer) were applied onto the grid and an excess of the sample was blotted with a filter paper. TEM images were acquired with a 200 kV FEI Tecnai G2 20 transmission electron microscope equipped with a 4 k × 4 k Eagle HS CCD and a 1 k × 1 k Olympus MegaView camera for overview images.

2.5. *Turbidimetry*

The transmittance of aqueous solutions of the PB-*b*-PDMAEMA diblock copolymer or aqueous dispersions of the AgNPs/PB-*b*-PDMAEMA metallopolymer hybrids with a concentration of 0.2 g/L (pH 9) was measured on a Shimadzu UV-1800 double-beam spectrophotometer (Shimadzu, Kyoto, Japan). The temperature-controlled sample holders were connected to a temperature controller allowing 6 simultaneous measurements. The samples were scanned at a fixed wavelength of 500 nm. Their light transmittance was recorded upon heating at a temperature variation rate of 0.4 °C/min. The inflection points on the transmittance–temperature curves, which correspond to the onset temperature of the transmittance decrease, are defined herein as the onset temperature of phase transition (T_{PT}).

2.6. Raman Spectroscopy

An SERS signal was measured using an innoRam Raman spectrometer (BWTek, Plainsboro, NJ, USA) with a laser wavelength of 532 nm and power of 40 mW. The SERS-active analyte, that is, 4-MPBA, was used as a reporter as this compound can covalently bind to the Ag surface due to its mercapto group. Aqueous solutions of 4-MPBA were prepared at the concentrations of 3, 6, 15, and 30 μM and then mixed with an AgNPs/PB-*b*-PDMAEMA sample in a 1:1 ratio. Afterwards, the prepared mixtures were left for 30 min. Then, a 10 μL aliquot of each mixture was applied onto aluminum foil. The SERS spectra were recorded in between 5–8 cycles of continuous laser exposition (each of 1 min duration). The acquisition time was 10 s. The characteristic peak of 4-MPBA at 1070 cm^{-1} was used as an analytical signal for the evaluation of the SERS efficiency.

3. Results and Discussion

3.1. Physico-Chemical Characterization of the AgNPs Templated on the PB-b-PDMAEMA Micelles

In the present study, we used a stimuli-responsive (pH- and thermoresponsive) PB-*b*-PDMAEMA diblock copolymer, which comprises both a hydrophobic poly(1,2-butadiene) (PB) block with a low glass transition temperature (T_g) of −15 °C and a hydrophilic (chargeable) poly(*N,N*-dimethylaminoethyl methacrylate) (PDMAEMA) block. The chemical structure of the PB-*b*-PDMAEMA diblock copolymer is shown in Figure S1A.

The stimuli-responsiveness of the PB-*b*-PDMAEMA diblock copolymer results from the presence of a PDMAEMA block. PDMAEMA is a weak polybase with each monomer unit containing a pendant tertiary amino group. A reversible protonation of such amino groups imparts pH sensitivity to the PB-*b*-PDMAEMA diblock copolymer. According to the published results [29,32], PDMAEMA undergoes a transition from the fully protonated (charged) state to the fully deprotonated (non-charged) state in a pH window of 4.0–8.5, and the pK_a^{app} (at $\alpha = 0.5$, where α is the degree of protonation of PDMAEMA) for the PB-*b*-PDMAEMA diblock copolymer was found to be 6.35. PDMAEMA is also related to thermosensitive polymers with cloud points of its aqueous solutions depending on the pH [23,24]. It is worth noting that the decreasing pH, which results in the charging of PDMAEMA due to the protonation of its monomer units, is accompanied by an increase in cloud points.

Being amphiphilic, the PB-*b*-PDMAEMA diblock copolymer self-assembles into micelles, which at the ambient temperature and neutral pH ($\alpha = 0.75$) possess an overall average diameter of 106 nm. The micellar structure of the diblock copolymer was confirmed by means of cryogenic TEM (Figure S1B). Within a micelle, a hydrophobic PB core has a mean diameter of 38 nm while a hydrophilic PDMAEMA corona has a thickness of >34 nm [29]. The latter grants the formed PB-*b*-PDMAEMA micelles sufficient colloidal stability in aqueous media.

The in situ synthesis of AgNPs was performed by the pre-complexation of Ag^+ ions with PDMAEMA blocks of the PB-*b*-PDMAEMA diblock copolymer, followed by their reduction to metallic Ag by sodium borohydride. The reduction was carried out at pH 9 by mixing an aqueous solution of $AgNO_3$ with an aqueous solution of the PB-*b*-PDMAEMA micelles. The molar ratio of amino groups of PDMAEMA to $AgNO_3$ (N:Ag) was kept at 20:1. According to our previously published results [25,29], all monomer units of PDMAEMA are uncharged ($\alpha = 0$) at pH 9. Hence, the formation of an ion-coordination bond between an Ag^+ ion and an uncharged N atom of a monomer unit of PDMAEMA was expected. The following reduction of Ag^+ ions by a 2 molar excess of sodium borohydride results in the immediate appearance of a yellow-orange color, which indicates the formation of the AgNPs.

Figure 1A,B shows the UV–Vis spectra of the AgNPs formed in situ in the presence and in the absence of the PB-*b*-PDMAEMA diblock copolymer. As one can see, both samples, when freshly prepared, exhibit a characteristic surface plasmon resonance band at approximately 410–420 nm. While its position weakly depends on the presence or the absence of the copolymer, a considerable stabilization effect of the PB-*b*-PDMAEMA micelles is obvious. A spontaneous aggregation of non-stabilized AgNPs in the control sample results in a fast absorbance intensity decrease and broadening of the plasmonic peak soon after the preparation of AgNPs (Figure 1B). In solutions of the PB-*b*-PDMAEMA micelles, the plasmonic peak of the AgNPs keeps its position as well as its intensity (Figure 1A). One only notes certain (up and down) absorbance intensity variations for the metallopolymer hybrid system, which takes place during the first few days (Figure 1A). Afterwards, its absorbance demonstrates nearly no changes over at least 6 months (data not shown). The well-known strong stabilization effect of the PDMAEMA [38] is expected to impart such high colloidal stability to the AgNPs/PB-*b*-PDMAEMA metallopolymer hybrids. Apparently, the micellar structure of the PB-*b*-PDMAEMA diblock copolymer (Figure S1) could additionally contribute to the enhanced colloidal stability of the AgNPs/PB-*b*-PDMAEMA samples.

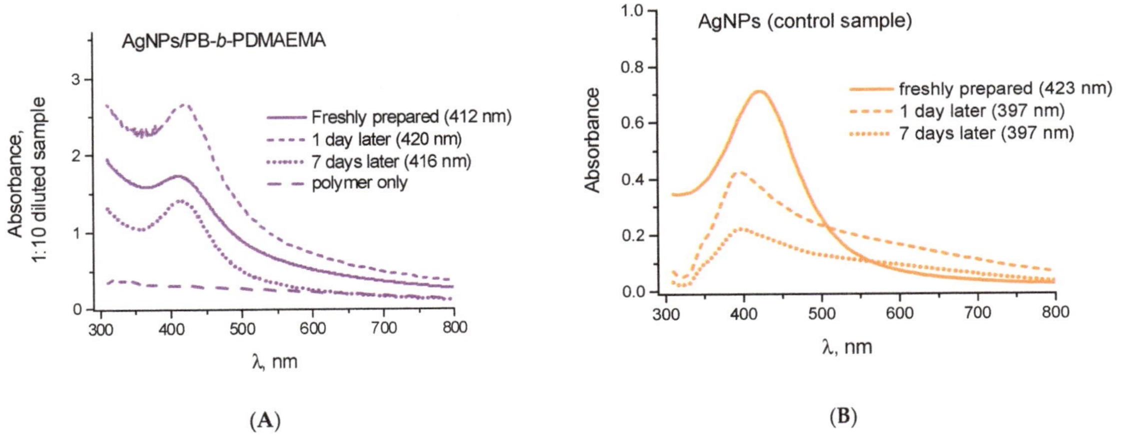

Figure 1. The time evolution of the UV–Vis spectra for the AgNPs prepared (**A**) in the presence and (**B**) in the absence of the PB-*b*-PDMAEMA micelles.

The TEM images of the AgNPs that were in situ templated on the PDMAEMA blocks of the PB-*b*-PDMAEMA micelles are demonstrated in Figure 2A. Two types of objects are clearly visible in the obtained images. Numerous small darker spots are undoubtedly attributed to the AgNPs while light grey round-shape structures, which are significantly weaker in contrast, are thought to represent PB-*b*-PDMAEMA micelles in a dry state. More TEM images of the prepared AgNPs/PB-*b*-PDMAEMA metallopolymer hybrids are given in Figure S2.

Furthermore, the AgNPs formed in the presence of the PB-*b*-PDMAEMA micelles were mostly spherical-like ones, which were separated from each other (Figure 2A). Although one cannot definitely see how the AgNPs are distributed within the PB-*b*-PDMAEMA micelles, they are most likely located in the external (periphery) part of the micelles, that is, in the micellar corona built up by PDMAEMA blocks. Importantly, the TEM image of the control sample of AgNPs, which were synthesized in the absence of the PB-*b*-PDMAEMA diblock copolymer, considerably differs from that taken for the AgNPs/PB-*b*-PDMAEMA sample. Indeed, only aggregated AgNPs are observed in this case, as shown in Figure S3. These results further confirm a considerable stabilization of the AgNPs in these metallopolymer hybrids.

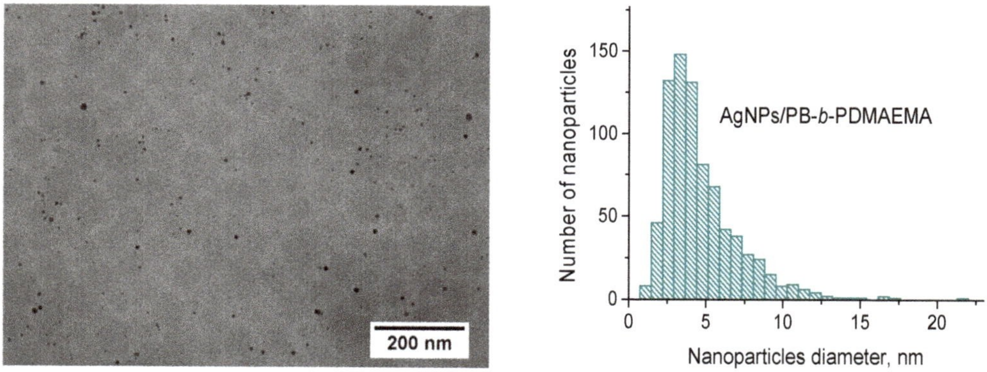

(**A**) (**B**)

Figure 2. (**A**) The TEM image and (**B**) the corresponding size distributions of the AgNPs prepared in the presence of PB-*b*-PDMAEMA micelles.

As follows from Figure 2B, the size distribution of the AgNPs formed in the presence of the PB-*b*-PDMAEMA micelles was found to be rather narrow. Their mean diameter was calculated as 4.6 ± 2.5 nm (n = 800). Assuming that all of them are isotropic and possess a crystalline structure with the closest-packed Ag atoms that form a *face-centered cubic lattice*, one can evaluate the mean number of Ag atoms per NP. Indeed, the volume of a lattice cell for a silver face-centered cubic lattice (a = b = c = 0.4086 nm) is V = 0.0682 nm^3 with the number of Ag atoms per lattice cell of 4. Hence, one AgNP with a mean diameter of 4.6 ± 2.5 nm contains approximately 2990 atoms. If we assume that Ag atoms form less-packed clusters, then one can calculate the same for a *simple cubic lattice* (a = b = c = 0.288 nm, where 0.288 nm is a diameter of Ag atom). In this case, the volume of the lattice cell is V = 0.0239 nm^3 with the number of Ag atoms per lattice cell of 1. Hence, one can evaluate approximately 2130 atoms per one AgNP with a mean diameter of 4.6 ± 2.5 nm. Thus, one can consider 2990 and 2130 values as an upper-bound estimate and a lower-bound estimate for the number of Ag atoms per one AgNP, respectively.

Keeping in mind that the molar ratio of N:Ag was set to 20:1 (see Section 2.2 for the conditions applied for the synthesis of the AgNPs), one can easily calculate the concentration of AgNPs in the AgNPs/PB-*b*-PDMAEMA system as 1.51×10^{17} particles/L (assuming a crystalline structure of the AgNPs) or 2.12×10^{17} particles/L (assuming a cluster structure of the AgNPs). Dividing these values by the concentration of the PB-*b*-PDMAEMA micelles (independently determined by means of nanoparticle tracking analysis as 3.7×10^{16} micelles/L [29]), one can further evaluate the mean number of the AgNPs per one micelle. Hence, one PB-*b*-PDMAEMA micelle apparently contains between 4 and 6 AgNPs. It is worth noting that these evaluations are in very good agreement with the analysis of the obtained TEM images (Figure 2A), wherein 5–6 AgNPs per one PB-*b*-PDMAEMA micelle were counted.

A thermoresponsive behavior of the PB-*b*-PDMAEMA micelles and the prepared AgNPs/PB-*b*-PDMAEMA metallopolymer hybrids was revealed in our work by means of turbidimetry. The measurements were performed for 0.2 g/L solutions of each sample at a heating rate of 0.4 °C/min. A temperature-induced hydrophobization of the PDMAEMA blocks and the subsequent aggregation of the PB-*b*-PDMAEMA micelles resulted in a change in the transmittance of the samples (Figure 3A). The inflection points in the transmittance–temperature curves, which correspond to the onset temperature of the transmittance decrease, are defined herein as the onset temperature of the phase transition T_{PT}. As shown in Figure 3A, the transmittance curves demonstrate temperature-induced phase transition for both the PB-*b*-PDMAEMA micelles and for the AgNPs/PB-*b*-PDMAEMA metallopolymer hybrids, although the intensity of the process and the temperature ranges are remarkably different. Indeed, the T_{PT} for the PB-*b*-PDMAEMA micelles was determined as 38 °C while the AgNPs/PB-*b*-PDMAEMA metallopolymer hybrids exhibit a higher T_{PT} of 52 °C (Figure 3A). Moreover, the phase transition in the latter case looks considerably less sharp and less pronounced. Apparently, the loading of PDMAEMA blocks with the AgNPs notably suppresses the temperature-induced phase transition. Furthermore, Figure 3B clearly demonstrates that the phase transition for the AgNPs/PB-*b*-PDMAEMA metallopolymer hybrids is completely reversible (at least, within the examined temperature cycle of 30 °C–60 °C–30 °C).

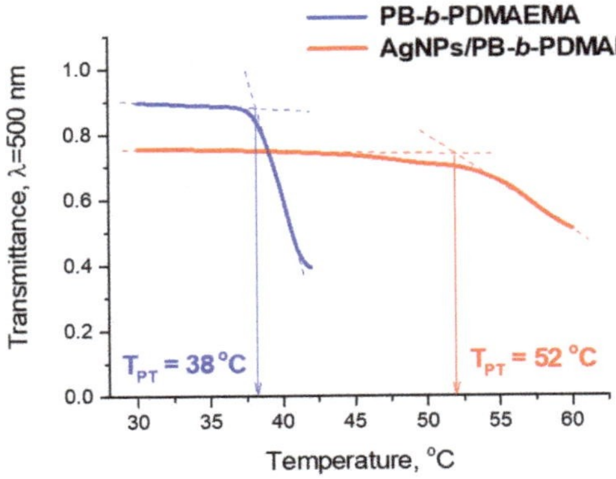

(A)

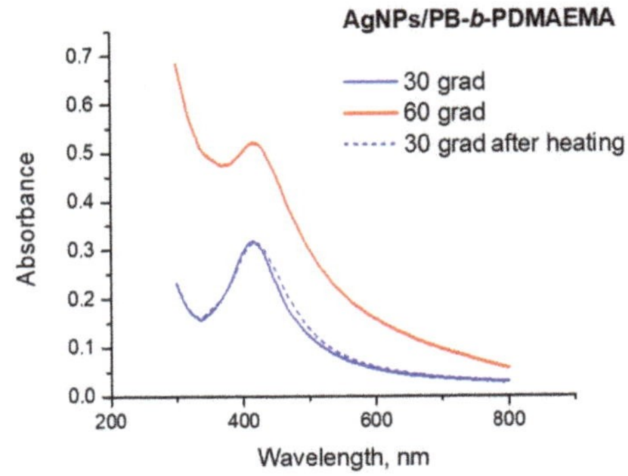

(B)

Figure 3. (**A**) Temperature-dependent changes in the transmittance of 0.2 g/L aqueous solutions of the PB-*b*-PDMAEMA micelles (blue line) and the AgNPs/PB-*b*-PDMAEMA hybrids (red line). (**B**) The UV–Vis spectra of 0.2 g/L aqueous solutions of the AgNPs/PB-*b*-PDMAEMA hybrids recorded at different temperatures. The pH values of all solutions were adjusted to pH 9 by 0.1 M NaOH.

Thus, a successful synthesis of the AgNPs on a PB-*b*-PDMAEMA micellar template was confirmed by means of UV–Vis spectroscopy and TEM, wherein the shape and size of the AgNPs were determined. We also demonstrated a reversible character of the temperature-induced phase separation for the AgNPs/PB-*b*-PDMAEMA metallopolymer hybrids. These findings allow us to expect a compulsory aggregation of the AgNPs upon a laser beam exposure of the AgNPs/PB-*b*-PDMAEMA samples. Indeed, the plasmonic heating of AgNPs is thought to result in a temperature-induced collapse of PDMAEMA blocks (coil-to-globule transition) and their aggregation, thereby initiating the approaching of the AgNPs and appearance of "hot spots". As consequence, the SERS effect is to be observed.

3.2. SERS for the AgNPs Templated on the PB-b-PDMAEMA Micelles

To reveal a potential of the AgNPs/PB-*b*-PDMAEMA metallopolymer hybrids to exhibit the SERS effect, the AgNPs were labeled with 4-MPBA. As it forms a strong Ag–S bond, this compound can covalently bind to AgNPs. The covalent bond formation also leads to the appearance of intense peaks in the SERS spectrum of 4-MPBA that renders it an outstanding reporter for SERS experiments with AgNPs. Moreover, boronic acid itself can covalently bind to polyols, including carbohydrates and glycated proteins, which can be advantageously exploited for the diagnosis of diabetes, thereby emphasizing the biochemical relevance of 4-MPBA [39].

The chemical structure and the exemplary SERS spectrum of 4-MPBA with characteristic vibrations is shown in Figure S4. The most intense peaks of 4-MPBA are at 414, 992, 1018, 1070, 1178, and 1570 cm^{-1}. The peak at 1070 cm^{-1} is related to the in-plane benzene ring breathing mode coupled with the C–S stretching mode [40]. In further SERS experiments, the amplitude of this peak was used as an analytical signal.

To demonstrate the SERS effect for the AgNPs/PB-*b*-PDMAEMA metallopolymer hybrids, a drop of the AgNPs/PB-*b*-PDMAEMA sample labeled with 4-MPBA was deposited on aluminum foil and then an initial SERS spectrum was acquired first. After that, the sample was continuously exposed to laser beam for 1 min. Then, a second SERS spectrum was acquired under the same conditions. This procedure was repeated eight times so that the total time of laser exposition was equal to 8 min. Figure 4 demonstrates that a continuous laser exposure leads to a notable increase in the intensity of the SERS spectra, the increase being proportional to the exposure time. This result is thought to be a consequence of the local "hot spot" generation through a laser-induced collapse of PDMAEMA blocks (coil-to-globule transition) and their aggregation, followed by AgNPs approaching.

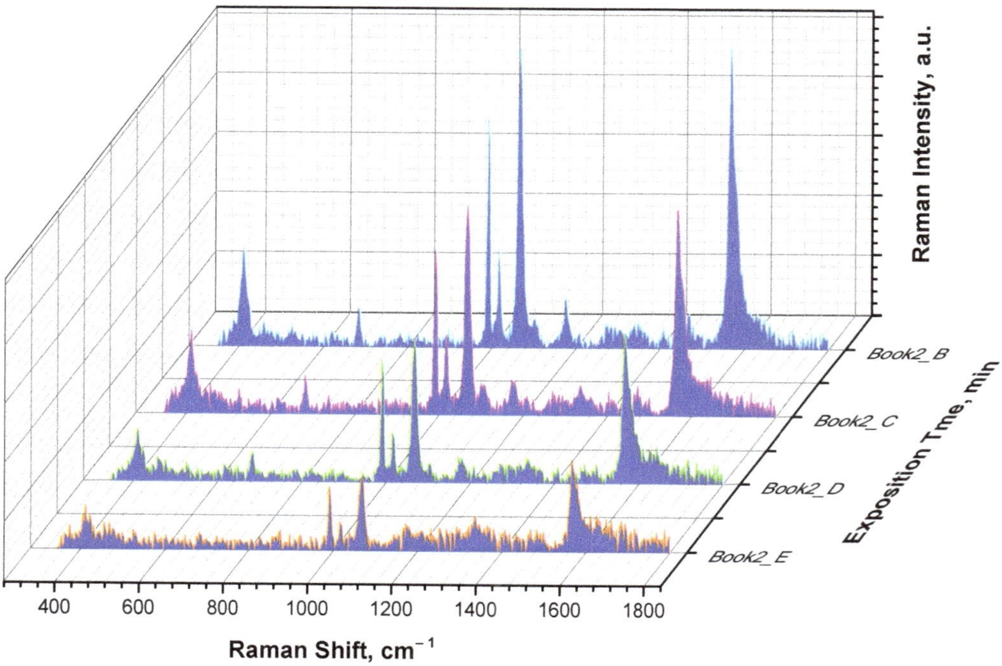

Figure 4. The evolution of the SERS spectra of the 10 μL drop of the AgNPs/PB-*b*-PDMAEMA hybrid with 30 μM of 4-MPBA at different exposure times: 0 min (orange line); 3 min (green line); 6 min (pink line); and 8 min (light blue line).

To prove whether the observed effect of the SERS intensity increase is due to a local laser exposure of the sample, a control experiment was performed, which started from recording the series of SERS spectra acquired in between five cycles of a 1 min laser exposure. Then, the laser was switched off, the drop of the sample was carefully mixed, and after that, one more SERS spectrum was recorded under the same conditions. As one can see from Figure 5, the intensity of the characteristic peak at 1070 cm^{-1} drops back to the initial value. Indeed, exchanging the exposed sample volume by a portion of the fresh sample via stirring leads to a decrease in the intensity of the SERS spectrum back to the start point, thus confirming that the local laser exposure plays a key role in the observed increase in the SERS signal.

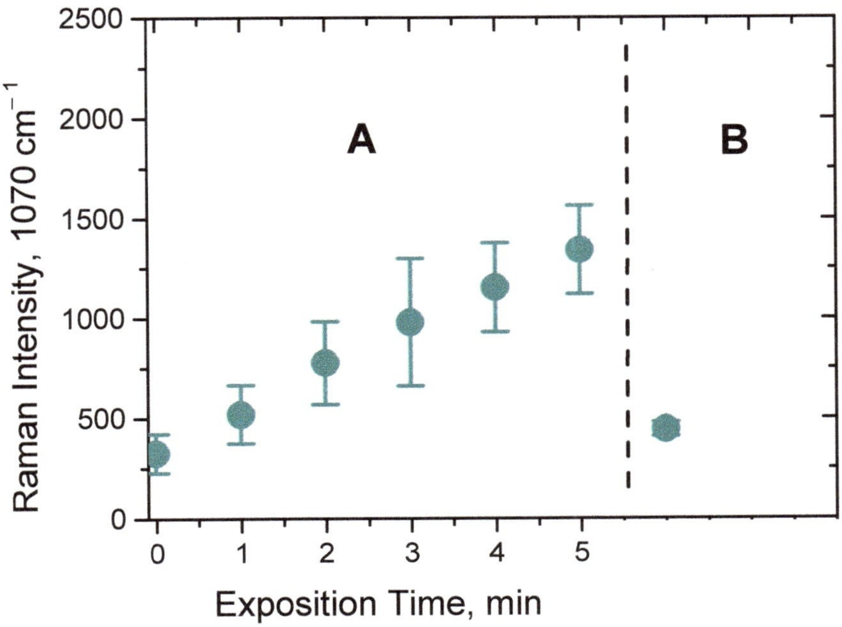

Figure 5. (**A**) The intensity increase in the 4-MPBA characteristic peak at 1070 cm^{-1} upon 5 cycles of a local laser exposure of a drop of the AgNPs/PB-*b*-PDMAEMA hybrid labeled with 4-MPBA; (**B**) The intensity of the 4-MPBA characteristic peak at 1070 cm^{-1} upon mixing a drop of the sample of the AgNPs/PB-*b*-PDMAEMA hybrid labeled with 4-MPBA. The data are presented as the mean ± SD for three independent experiments.

Furthermore, we recorded the SERS spectra in the same manner at different 4-MPBA concentrations and compared the intensities of the characteristic peak at 1070 cm^{-1} at different times of laser exposure. As one can see from Figure 6B–E, eight 1 min cycles of the continuous laser exposure lead to a notable increase in the start point, the endpoint, and the signal growth rate of the SERS signal intensity, the increase being proportional to the time of the laser exposure and to the concentration of 4-MPBA. A close to linear signal growth with the laser exposure is evident, although some deviations from linearity are found at the high concentrations of 4-MPBA. One can also note a concentration dependence of the initial SERS signal in the start point. This might be a consequence of a certain initial aggregation of the AgNPs, which could be induced by 4-MPBA upon its mixing with the AgNPs/PB-*b*-PDMAEMA sample. A control sample, which was measured in the absence of 4-MPBA, demonstrates no SERS signal (Figure 6A). One more control sample measured at 30 μM of 4-MPBA in the absence of the AgNPs/PB-*b*-PDMAEMA metallopolymer hybrids shows no SERS signal either (data not shown). Here, the time of the laser exposure was limited to 8 min to avoid the significant evaporation of the sample. The evaporation of a 10 μL drop was followed in a separate experiment and was found to be not exceeding 20% of the drop volume.

The endpoint intensities of the characteristic peak at 1070 cm^{-1} were plotted against the concentration of 4-MPBA (Figure 6F). By this experiment, we demonstrated a potential benefit of such local triggering effects for the AgNPs/PB-*b*-PDMAEMA metallopolymer hybrids in the context of the SERS sensing applications. We think that the new approach considered herein could significantly broaden the applied range of SERS sensing, which could have possible applications in areas such as nanosensors for biologically relevant molecules.

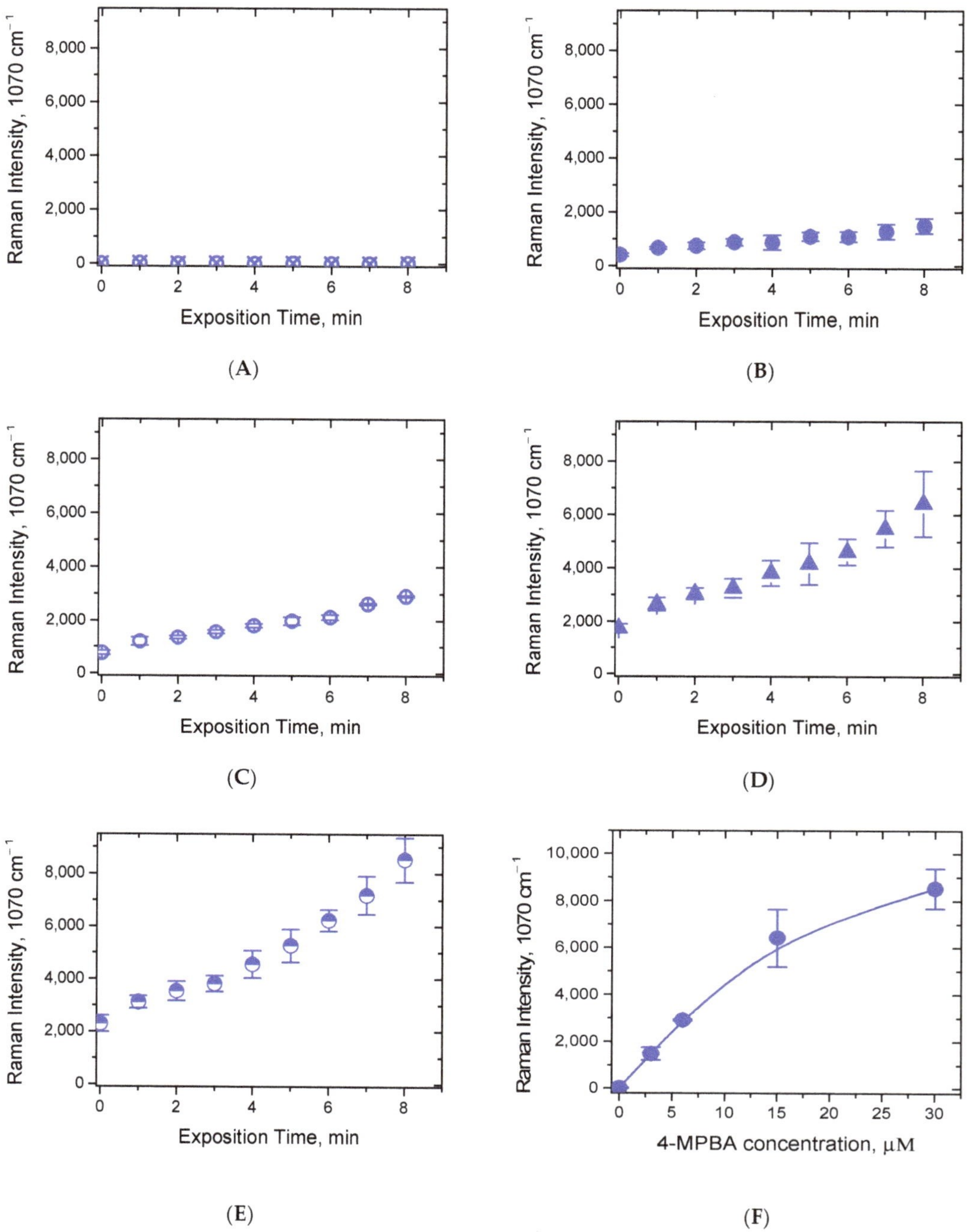

Figure 6. (**A–E**) The intensity increase in the 4-MPBA characteristic peak at 1070 cm^{-1} upon 8 cycles of local laser exposure of a drop of the AgNPs/PB-*b*-PDMAEMA hybrid labeled with 4-MPBA at different 4-MPBA concentrations (**A**) 0 μM (control with water), (**B**) 3 μM, (**C**) 6 μM, (**D**) 15 μM, and (**E**) 30 μM; (**F**) the dependence of the intensity of characteristic peak at 1070 cm^{-1} on 4-MPBA concentration plotted for the laser exposure time equal to 8 min. The data are presented as the mean ± SD for three independent experiments.

3.3. Theoretical Simulation of Sample Heating by a Laser Beam in SERS Experiments

To confirm that a laser exposure of the AgNPs/PB-*b*-PDMAEMA sample in the SERS experiments leads to heating that is sufficient for the temperature-induced collapse and aggregation of PDMAEMA blocks, a theoretical consideration of sample heating by a laser beam was carried out. The temperature distribution in a sample drop was simulated, based on the volume-specific heat source distribution resulting from the absorption of the laser light by AgNPs. The absorbance spectra (Figure 1A) of the diblock copolymer solutions with and without the AgNPs were used to determine the light absorption per unit volume. It was assumed that the extinction of a solution without the AgNPs resulted from the light scattering by PB-*b*-PDMAEMA micelles. Then, the difference between the extinctions of solutions with and without the AgNPs is expected to be exclusively due to light absorption in the AgNPs. Here, we neglected the own light scattering of the AgNPs since for the AgNPs with a size distribution as in our samples (Figure 2B), the average absorption cross-section is approximately an order of magnitude larger than the average scattering cross-section at $\lambda = 532$ nm. To determine the volume-specific heat source distribution, the absorption coefficient of the AgNPs/PB-*b*-PDMAEMA sample (determined from the difference between the extinction spectra of the samples with and without the AgNPs) and the exponential attenuation of the laser intensity (determined from the entire extinction of the sample with the AgNPs) were taken into account.

Figure 7 shows the temperature increase distribution (relative to the room temperature) in a sample drop lying on an aluminum foil and in the surrounding media in the steady state. The maximum temperature increase in the sample was assessed as 37.7 K with respect to room temperature, which is sufficient for PDMAEMA blocks to collapse and aggregate as experimentally shown (Figure 3A). We assume that the SERS signal increase could be explained by the gradual temperature-induced transformation of the polymer in the steady temperature distribution (according to [19,20], the temperature-induced collapse time of thermoresponsive polymers might take approximately hundreds of seconds).

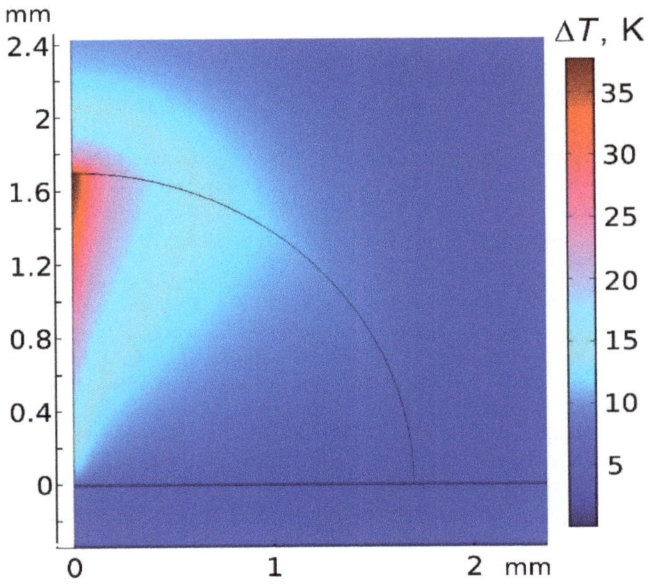

Figure 7. The simulated temperature increase distribution (with respect to the room temperature) resulting from laser heating in a sample drop (by a volume of 10 µL) on a foil and in the surrounding air (in the axisymmetric geometry). The horizontal direction corresponds to the radial coordinate (the symmetry axis is perpendicular to the foil plane and coincides with the laser beam axis).

4. Conclusions

In summary, we synthesized the AgNPs templated on the thermoresponsive PB-*b*-PDMAEMA micelles and showed a temperature-induced reversible phase transition of the prepared AgNPs/PB-*b*-PDMAEMA metallopolymer hybrids. With a SERS reporting probe (4-MPBA), we further revealed a laser-induced enhancement of a SERS signal observed under a constant local laser irradiation. The triggered SERS effect depends on the time of the laser exposure and on the concentration of 4-MPBA. The local heating of the AgNPs/PB-*b*-PDMAEMA sample in a laser spot is in good agreement with a performed theoretical consideration pointing to a temperature increase up to approximately 38 °C. A plasmonic heating of the AgNPs during the local laser exposure of the AgNPs/PB-*b*-PDMAEMA sample is thought to be responsible for the triggered SERS effect. Indeed, the plasmonic heating initiates a collapse of the PDMAEMA blocks and their gradual aggregation, which leads to the approaching of the AgNPs and generation of "hot spots". However, an optical trapping effect (optical tweezers) that might appear as a result of a decreasing distance among the AgNPs as reported in [21] could also be a reasonable explanation.

We think that the smart stimuli-responsive metallopolymer hybrid materials reported herein could have promising applications in areas such as SERS nanosensors for the detection and quantification of biologically relevant molecules. At the same time, further optimization of the technique and the experimental conditions appears to be necessary to improve the approach and to increase the sensitivity, both of which will be subjects of future work.

Supplementary Materials: The following supporting information can be downloaded at: https://www.mdpi.com/article/10.3390/bios12080628/s1, Figure S1: The chemical structure of the PB-b-PDMAEMA (A) and the cryo-TEM micrograph of the PB-b-PDMAEMA micelles (B) in 50 mM sodium phosphate (c = 2.5 g/L) at pH 7.0. Figure S2: The TEM images of the AgNPs/PB-b-PDMAEMA hybrids taken from different places of the sample at different magnification. Figure S3: The TEM image of the control AgNPs sample that was prepared under the same conditions in the absence of the PB-b-PDMAEMA micelles. Figure S4: The typical SERS spectrum of 4-MPBA with characteristic vibrations.

Author Contributions: Conceptualization, L.V.S.; methodology, L.V.S., N.L.N., I.N.K., A.M.M., A.I.I. and D.V.P.; validation, N.L.N., L.Y.F., T.Z.S., A.I.I. and D.V.P.; formal analysis, N.L.N., T.Z.S., L.Y.F., A.I.I. and D.V.P.; investigation, J.E., N.L.N. and L.Y.F.; resources, I.N.K. and F.H.S.; data curation, L.V.S., I.N.K. and A.M.M.; writing—original draft preparation, L.V.S., N.L.N. and A.I.I.; writing—review and editing, L.V.S., D.V.P. and F.H.S.; visualization, L.V.S., N.L.N., A.I.I. and T.Z.S.; supervision, L.V.S., D.V.P. and I.N.K.; project administration, L.V.S., F.H.S. and I.N.K.; funding acquisition, L.V.S., F.H.S. and I.N.K. All authors have read and agreed to the published version of the manuscript.

Funding: The synthesis of the AgNPs/PB-*b*-PDMAEMA hybrids used in this research was supported by the Russian Science Foundation (RSF, project no. 18-44-04011) and the Deutsche Forschungsgemeinschaft (DFG, SCHA1640/18-1) within a joint RSF-DFG grant. The TEM facilities of the Jena Center for Soft Matter (JCSM) were established with a grant from the German Research Council (DFG) and the European Fonds for Regional Development (EFRE). The physico-chemical characterization of the PB-*b*-PDMAEMA diblock copolymer and the AgNPs/PB-*b*-PDMAEMA hybrids was supported by the State Registration Topic "Modern Problems in Chemistry and Physical Chemistry of High Molecular Compounds" for 2021–2024 (no. AAAA-A21-121011990022-4). The SERS experiments were supported by the State Registration Topics "Molecular Design, Structure-Function Analysis and Regulation of Enzyme Systems, Cellular Structures, Bionanomaterials: Fundamental Basis and Applications in Technology, Medicine, Environmental Protection" for 2021–2024 (no. 121041500039-8), "Physical Chemistry of Bioanalytical Processes and Development of New Sensor Materials" (no. AAAA-A19-119071890024-8), and the M.V. Lomonosov Moscow State University Program of Development.

Institutional Review Board Statement: Not applicable.

Informed Consent Statement: Not applicable.

Data Availability Statement: Research data are not shared.

Acknowledgments: The authors thank Johannes B. Max for the technical assistance with TEM experiments.

Conflicts of Interest: The authors declare no conflict of interest.

References

1. Cialla, D.; März, A.; Böhme, R.; Theil, F.; Weber, K.; Schmitt, M.; Popp, J. Surface-enhanced Raman spectroscopy (SERS): Progress and trends. *Anal. Bioanal. Chem.* **2012**, *403*, 27–54. [CrossRef] [PubMed]
2. Haynes, C.L.; McFarland, A.D.; Van Duyne, R.P. Surface-enhanced Raman spectroscopy. *Anal. Chem.* **2005**, *77*, 339A–346A. [CrossRef]
3. Asiala, S.M.; Schultz, Z.D. Characterization of hotspots in a highly enhancing SERS substrate. *Analyst* **2011**, *136*, 4472–4479. [CrossRef] [PubMed]
4. Brouers, F.; Blacher, S.; Lagarkov, A.; Sarychev, A.K. Theory of giant Raman scattering from semicontinuous metal films. *Phys. Rev. B—Condens. Matter Mater. Phys.* **1997**, *55*, 13234–13245. [CrossRef]
5. Guerrini, L.; Graham, D. Molecularly-mediated assemblies of plasmonic nanoparticles for Surface-Enhanced Raman Spectroscopy applications. *Chem. Soc. Rev.* **2012**, *41*, 7085–7107. [CrossRef]
6. Bell, S.E.J.; Sirimuthu, N.M.S. Surface-Enhanced Raman Spectroscopy as a Probe of Competitive Binding by Anions to Citrate-Reduced Silver Colloids. *J. Phys. Chem. A* **2005**, *109*, 7405–7410. [CrossRef]
7. Radziuk, D.; Grigoriev, D.; Zhang, W.; Su, D.; Möhwald, H.; Shchukin, D. Ultrasound-Assisted Fusion of Preformed Gold Nanoparticles. *J. Phys. Chem. C* **2010**, *114*, 1835–1843. [CrossRef]
8. Podoynitsyn, S.N.; Sorokina, O.N.; Nechaeva, N.L.; Yanovich, S.V.; Kurochkin, I.N. Surface-enhanced Raman spectroscopy in tandem with a gradient electric field from 4-mercaptophenylboronic acid on silver nanoparticles. *Microchim. Acta* **2020**, *187*, 566. [CrossRef]
9. Ding, T.; Rudrum, A.W.; Herrmann, L.O.; Turek, V.; Baumberg, J.J. Polymer-assisted self-assembly of gold nanoparticle monolayers and their dynamical switching. *Nanoscale* **2016**, *8*, 15864–15869. [CrossRef]
10. Manikas, A.C.; Aliberti, A.; Causa, F.; Battista, E.; Netti, P.A. Thermoresponsive PNIPAAm hydrogel scaffolds with encapsulated AuNPs show high analyte-trapping ability and tailored plasmonic properties for high sensing efficiency. *J. Mater. Chem. B* **2015**, *3*, 53–58. [CrossRef]
11. Nguyen, M.; Kanaev, A.; Sun, X.; Lacaze, E.; Lau-Truong, S.; Lamouri, A.; Aubard, J.; Felidj, N.; Mangeney, C. Tunable Electromagnetic Coupling in Plasmonic Nanostructures Mediated by Thermoresponsive Polymer Brushes. *Langmuir* **2015**, *31*, 12830–12837. [CrossRef] [PubMed]
12. Schild, H.G. Poly (N-isopropylacrylamide): Experiment, theory and application. *Prog. Polym. Sci.* **1992**, *17*, 163–249. [CrossRef]
13. Yin, P.G.; Chen, Y.; Jiang, L.; You, T.T.; Lu, X.Y.; Guo, L.; Yang, S. Controlled dispersion of silver nanoparticles into the bulk of thermosensitive polymer microspheres: Tunable plasmonic coupling by temperature detected by surface enhanced Raman scattering. *Macromol. Rapid Commun.* **2011**, *32*, 1000–1006. [CrossRef]
14. Dong, X.; Chen, S.; Zhou, J.; Wang, L.; Zha, L. Self-assembly of monodisperse composite microgels with bimetallic nanorods as core and PNIPAM as shell into close-packed monolayers and SERS efficiency. *Mater. Des.* **2016**, *104*, 303–311. [CrossRef]
15. Dong, X.; Zou, X.; Liu, X.; Lu, P.; Yang, J.; Lin, D.; Zhang, L.; Zha, L. Temperature-tunable plasmonic property and SERS activity of the monodisperse thermo-responsive composite microgels with core-shell structure based on gold nanorod as core. *Colloids Surf. A Physicochem. Eng. Asp.* **2014**, *452*, 46–50. [CrossRef]
16. Zheng, Y.; Soeriyadi, A.H.; Rosa, L.; Ng, S.H.; Bach, U.; Justin, J. Gooding, Reversible gating of smart plasmonic molecular traps using thermoresponsive polymers for single-molecule detection. *Nat. Commun.* **2015**, *6*, 8797. [CrossRef]
17. Elashnikov, R.; Mares, D.; Podzimek, T.; Švorčík, V.; Lyutakov, O. Sandwiched gold/PNIPAm/gold microstructures for smart plasmonics application: Towards the high detection limit and Raman quantitative measurements. *Analyst* **2017**, *142*, 2974–2981. [CrossRef]
18. Wu, Y.; Zhou, F.; Yang, L.; Liu, J. A shrinking strategy for creating dynamic SERS hot spots on the surface of thermosensitive polymer nanospheres. *Chem. Commun.* **2013**, *49*, 5025–5027. [CrossRef]
19. Hofkens, J.; Hotta, J.; Sasaki, K.; Masuhara, H.; Iwai, K. Molecular assembling by the radiation pressure of a focused laser beam: Poly(N-isopropylacrylamide) in aqueous solution. *Langmuir* **1997**, *13*, 414–419. [CrossRef]
20. Tsuboi, Y.; Nishino, M.; Sasaki, T.; Kitamura, N. Poly (*N*-Isopropylacrylamide) Microparticles Produced by Radiation Pressure of a Focused Laser Beam: A Structural Analysis by Confocal Raman Microspectroscopy Combined with a Laser-Trapping Technique. *J. Phys. Chem. B* **2005**, *109*, 7033–7039. [CrossRef]
21. Aibara, I.; Mukai, S.; Hashimoto, S. Plasmonic-Heating-Induced Nanoscale Phase Separation of Free Poly(N-isopropylacrylamide) Molecules. *J. Phys. Chem. C* **2016**, *120*, 17745–17752. [CrossRef]
22. Yeshchenko, O.A.; Bartenev, A.O.; Naumenko, A.P.; Kutsevol, N.V.; Harahuts, I.I.; Marinin, A.I. Laser-driven aggregation in dextran-graft-PNIPAM/silver nanoparticles hybride nanosystem: Plasmonic effects. *Ukr. J. Phys.* **2020**, *65*, 254–267. [CrossRef]

23. Plamper, F.A.; Schmalz, A.; Ballauff, M.; Müller, A.H.E. Tuning the Thermoresponsiveness of Weak Polyelectrolytes by pH and Light: Lower and Upper Critical-Solution Temperature of Poly(N,N-dimethylaminoethyl methacrylate). *J. Am. Chem. Soc.* **2007**, *129*, 14538–14539. [CrossRef]
24. Plamper, F.A.; Ruppel, M.; Schmalz, A.; Borisov, O.; Ballauff, M.; Müller, A.H.E. Tuning the Thermoresponsive Properties of Weak Polyelectrolytes: Aqueous Solutions of Star-Shaped and Linear Poly(N,N-dimethylaminoethyl Methacrylate). *Macromolecules* **2007**, *40*, 8361–8366. [CrossRef]
25. Sigolaeva, L.V.; Günther, U.; Pergushov, D.V.; Gladyr, S.Y.; Kurochkin, I.N.; Schacher, F.H. Sequential pH-dependent adsorption of ionic amphiphilic diblock copolymer micelles and choline oxidase onto conductive substrates: Toward the design of biosensors. *Macromol. Biosci.* **2014**, *14*, 1039–1051. [CrossRef] [PubMed]
26. Sigolaeva, L.V.; Konyakhina, A.Y.; Pergushov, D.V.; Kurochkin, I.N. Electrochemical Biosensor Based on Nanosized Polymer-Enzyme Films Composed of Linear Poly(*N,N*-Dimethylaminoethyl Methacrylate) and Choline Oxidase. *Moscow Univ. Chem. Bull.* **2021**, *76*, 334–342. [CrossRef]
27. Sigolaeva, L.V.; Pergushov, D.V.; Synatschke, C.V.; Wolf, A.; Dewald, I.; Kurochkin, I.N.; Fery, A.; Müller, A.H.E. Co-assemblies of micelle-forming diblock copolymers and enzymes on graphite substrate for an improved design of biosensor systems. *Soft Matter* **2013**, *9*, 2858–2868. [CrossRef]
28. Makhaeva, G.F.; Rudakova, E.V.; Sigolaeva, L.V.; Kurochkin, I.N.; Richardson, R.J. Neuropathy target esterase in mouse whole blood as a biomarker of exposure to neuropathic organophosphorus compounds. *J. Appl. Toxicol.* **2016**, *36*, 1468–1475. [CrossRef]
29. Shumyantseva, V.V.; Sigolaeva, L.V.; Agafonova, L.E.; Bulko, T.V.; Pergushov, D.V.; Schacher, F.H.; Archakov, A.I. Facilitated biosensing via direct electron transfer of myoglobin integrated into diblock copolymer/multi-walled carbon nanotube nanocomposites. *J. Mater. Chem. B* **2015**, *3*, 5467–5477. [CrossRef]
30. Sigolaeva, L.V.; Bulko, T.V.; Konyakhina, A.Y.; Kuzikov, A.V.; Masamrekh, R.A.; Max, J.B.; Köhler, M.; Schacher, F.H.; Pergushov, D.V.; Shumyantseva, V.V. Rational design of amphiphilic diblock copolymer/MWCNT surface modifiers and their application for direct electrochemical sensing of DNA. *Polymers* **2020**, *12*, 1514. [CrossRef]
31. Shumyantseva, V.V.; Agafonova, L.E.; Bulko, T.V.; Kuzikov, A.V.; Masamrekh, R.A.; Yuan, J.; Pergushov, D.V.; Sigolaeva, L.V. Electroanalysis of Biomolecules: Rational Selection of Sensor Construction. *Biochemistry* **2021**, *86*, S140–S151. [CrossRef] [PubMed]
32. Shumyantseva, V.V.; Bulko, T.V.; Kuzikov, A.V.; Masamrekh, R.A.; Konyakhina, A.Y.; Romanenko, I.; Max, J.B.; Köhler, M.; Gilep, A.A.; Usanov, S.A.; et al. All-electrochemical nanocomposite two-electrode setup for quantification of drugs and study of their electrocatalytical conversion by cytochromes P450. *Electrochim. Acta* **2020**, *336*, 135579. [CrossRef]
33. Shumyantseva, V.V.; Bulko, T.V.; Tikhonova, E.G.; Sanzhakov, M.A.; Kuzikov, A.V.; Masamrekh, R.A.; Pergushov, D.V.; Schacher, F.H.; Sigolaeva, L.V. Electrochemical studies of the interaction of rifampicin and nanosome/rifampicin with dsDNA. *Bioelectrochemistry* **2021**, *140*, 107736. [CrossRef] [PubMed]
34. Schallon, A.; Synatschke, C.V.; Pergushov, D.V.; Jérôme, V.; Müller, A.H.E.; Freitag, R. DNA Melting Temperature Assay for Assessing the Stability of DNA Polyplexes Intended for Nonviral Gene Delivery. *Langmuir* **2011**, *27*, 12042–12051. [CrossRef]
35. Raup, A.; Wang, H.; Synatschke, C.V.; Jérôme, V.; Agarwal, S.; Pergushov, D.V.; Müller, A.H.E.; Freitag, R. Compaction and transmembrane delivery of pDNA: Differences between l-PEI and two types of amphiphilic block copolymers. *Biomacromolecules* **2017**, *18*, 808–818. [CrossRef] [PubMed]
36. Ou, Y.; Wang, L.Y.; Zhu, L.W.; Wan, L.S.; Xu, Z.K. In-situ immobilization of silver nanoparticles on self-assembled honeycomb-patterned films enables surface-enhanced raman scattering (SERS) substrates. *J. Phys. Chem. C* **2014**, *118*, 11478–11484. [CrossRef]
37. Zhou, J.; Zha, X.; Chen, X.; Ma, J. β-Cyclodextrin-g-Poly(2-(dimethylamino) ethyl methacrylate) as the Stabilizer and Reductant to Prepare Colloid Silver Nanoparticles in situ. *Indian J. Pharm. Educ. Res.* **2017**, *51*, 543–550. [CrossRef]
38. Sun, H.; Gao, Z.; Yang, L.; Gao, L.; Lv, X. Synthesis and characterization of novel four-arm star PDMAEMA-stabilized colloidal silver nanoparticles. *Colloid Polym. Sci.* **2010**, *288*, 1713–1722. [CrossRef]
39. Nechaeva, N.L.; Boginskaya, I.A.; Ivanov, A.V.; Sarychev, A.K.; Eremenko, A.V.; Ryzhikov, I.A.; Lagarkov, A.N.; Kurochkin, I.N. Multiscale flaked silver SERS-substrate for glycated human albumin biosensing. *Anal. Chim. Acta* **2020**, *1100*, 250–257. [CrossRef]
40. Sun, F.; Bai, T.; Zhang, L.; Ella-Menye, J.-R.; Liu, S.; Nowinski, A.K.; Jiang, S.; Yu, Q. Sensitive and Fast Detection of Fructose in Complex Media via Symmetry Breaking and Signal Amplification Using Surface-Enhanced Raman Spectroscopy. *Anal. Chem.* **2014**, *86*, 2387–2394. [CrossRef]

biosensors

Article

Investigation of the "Antigen Hook Effect" in Lateral Flow Sandwich Immunoassay: The Case of Lumpy Skin Disease Virus Detection

Simone Cavalera [1], Giulia Pezzoni [2], Santina Grazioli [2], Emiliana Brocchi [2], Stefano Baselli [2], Davide Lelli [2], Barbara Colitti [3], Thea Serra [1], Fabio Di Nardo [1], Matteo Chiarello [1], Valentina Testa [1], Sergio Rosati [3], Claudio Baggiani [1] and Laura Anfossi [1,*]

[1] Department of Chemistry, University of Turin, Via Pietro Giuria 7, 10137 Turin, TO, Italy
[2] Istituto Zooprofilattico Sperimentale della Lombardia e dell'Emilia-Romagna, National/OIE/FAO Reference Centre for FMD and SVD, Via Antonio Bianchi 7, 25124 Brescia, BS, Italy
[3] Department of Veterinary Science, University of Turin, Largo Paolo Braccini 2, 10095 Grugliasco, TO, Italy
* Correspondence: laura.anfossi@unito.it

Abstract: Lumpy skin disease (LSD) is an infectious disease affecting bovine with severe symptomatology. The implementation of effective control strategies to prevent infection outbreak requires rapid diagnostic tools. Two monoclonal antibodies (mAbs), targeting different epitopes of the LSDV structural protein p32, and gold nanoparticles (AuNPs) were used to set up a colorimetric sandwich-type lateral flow immunoassay (LFIA). Combinations including one or two mAbs, used either as the capture or detection reagent, were explored to investigate the hook effect due to antigen saturation by the detector antibody. The mAb-AuNP preparations were optimized by a full-factorial design of experiment to achieve maximum sensitivity. Opposite optimal conditions were selected when one Mab was used for capture and detection instead of two mAbs; thus, two rational routes for developing a highly sensitive LFIA according to Mab availability were outlined. The optimal LFIA for LSDV showed a low limit of detection ($10^{3.4}$ $TCID_{50}$/mL), high inter- and intra-assay repeatability (CV% < 5.3%), and specificity (no cross-reaction towards 12 other viruses was observed), thus proving to be a good candidate as a useful tool for the point-of-need diagnosis of LSD.

Keywords: point-of-care test; rapid diagnosis; infectious diseases; single-epitope sandwich; double-epitope sandwich; lumpy skin disease

Citation: Cavalera, S.; Pezzoni, G.; Grazioli, S.; Brocchi, E.; Baselli, S.; Lelli, D.; Colitti, B.; Serra, T.; Nardo, F.D.; Chiarello, M.; et al. Investigation of the "Antigen Hook Effect" in Lateral Flow Sandwich Immunoassay: The Case of Lumpy Skin Disease Virus Detection. *Biosensors* **2022**, *12*, 739. https://doi.org/10.3390/bios12090739

Received: 29 July 2022
Accepted: 5 September 2022
Published: 8 September 2022

1. Introduction

In the recent years, more than ever, screening tests used for monitoring infectious diseases have been potentiated to counteract pathogenic local outbreaks, as well as pandemics [1–3]. The World Health Organization (WHO) has suggested increasing the use of the point-of-care tests (POCTs), as they optimally fit with criteria gathered under the acronym of (RE)ASSURED (Real-time connectivity, Easy of specimen collection, Affordability, Sensitivity, Specificity, User-friendliness, Rapidness/Robustness, Equipment-free, Deliverability) [4]. Among analytical strategies involved in POCTs, the most diffuse and successful technique is the immunochromatographic strip test, also known as the lateral flow immunoassay (LFIA). Many POCTs in the LFIA format are routinely employed to monitor infectious diseases by the Centers for Disease Control and Prevention (CDC) [5,6]. As well as any kind of immunoassay, the LFIA exploits the affinity of specific antibodies to detect antigens from viruses, fungi, or bacteria, thus enabling the direct diagnosis of the infection. The LFIA for antigen detection is based on the formation of the immunocomplex between a labeled specific antibody (detection antibody) and the target antigen, and its capture by another specific antibody (capture antibody) anchored to a support to create a reactive zone (test line). The signal is generated by the accumulation of the labeled antibody and antigen immunocomplex in correspondence to the test line. For colorimetric LFIAs, the label is a colored nanomaterial, such as metal (e.g., Au, Ag), latex, or carbon nanoparticles [7,8]. As a widespread and impactful infectious disease affecting bovines, lumpy skin disease (LSD) needs an effective control strategy to limit the spread of the infection. The etiological agent is the lumpy skin disease virus (LSDV), a large double-stranded DNA virus with an oval or brick-shaped morphology, and a genome of about 150 kDa. LSDVs belong to the *Capripoxvirus genus*, of the *Poxviridae* family. LSD causes economic damage, such as loss in milk production, besides limping and more severe symptomatology [9,10]. During the 2012–2018 Eurasian LSD epidemic, the morbidity and mortality associated with the epidemic were reported at 9% to 26% and 0.5% to 2%, respectively [11]. Though characteristic clinical signs of LSD enable a presumptive diagnosis, laboratory confirmation is necessary for the successful control and eradication of LSD, particularly in cases of mild disease. Laboratory confirmation of LSD relies on electron microscopy examination, virus isolation (VI), serological tests, and polymerase chain reaction (PCR). Viral isolation is the gold standard for LSDV diagnosis; however, it requires several weeks to isolate the virus. Molecular methods, including conventional or real-time polymerase chain reaction and loop-mediated isothermal amplification, are the most sensitive methods for detecting LSDV [12–19]. As far as immunodiagnostics is concerned, serological assays are available to indirectly diagnose LSDV by detecting the immune response to the virus in the blood of infected animals, while assays aimed at detecting viral antigens have not been reported yet, with the exception of the immunohistochemical method for detecting the LSDV antigens in skin nodules of infected cattle [20]. However, there is a lack of simpler virological tests, such as ELISA, for antigen detection or rapid tests enabling the point-of-need diagnosis of LSD, which would be particularly useful in endemic countries. In a previous study, monoclonal antibodies (mAbs) raised against the Neethling strain of LSDV and recognizing different epitopes of a 35 kDa viral protein (LSDV p32) were shown to detect LSDV isolates when used as coating and peroxidase-conjugated antibodies in a sandwich ELISA format [21]. Therefore, taking advantage of these well-characterized pairs of mAbs, here we aimed at developing a sensitive colorimetric LFIA for the on-field and visual detection of the LSD virus in clinical samples. Though apparently antibodies can be employed for capture or detection of target antigens, optimizing the role of each bioreagent modulated the performance of the LFIA device [22]. In addition, other parameters, such as the antibody-to-label ratio, the amount of the labeled antibody, the concentration of the capture antibody, and the position of the test line play a crucial role in developing effective LFIA devices [22,23]. Typical sandwich-type immunoassays rely on the use of two antibodies, which specifically target different epitopes of the antigen. In the eventuality that the pathogen shows repeated copies of the same antigen or that the antigen is composed of repeated molecular domains, a single antibody targeting the repeated epitope can be employed successfully to develop a sandwich-type immunoassay. In a previous work, we evidenced a different behavior between resorting to a "single epitope" (SE) targeting sandwich assay, i.e., one antibody for capture and detection, and using a "double epitope" (DE) targeting sandwich assay, where the capture and detection antibodies differed [23]. Indeed, we observed that for DE assays, the increase in the antibodies led to an increase in the analytical signal. For the SE assays, the observed effect was the opposite; i.e., when increasing the amount of the antibodies, the signal decreased. In this work, we deeply investigated the phenomenon to possibly establish if the evidence could be generalized or if it was due to that specific antigen (e.g., to the particular number and distancing of the epitopes). Therefore, two lateral flow devices were designed, using gold nanoparticles (AuNPs) as the optical signal reporter (Figure 1) and passive adsorption for antibody attachment. Although covalent coupling of antibodies to AuNPs ensures the best orientation of the ligand for binding to the antigen [24], the adsorption strategy is universally employed due to its simplicity, cost efficiency, and preservation of antibody structure and binding ability [24,25]. One system employed a traditional sandwich-type immunoassay, in which two antibodies, which recognize two different epitopes of the LSDV p32, were used (DE, Figure 1b). To this end, these two mAbs (identified as #2C6 and #2F10) were either linked to the AuNPs or immobilized onto the membrane to explore their performance as capture and detection ligands. The optimal capture–detection combination was defined according to the stability of the resulting mAb_AuNP conjugate and to the highest.

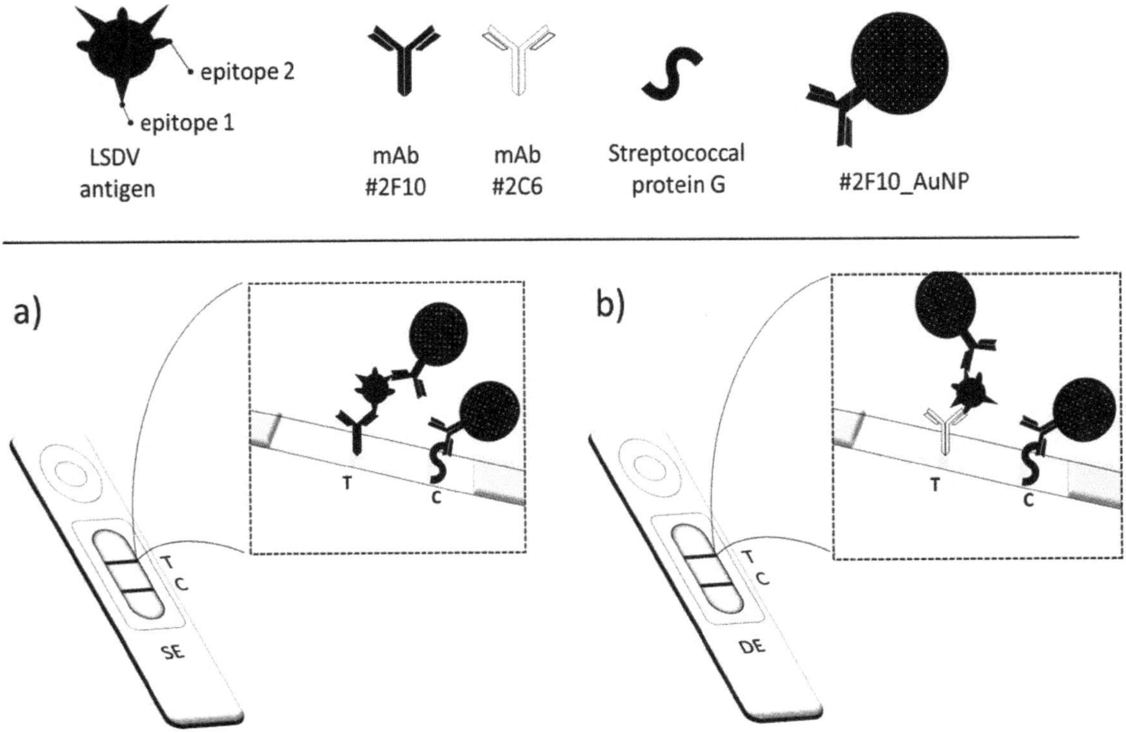

Figure 1. Schematic of the two sandwich-type immunoassays using: one antibody (single epitope, (**a**)) and two different antibodies (double epitope, (**b**)) as capture and detection.

Signal-to-noise ratio (i.e., the ratio between the color developed by applying the supernatant of an inactivated virus suspension and the color developed by applying the sample diluent). The second device was designed by using each of the mAb as the capture and detection ligand in combination with itself (SE, Figure 1a). We studied the effect of varying the characteristics of the mAb_AuNP probes (such as varying the optical density of the probe, the mAb-to-AuNP ratio, and the size of AuNPs) on assay sensitivity. The probes were characterized by UV-visible spectroscopy, zeta-potential, and high-resolution transmission electron microscopy (HR-TEM), and their effectiveness in the LFIA was studied by a full-factorial design of experiment (FF-DoE), mapping the intensity of the color obtained by a "positive control" (inactivated LSDV Neethling strain). Experiments were replicated for the SE system, using one mAb as the capture and detection ligand, and for the DE system, which employed the two mAbs. The two systems were optimized and compared to confirm whether the atypical hook effect, due to the saturation of antigen epitopes by the detection ligand, could be generalized and occurs for sandwich-type LFIA based on one mAb. The DE-based assay, which showed the best performance, was further characterized by evaluating the analytical sensitivity, the inter- and intra-assay repeatability, and the specificity towards other viruses of the *lentivirus* and *parapoxvirus* genii.

2. Materials and Methods

2.1. Synthesis and Characterization of the Gold Nanoparticles (AuNPs)

AuNPs with different sizes (24, 30, 36 nm) were prepared by tetra chloroauric acid reduction with different amounts of sodium citrate as previously described [26]. The colloidal gold solutions were spectroscopically characterized by means of a Cary 60 UV-Visible spectrophotometer (Agilent, Santa Clara, CA, USA) in the UV-visible range, and hydrodynamic diameters were acquired by using a Z-view^R Nanoparticle Tracking Analyzer PMX 120 (Particle Metrix Gmbh, Inning, Germany). The 32 nm AuNPs and conjugates were also characterized by high-resolution transmission electron microscopy (HR-TEM) (Jeol JEM 3010-UHR, Tokyo, Japan).

2.2. Synthesis and Characterization of the mAb_AuNP Conjugates

Signal reporters used in the LFIAs were prepared by passively adsorbing the mAbs onto the surface of the AuNPs, followed by overcoating of the AuNP-free surface with BSA [24]. In detail, the pH of 10 mL of AuNP solution was adjusted to 8 with carbonate buffer (0.05 M, pH 9.6) and added to 1 mL of borate buffer (0.02 M, pH 8). The mAb solution was gently mixed with the AuNPs and incubated for 40 min at 37 °C at room temperature. Then, 1 mL of BSA (1% *w*/*v* in borate buffer) was added and reacted for 10 min. The mAb_AuNP conjugates were recovered by centrifugation (10,000 rpm, 15 min) and washed twice with borate buffer supplemented with 0.1% *w*/*v* BSA. Finally, mAb_AuNPs were re-suspended in AuNP storage buffer (borate buffer with 1% *w*/*v* BSA, 0.25% *v*/*v* Tween 20, *w*/*v* 2% sucrose, and 0.02% *w*/*v* sodium azide) and stored at 4 °C until use. The #2F10_AuNP conjugates made from the 32 nm AuNPs were characterized by their visible spectra], Z-potential, and dynamic diameter, and HR-TEM imaging. For these experiments, the mAb_AuNP were not overcoated with BSA.

2.3. Production of the LFIA Strips

The capture mAbs (#2F10 and #2C6) used for drawing test lines and the Streptococcal protein G for the control line of the LFIA devices were diluted in phosphate buffer (0.02 M pH 7.4) and applied at 1 μL/cm onto the nitrocellulose membrane by means of a XYZ3050 platform (Biodot, Irvine, CA, USA) equipped with a BioJetQuanti 3000 Line Dispenser for non-contact dispensing. The mAb concentration used for the experimental design and in the final device was 1 mg/mL and 2 mg/mL, respectively. Strips were composed as follows: sample pad, conjugate pad, membrane, and adsorbent pad and were cut to 4.2 mm in width) by means of a CM4000 guillotine (Biodot, Irvine, CA, USA). Finally, strips were inserted into plastic cassettes (Kinbio, Bejing, China) to fabricate the ready-to-use LFIA device. Cassettes were stored in the dark in plastic bags containing silica at room temperature until use.

2.4. Experimental Design for the SE and for the DE

Three AuNPs preparations differing in size were used to produce mAb_AuNP conjugates. For each AuNP preparation, we changed, independently, (i) 4 levels of optical density for the SE (Table 1) and 3 for the DE (Table 2), and (ii) 4 levels of the mAb-to-AuNP ratio. The combinations were tested in duplicate by using a positive and a negative sample, for a total of 336 experiments. The probes were diluted with the AuNP dilution buffer (borate buffer with 0.25% Tween 20 *v*/*v*, 2% *w*/*v* sucrose and 0.02% *w*/*v* sodium azide) and adsorbed onto a pre-saturated glass fiber conjugate pad. Pads were dried for 4 h at room temperature. The format and condition giving the highest color intensity of the test line was extracted from the FF-DoE and the optimized device (LSD_LFIA).

Table 1. The FF-DoE scheme for SE optimization. Each experiment was repeated twice.

mAb-to-AuNP (× 10 μg/mL) [a]	OD	Size of AuNP (nm)								
			24			32			36	
0.5×	0.5	−2	−2	−1	−2	−2	0	−2	−2	1
	1	−2	−1	−1	−2	−1	0	−2	−1	1
	1.5	−2	1	−1	−2	1	0	−2	1	1
	2	−2	2	−1	−2	2	0	−2	2	1
1.0×	0.5	−1	−2	−1	−1	−2	0	−1	−2	1
	1	−1	−1	−1	−1	−1	0	−1	−1	1
	1.5	−1	1	−1	−1	1	0	−1	1	1
	2	−1	2	−1	−1	2	0	−1	2	1
1.5×	0.5	1	−2	−1	1	−2	0	1	−2	1
	1	1	−1	−1	1	−1	0	1	−1	1
	1.5	1	1	−1	1	1	0	1	1	1
	2	1	2	−1	1	2	0	1	2	1
2.0×	0.5	2	−2	−1	2	−2	0	2	−2	1
	1	2	−1	−1	2	−1	0	2	−1	1
	1.5	2	1	−1	2	1	0	2	1	1
	2	2	2	−1	2	2	0	2	2	1

[a] Defined as a n-fold multiple of the minimum stabilizing mAb-to-AuNP ratio.

Table 2. The FF-DoE scheme for DE optimization. Each experiment was repeated twice.

mAb-to-AuNP (× 10 μg/mL) [a]	OD	Size of AuNP (nm)								
			24			32			36	
0.5×	2	−2	−1	−1	−2	−1	0	−2	−1	1
	3	−2	0	−1	−2	0	0	−2	0	1
	4	−2	1	−1	−2	1	0	−2	1	1
1.0×	2	−1	−1	−1	−1	−1	0	−1	−1	1
	3	−1	0	−1	−1	0	0	−1	0	1
	4	−1	1	−1	−1	1	0	−1	1	1
1.5×	2	1	−1	−1	1	−1	0	1	−1	1
	3	1	0	−1	1	0	0	1	0	1
	4	1	1	−1	1	1	0	1	1	1
2.0×	2	2	−1	−1	2	−1	0	2	−1	1
	3	2	0	−1	2	0	0	2	0	1
	4	2	1	−1	2	1	0	2	1	1

[a] Defined as a n-fold multiple of the minimum stabilizing mAb-to-AuNP ratio.

2.5. Viruses

The virulent LSD virus, the Neethling strain, was grown in the ovine testis cell line (OA3.Ts). The supernatant of infected cells was harvested when the cytopathic effect was at its maximum and combined with the supernatant recovered from cell debris subjected to freeze–thaw cycles. The infectious virus titer was expressed as Tissue Culture Infective Dose 50 (TCID50) calculated by the Reed–Muench method [27]. For inactivation, the virus suspension was clarified by centrifugation at 5000 rpm for 20 min, followed by filtration through 0.2 μm filters, and then β-Propiolactone was added at a concentration of 0.01% and kept for 48 h at 4 °C. Inactivation was confirmed by repeated passages in the OA3.Ts cell line. A series of virus isolates belonging to the family Poxviridae, as detailed in Table 3, were additionally used for LSD-LFIA initial validation. The sole running buffer was used as the negative control.

Table 3. Spectroscopic and dimensional characteristics of conjugates as a function of different amounts of mAb adsorbed on the surface of the 32 nm AuNP.

Amount of mAb for mL of AuNP (OD1)	λmax of LSPR	Hydrodynamic Diameter [a]	Δ [b]	Z-Potential
(μg)	(nm)	(nm)	(nm)	(mV)
0	525.5	46.4 ± 0.3		−41.7 ± 0.4
5	536.0	89.8 ± 3.8	(21.7) [c]	−34.5 ± 0.7
10	529.5	50.6 ± 1.5	2.1	−24.2 ± 0.3
15	530.0	51.6 ± 0.2	2.6	−23.5 ± 0.5
20	530.5	52.7 ± 1.7	3.2	−29.7 ± 1.2

[a] as measured by the Z-view[R] Nanoparticle Tracking Analyzer. [b] (hydrodynamic diameter of the conjugate—hydrodynamic diameter of the bare AuNP)/2. [c] the value represents the effect of aggregation rather than that of mAb adsorption.

2.6. In-House Validation of the LSD-LFIA Device

The visual cut-off was defined as the LSDV level corresponding to the complete disappearance of the test line (visual LOD, vLOD). The specificity was evaluated by testing viruses belonging to the Poxviridae family, which LSDV also belongs to, and some viruses infecting bovine. The LSD-LFIA imprecision was estimated by measuring between- and within-assay reproducibility and was calculated by analyzing three dilutions of the inactivated virus suspension (4×, 1×, and 0.5× of the vLOD) in three replicates each, three times within the same day, and on three days. The overall assay variability was calculated as the mean of coefficients of variation (CV%) from all sessions for each level (n = 18). The within-assay variability was estimated based on the mean of the CV% within days (n = 9) and between days (n = 3). The shelf life of the device at 4 °C, room temperature, and 37 °C was explored after 7, 14, 30, and 90 days. The analytical sensitivity was evaluated in parallel with a sandwich ELISA using live LSDV Neethling strain growth on the ovine testis cell line (OA3.Ts), titrated by the Reed–Muench method, expressed as 50% Tissue Culture Infective Dose (TCID50), and tested in serial dilutions [28]. For the sandwich ELISA, the two selected mAbs (#2F10 and #2C6) were preliminarily evaluated for their capability in different combinations of adsorbed and HRP-conjugated MAb, to efficiently bind and detect the LSDV antigen [28]; thus, the combination 2F10 as adsorbed mAb and #2C6 as HRPO-conjugated mAb was selected. The 2F10 mAb was adsorbed to the microtiter plates (NUNC, Maxisorp, Roskilde, Denmark) at a saturating concentration of 2 μg/mL in a carbonate–bicarbonate buffer (pH 9.6), and after overnight incubation at 4 °C, three washes with PBS containing 0.05% Tween 20 were performed. The viral antigen was incubated for 1 h at 37 °C, at three-fold serial dilution. After three washes with the same buffer, the #2C6 HRP-conjugated mAb was added at serial checkerboard dilutions with respect to the viral antigen, for a further 1 h of incubation at 37 °C. Following three final washes with the same PBS–Tween buffer, 0.5 mg/mL of OPD diluted in phosphate–citrate buffer (pH 5.6) and supplemented with 0.02% H_2O_2 was added for 10 min of incubation at room temperature; the reaction was stopped with H_2SO_4 (1 M). Plates were analyzed using a Multiscan Ascent spectrophotometer (Thermo Fisher Scientific, Waltham, MA, USA) at 492 nm wavelength. The results are expressed as net optical density (OD) obtained by the OD of the reaction with the viral antigen subtracted by the OD value of the reaction without the viral antigen.

3. Results and Discussion

3.1. Preparation and Spectroscopic Characterization of the mAb_AuNP Conjugates

AuNPs with different localized surface plasmon resonance (LSPR) maximum wavelengths (523.0, 525.5, 527.0 nm) were prepared by reducing tetra chloroauric acid with different amounts of trisodium citrate. According to the model proposed by Khlebetsov et al., the mean diameters of the AuNPs were calculated as 24, 30, and 36 nm, respectively [29]. The amount of the antibody needed to stabilize AuNPs was defined by a salt-induced aggregation test and was identified as the one providing a 540/620 absorbance ratio (corresponding to the non-aggregated/aggregated nanoparticles) over 3, which represents the initial value shown by the unperturbed colloidal gold suspension. The salt-induced aggregation tests showed the typical behavior, with an increasing stabilization due to the increasing quantity of the antibody, which passivated the surface of the AuNPs and protected them from the aggregation (Figure S1). Interestingly, although the AuNPs involved in the study differed in size, the quantity of antibodies needed for the stabilization was apparently the same. The smallest AuNPs showed lower absorbance, and largest AuNPs were more prone to aggregation. As the 32 nm AuNPs are the typically employed probes in the LFIA, we continued the study on these AuNPs. The minimum amount of mAb required for stabilizing 1 mL of AuNP diluted at a level giving OD1 was defined as 6 μg for the mAb #2C6, and 10 μg for the mAb #2F10.

However, with the aim of maximizing the sensitivity of the LFIA for LSD diagnosis and, concurrently, to investigate the effect of some key parameters on assay performance, we synthesized several #2F10-AuNP conjugates by varying the mAb-to-AuNP ratio as a fraction or a multiple of the stabilizing amount (Figure S2a). The #2F10_AuNP conjugates made from the 32 nm AuNPs were also characterized by their visible spectra, Z-potential, and hydrodynamic diameter (Table 2), and HR-TEM imaging (Figure S2b–f). The Vis spectra in Figure S2g and their first derivative (Figure S2h) show a slight blue shift of the maximum of the LSPR corresponding to the addition of 10 μg, 15 μg, and 20 μg of the mAb. On the contrary, the addition of 5 μg caused a modification of the peak profile, which can be ascribed to the conjugate instability and aggregation phenomena. This is consistent with the flocculation stress test, in which the minimum quantity of the mAb needed for stabilization was set at 10 μg. The band shift of 4 nm observed upon the addition of 10 μg of mAb was consistent with the addition of a layer of bioreagents on the AuNP surface (Table 3).

Interestingly, the addition of higher amounts of the mAb induced a further limited increase in the λ_{max} (0.5 nm each for 15 and 20 μg of the mAb, respectively). This result suggests that the quantity of mAb adsorbed onto the AuNP surface varied, though not to the extent that it was compatible with the formation of a second mAb layer. We hypothesized that increasing the mAb put into contact with AuNPs over the stabilizing value resulted in a larger amount of the bioligand adsorbed. This was also supported by the measurement of the hydrodynamic diameter and Z-potential (Table 3). The thickness of the bioligand layer adsorbed onto the AuNP was estimated as the difference between hydrodynamic diameters of the mAb_AuNP conjugate and the one of the pristine AuNP, divided by two [29]. According to the calculation, the protein layer was 2.1 nm-thick when the stabilizing amount of mAb was adsorbed and increased to 3.2 nm when the amount was doubled. Two possible phenomena can explain the increase: (1) the mAb added formed a second layer of bioligands with a random orientation, or (2) the addition of the mAb induced the formation of an increasingly bundled monolayer of the bioligand, forcing the orientation of the antibody towards head-on/tail-on, in such a way that the Fab was more exposed for the binding (Figure 2) [25].

Though both reasonable, only the second hypothesis was also able to explain the parallel increase in the binding ability with the increasing quantity of mAb adsorbed. All mAb_AuNP conjugates showed a largely negative z-potential, which varied from the −41.7 mV value of the bare AuNP to −24–29 mV for the conjugates prepared by adsorbing the stabilizing amount of the mAb or exceeding this value. The z-potential for these conjugates was comparable to the one of mAb at pH 8.0, thus confirming the formation of the antibody layer on the AuNP surface when 10 μg of the mAb was adsorbed [30]. Coherently with the use of an insufficient amount of antibody to cover the AuNP surface, the conjugate prepared by adsorbing 5 μg of the mAb showed a more negative z-potential between the one of citrate-capped AuNPs and the one corresponding to AuNP completely shielded by the mAb layer. According to the suggested aggregation, the 5 μg conjugate showed an apparent diameter of about 90 nm.

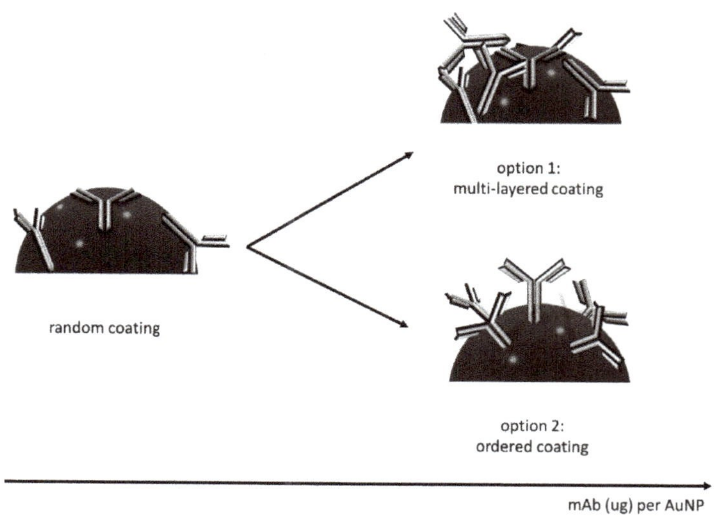

Figure 2. Scheme of the hypothesized increase in mAb adsorbed onto AuNP: (**1**) absorption of a second layer of randomly oriented mAb; (**2**) the additional mAb forced the head-on/tail-on orientation.

3.2. *Development of the LSDV_LFIA Based on the Combination #2F10_AuNP/#2F10 (SE)*

Combining two mAbs conjugated to AuNPs (detectors, #2F10* and #2C6*) and the same two used as capturing bioreagents (#2F10c and #2C6c), four combinations were possible and were investigated: two DEs (#2F10*/#2C6c, #2C6*/2F10c), and two SEs (#2F10*/#2F10c and #2C6*/#2C6c). Preliminarily, we excluded non-specific bindings in the explored combinations. DEs and SEs employing #2F10* showed the absence of non-specific signals, whereas the ones with #2C6* showed a strong non-specific signal. Some attempts were made to mitigate this non-specific binding by modifying: (i) the running buffer, (ii) the protein used to saturate the AuNP-free surface, and (iii) the mAb_AuNP storage buffer used for the dilution of the gold conjugate. In brief, we changed the type (BSA and casein) and concentration (1, 0.25, 0.1% w/v) of the protein and the ionic strength of the running buffer by adding different amounts (30, 80, 130, 180 mM) of sodium chloride. The casein was used for its ability to interfere with protein–protein non-specific interactions, while the ionic strength was used to reduce electrostatic interactions. The addition of 0.1% w/v casein and 80 mM sodium chloride to the running buffer reduced the non-specific binding when 6 µg of #2C6 per mL of AuNPs was adsorbed. However, it was ineffective for higher mAb-to-AuNP ratios (Table S1). As the attempts made to reduce non-specific interaction of the #2C6* were unfruitful, the combinations involving this probe were no further studied. The optimal conditions for detecting LSDV by the LFIA including the same mAbs as the capture and detection reagents were defined by exploiting an FF-DoE. This was made by mapping the intensity of the test line as a function of the mAb-to-AuNP ratio and optical density. As previously observed for a similar LFIA for FMDV detection, the SE showed a saturation behavior, where the intensity of the test line decreased by increasing the 2F10* optical density (Figure S3 and Table S2) [23]. The decrease in the signal, while increasing antigen concentration, occurs for sandwich-type immunoassays when the antigen amount exceeds the capture and/or detection antibodies. The effect, named the "hook effect", has been explained as the overwhelming inhibition by antigen-saturated capture and detection antibodies of the formation of the capture–antigen–detection "sandwich" complex that is responsible for the antigen-related signal. To limit the hook effect and increase the dynamic range of the assay, typically, the quantity of the antibodies should be augmented. In the LFIA platform, the equilibria between the antigen and capture and detection antibodies occur separately, so the effect is just apparently similar to the typical hook effect. In this case, the observed decrease in the signal was not due to the antigen but to the excess of the detection antibody. The effect was described as an "antigen saturation hook effect" (asHE) (Figure 3a) where the capture and the detection antibodies competed for the same epitope of the antigen.

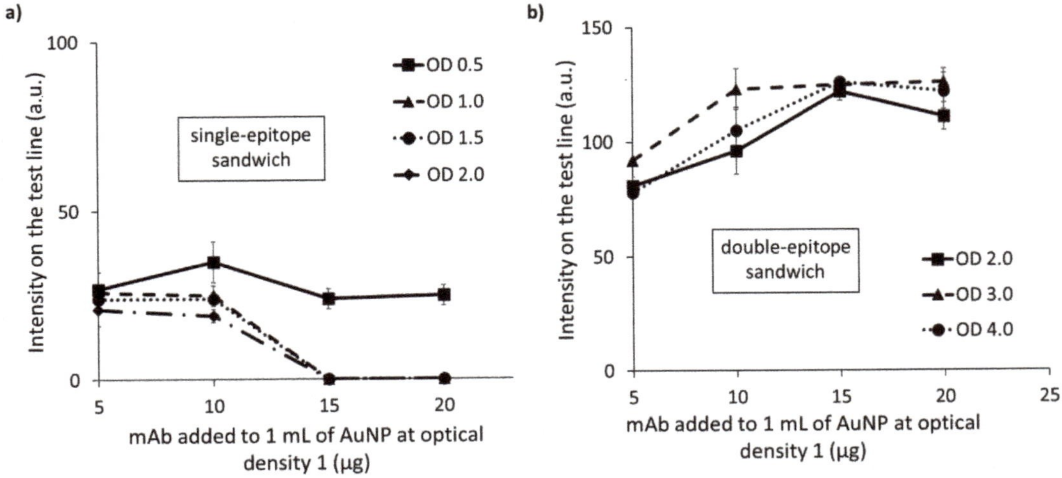

Figure 3. Intensity map for the 32 nm AuNPs with increasing amounts of #2F10_AuNP (optical density 0.5-1-1.5-2 for the SE (**a**) and 2-3-4 for the DE (**b**)) and antibody adsorbed to the surface of the AuNPs (5-10-15-20 µg for each mL of AuNP at optical density 1).

In this competition, the detection antibody was favored considering that it was mixed with the antigen in the solution and that it had much more contact time with the antigen compared to the capture antibody. The saturation of the antigen by the detection antibody inhibited its further binding to the capture antibody at the test line and resulted in a lowering of the test line color. The same behavior happened when increasing both the quantity of mAb-AuNP and the mAb-to-AuNP ratio. The experimental design was repeated with AuNPs of various sizes. From the data, we always observed the asHE phenomenon, regardless of the particle size (Figure S3).

3.3. Development of the LSDV_LFIA Based on the Combination #2F10_AuNP/#2C6 (DE)

When the DE was developed, we observed the opposite effect than that seen for the SE, and the pattern of color intensities followed the typical trend expected for non-competitive immunoassays: the intensity of the test line (reported as arbitrary units, a.u., from the image processing) increased with the increase in the antibody amount, both considered as the OD of the conjugate and the mAb-to-AuNP ratio parameters (Figure 3b). As for the SE, we completed the experimental design with various sizes of AuNPs (Figure S4 and Table S3). From the intensity maps, we determined the experimental conditions leading to the maximal signal for both the SE and DE approaches. The top performing condition for the SE approach was characterized by 36 nm AuNPs functionalized with 5 µg for each milliliter of AuNP of optical density 1, and used at an optical density of 0.5. As an alternative, the 32 nm AuNPs were also effective when functionalized with 10 µg for each milliliter of AuNP of optical density 1 and used at an optical density of 0.5. Both the conditions provided a signal between 35 and 40 a.u. from the digitalizing processing of the test line color for the inactivated virus suspension. Much more intense signals were achieved with the DE, for which the top performing condition was obtained from 32 nm AuNPs functionalized with 20 µg for each milliliter of AuNP of optical density 1 and used at an optical density of 3. This condition provided a signal of 127 a.u., overcoming the top performers of the SE approach by a factor of 3.6 and 3.2, respectively. Therefore, the optimized DE (32 nm, 20 µg/mL, OD3) was selected for the LSDV_LFIA development.

3.4. The LFIA Device for the Rapid LSD Antigen Detection

Further refining of the assay sensitivity was realized by modulating additives used for AuNP conjugate preparation and for the running buffer, until reaching of a substantial gain in the color intensity (Table 3). In particular, by combining the use of a hydrogen carbonate buffer (26 mM pH 7.9 supplemented with 1% *v*/*v* tween 20 and 0.02% *w*/*v* sodium azide, without any additional proteins) as the running buffer, and the use of a borate buffer (20 mM pH 8 supplemented with 2% *w*/*v* sucrose, 0.25% *v*/*v* Tween 20, 0.02% *w*/*v* sodium azide, 0.8% *w*/*v* BSA, and 0.1% *w*/*v* casein) as the mAb_AuNP dilution buffer, we obtained an increment in the signal-to-noise ratio by a factor of three compared to the use of the typical phosphate buffer.

Furthermore, by doubling the concentration of the capture antibody, the signal increased again by a factor of about three (Table S5).

The optimized LSDV-LFIA was assessed with serial dilutions of the live LSDV Neethling strain in parallel with the sandwich ELISA test (Figure 4). The test line was still visible at 400-fold dilution of the viral antigen. In the sandwich ELISA, the conditions with the best signal-to-noise ratio were considered, and the cut-off was set at 1/250 (net OD ranged between 0.1 and 0.15). In accordance with the virus titration of 10^6 $TCID_{50}$/mL, the visual limit of detection (vLOD) for the LSDV-LFIA resulted in $10^{3.4}$ $TCID_{50}$/mL, which is comparable with the value obtained in the sandwich ELISA ($10^{3.6}$ $TCID_{50}$/mL). Although ELISA is an accurate and reproducible analytical method and it is largely employed for the diagnosis of infectious diseases, its applicability is confined to the laboratory, and it is not suited for timely and onsite testing. It is a general opinion that the LFIA possesses lower sensitivity, specificity, and higher variability in results than laboratory-based immunoassays, such as ELISA. In this work, we showed that the optimization of the LFIA allows for obtaining comparable performance, while the same antibodies are employed.

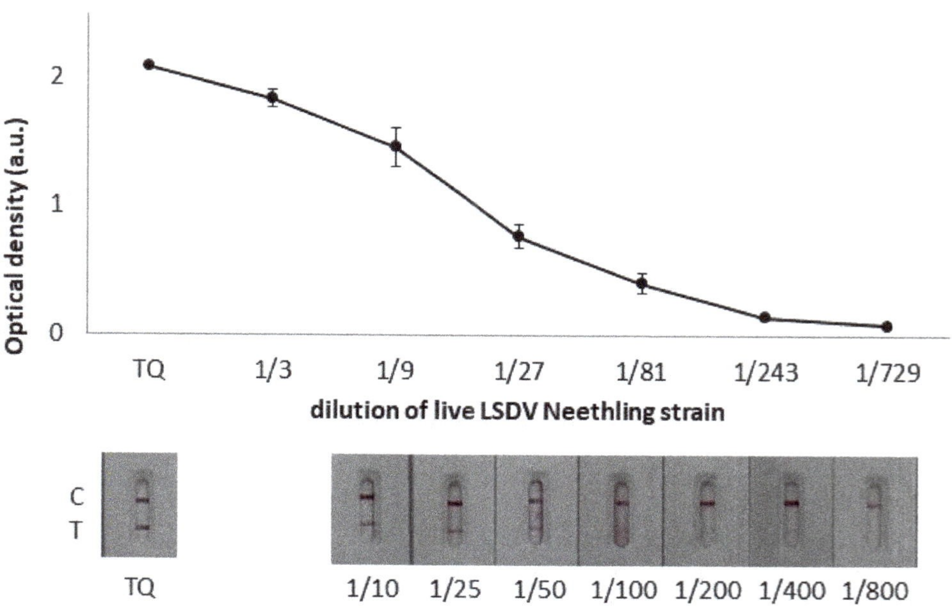

Figure 4. LSD_LFIA sensitivity assessment by testing serial dilution of live LSDV Neethling strain titrated as 10^6 $TCID_{50}$/mL in parallel with the ELISA.

As for the imprecision assessment, we selected three levels corresponding to 4×, 1×, and 0.5× of the vLOD and repeated the analysis in triplicate in each experimental session (same sample preparation, same batch of strips) for three experimental sessions (new sample preparation, different strip batch) each day for three days. No false positives were reported considering the total number of experiments for the inactivated virus suspension diluted below the vLOD (n = 18). Moreover, no false-negative results were observed considering the experiments including the inactivated virus suspension at levels above the vLOD (n = 36). As summarized in Table 4, the assay showed very low CV% for both within- and between-assay variability for the 4x vLOD level (5.3% and 4.4%), which increased for the 1x vLOD level (20.3% and 26.2%). However, these values were still considered to be acceptable as they were measured in correspondence of the limit of detection of the method.

Table 4. Imprecision of the LFD-LFIA.

	Within-Assay			Between-Assay		
Level	**Within-Day** (n = 3)	**Between-Day** (n = 3)	**Overall** (n = 9)	**Within-Day** (n = 3)	**Between-Day** (n = 3)	**Overall** [a] (n = 9)
2× vLOD	3, 7	7, 1	5, 3	0, 6	8, 3	4, 4
1× vLOD	19, 1	21,2	20, 3	27, 5	25, 7	26,2

[a] calculated as mean coefficient of variation % (CV%).

The device was evaluated for its specificity towards one lentivirus infecting bovine and viruses belonging to Poxviridae family, namely five orf virus isolates, four bovine popular stomatitis viruses, and two pseudocowpox viruses; no false-positive results were observed, proving that the assay was specific for the LSDV (Table 5).

Table 5. List of the viruses tested to evaluate the specificity of LSDV_LFIA.

Sample (#)	Genus	Species	Sampling (m/y)	LSD-LFIA
1	lentivirus	...	...	NEG
2		Orf virus	...	NEG
3		Orf virus	...	NEG
4		Orf virus	...	NEG
5		Orf virus	...	NEG
6		Orf virus	...	NEG
7	parapoxvirus	Bovine papular stomatitis virus	8/2008	NEG
8		Bovine papular stomatitis virus	2/2008	NEG
9		Bovine papular stomatitis virus	8/1998	NEG
10		Bovine papular stomatitis virus	4/2020	NEG
11		Pseudocowpox	5/2021	NEG
12		Pseudocowpox	5/2009	NEG

The stability of the devices was verified after 90 days from the production by applying the running buffer and a positive sample (inactivated virus suspension) to LSD_LFIA devices stored at 4, 25, and 37 °C. No appreciable loss in color intensity was observed in any of the three conditions, and no false positive/false negatives were obtained.

4. Conclusions

We had previously observed a peculiar hook effect in the LFIA [23] and made the hypothesis that it was due to the antigen saturation caused by the excess of the labeled antibody. Therefore, we compared the cases of one antibody used as a capture and detector ligand (i.e., one epitope of the antigen is involved in the binding with both ligands, so that the saturation of the antigen by the detector impedes its subsequent capture by the reagent deposed on the test line) and two antibodies targeting different epitopes (where we expected that the two would not interfere reciprocally and the capture would not be hampered by the excess of the detector). We suggested that antigen saturation occurred in the time frame between detector resuspension by the sample and reaching the capture ligand (test line), which encompassed a few seconds. Although the contact time between the detector and the antigen was quite low, the large excess of the detector antibody generally used to increase the sensitivity of the sandwich-type LFIA caused the partial or complete saturation of the available epitopes of the antigen. In conclusion, the modulation of the reagents should be considered very carefully to balance detectability with saturation. In this work, we demonstrated that the variables influencing the saturation phenomena are different (AuNP size, mAb-to-AuNP ratio, optical density). To limit the number of experiments (and the consumption of precious reagents), we applied a design of experiment strategy to optimize the variables, instead of the traditional checkerboard titration, which allows for optimizing just two variables at one time. The signal intensity (and, therefore, assay sensitivity) was used to define optimal conditions.

To investigate the peculiar hook effect, we used a model case, i.e., the detection of the lumpy skin disease virus. By employing two monoclonal antibodies specific for LSDV, we traced the route for the optimization of the LFIA in the cases of using both one and two antibodies for the viral antigen detection. Indeed, even if using different antibodies is certainly more convenient and also proved to lead to higher sensitivity in our work, is not uncommon to have just one efficient antibody available (or to have more than one but it binds to the same or adjacent epitopes). As previously observed for the competitive LFIA, as well as for the sandwich LFIA employing one antibody, involving the competition between the capture and detector ligand, the best approach follows the rule "less is more" [31]: the lower the amount of the detector antibody, the higher the sensitivity. In particular, the effect of the optical density, and even more so, the mAb-to-AuNp ratio, should be low to provide better performances.

Moreover, to the best of our knowledge, no LFIA has been reported in the literature yet for the on-field rapid detection of the LSDV antigen. In the panorama of LSDV diagnosis, molecular methods are largely predominant, and immunoassays are used for serological indirect diagnosis and immunohistochemistry. We presented the very first lateral flow test for the diagnosis of LSDV antigen, which was confirmed to be as sensitive as the ELISA when exploiting the same antibody pair, reproducible, specific, and robust. The LSD_LFIA was confirmed to be as sensitive as the ELISA when exploiting the same antibody pair, reproducible, specific, and robust. So, the LSD_LFIA candidate itself has proved to be a useful tool for the on-field rapid diagnosis of LSD considering all the advantages of a rapid on-field testing method. The clinical validation of the assay through analyzing samples belonging to infected animals is ongoing and will reinforce its application for helping develop strategies to control LSD transmission and spread.

Supplementary Materials: The following supporting information can be downloaded at: https://www.mdpi.com/article/10.3390/bios12090739/s1. Chemicals, AuNP synthesis, flocculation test [32]. Figure S1: Flocculation stress test for the mAb #2C6 (a) and for the mAb #2F10 (b). The ratio between Absorbance at 540 nm and Absorbance at 620 nm versus the amount of mAb used for each millilitre of AuNP with different size of AuNPs is plotted. The stabilizing amounts of the mAbs #2C6 and #2F10 are highlighted with a circle. Figure S2: Characterization of the 32 nm AuNPs and its conjugates with #2F10 mAb. (a) The flocculation stress test carried out by adsorbing 2–4–6–8–10 µg of the mAb to 1 mL of AuNP (OD1) followed by saline shock with aqueous 10% w/v NaCl: the absorbance at 540 nm and 620 nm are due to non-aggregated and aggregated fractions of AuNP, so increasing the stabilization increases the 540/620 absorbance ratio; (b–f) HR-TEM images of the mAb_AuNP from figure 2a, (g) visible spectra of the AuNP and conjugates obtained by increasing the amount of mAb adsorbed (0–5–10–15–20 µg per ml of AuNP) and (h) their first derivative. Figure S3: Response surface from the FF-DoE for the SE based on the mAb #2F10 and using different sizes of AuNP: 24nm (a), 32 nm (b), and 36 nm (c). Data were processed by means of the Software CAT. Figure S4: Response surface from the FF-DoE for the DE based on the mAb #2F10 as the detection and #2C6 as the capture ligand and using different sizes of AuNP: 24 nm (a), 32 nm (b), and 36 nm (c). Data were processed by means of the Software CAT. Table S1: Attempts to eliminate the background signal observed when using the #2C6_AuNP gold conjugate the mAb-to-AuNP ratio was 12 µg (2.0×) per mL of AuNP and the probe was diluted to optical density equal to 4. A solution of the inactivated viral culture

(+) and the sample diluent (−) were applied and the colour intensity at the test line quantified. The condition selected is highlighted in bold. Table S2: Results from the design of experiment for the optimization of the SE based on the mAb #2F10. A 1 + 3 dilution solution of the inactivated viral culture in the sample diluent was used as the positive control and the colour intensity at the test line quantified. Table S3: Results from the design of experiment made on the DE based on the mAb #2F10 as the detection and #2C6 as the capture ligand. A 1 + 3 dilution solution of the inactivated viral culture in the sample diluent was used as the positive control and the colour intensity at the test line quantified. Table S4: Effects of the composition of the buffer used as running buffer and #2F10_AuNP dilution buffer. Colour of the test line was measured upon application of a solution of the inactivated viral culture (+) and of the running buffer (−). The condition selected is highlighted in bold. Table S5: Study of the optical density of the #2F10-AuNP probe (mAb-to-AuNP 20 µg per mL) for the device including the test line at 2 mg/mL. The condition selected is highlighted in bold.

Author Contributions: S.C.: Conceptualization, Investigation, Visualization, Original Draft. S.G.: Conceptualization, Investigation, Visualization, Validation. S.B.: Review and Editing. D.L.: Validation. B.C.: Investigation, Visualization. S.R.: Conceptualization, Supervision, Resources. F.D.N.: Methodology, Formal Analysis. T.S.: Formal Analysis. V.T.: Formal Analysis, M.C.: Formal Analysis. C.B.: Supervision, Resources. G.P.: Investigation, Visualization, Review and Editing. E.B.: Conceptualization, Supervision, Writing—Review and Editing. L.A.: Conceptualization, Supervision, Writing—Review and Editing. All authors have read and agreed to the published version of the manuscript.

Funding: This research was funded by the Italian Ministry of Health (IZSLER 02/20—PRC2020002) and University of Torino (Ricerca locale 2019-A).

Institutional Review Board Statement: Not applicable.

Informed Consent Statement: Not applicable.

Data Availability Statement: Not applicable.

Conflicts of Interest: The authors declare that they have no known competing financial interests or personal relationships that could have appeared to influence the work reported in this paper.

References

1. Briggs, N.; Campbell, S.; Gupta, S. Advances in Rapid Diagnostics for Bloodstream Infections. *Diagn. Microbiol. Infect. Dis.* **2021**, *99*, 115219. [CrossRef] [PubMed]
2. A Review of Antigen-Detecting Rapid Diagnostic Tests for COVID-19. Available online: https://www.news-medical.net/news/20211108/A-review-of-antigen-detecting-rapid-diagnostic-tests-for-COVID-19.aspx (accessed on 24 November 2021).
3. Guidance for SARS-CoV-2 Point-of-Care and Rapid Testing CDC. Available online: https://www.cdc.gov/coronavirus/2019-ncov/lab/point-of-care-testing.html (accessed on 24 November 2021).
4. Land, K.J.; Boeras, D.I.; Chen, X.S.; Ramsay, A.R.; Peeling, R.W. REASSURED Diagnostics to Inform Disease Control Strategies, Strengthen Health Systems and Improve Patient Outcomes. *Nat. Microbiol.* **2019**, *4*, 46–54. [CrossRef]
5. Rapid Diagnostic Tests for Infectious Diseases—Chapter 11—2020 Yellow Book Travelers' Health CDC. Available online: https://wwwnc.cdc.gov/travel/yellowbook/2020/posttravel-evaluation/rapid-diagnostic-tests-for-infectious-diseases (accessed on 27 July 2022).
6. Boehringer, H.R.; O'Farrell, B.J. Lateral Flow Assays in Infectious Disease Diagnosis. *Clin. Chem.* **2021**, *68*, 52–58. [CrossRef]
7. Sajid, M.; Daud, M. Designs, Formats and Applications of Lateral Flow Assay: A Literature Review. *J. Saudi Chem. Soc.* **2015**, *19*, 689–705. [CrossRef]
8. Di Nardo, F.; Chiarello, M.; Cavalera, S.; Baggiani, C.; Anfossi, L. Ten Years of Lateral Flow Immunoassay Technique Applications: Trends, Challenges and Future Perspectives. *Sensors* **2021**, *21*, 5185. [CrossRef]
9. Dilaveris, D.; European Commission Directorate-General for Health and Food Safety G3-Crisis Management in Food, Animals and Plants Lumpy Skin Disease (LSD). Exit Strategy in South East Europe Possibilities and Challenges Standing Group of Experts on Lumpy Skin Disease in the South East Europe Region under the GF-TADs Umbrella Seventh Meeting (SGE LSD7). Available online: https://rr-europe.woah.org/wp-content/uploads/2019/09/5_sge-lsd7_commission-exit-strategy.pdf9 (accessed on 20 July 2022).
10. Klement, E. Economic Impact of Lumpy Skin Disease. In *Lumpy Skin Disease*; Springer: Cham, Switzerland, 2018. [CrossRef]
11. Molla, W.; de Jong, M.C.M.; Gari, G.; Frankena, K. Economic Impact of Lumpy Skin Disease and Cost Effectiveness of Vaccination for the Control of Outbreaks in Ethiopia. *Prev. Vet. Med.* **2017**, *147*, 100–107. [CrossRef] [PubMed]
12. Babiuk, S. Diagnostic Tools. In *Lumpy Skin Disease*; Springer: Cham, Switzerland, 2018. [CrossRef]

13. Tuppurainen, E.S.M.; Venter, E.H.; Coetzer, J.A.W. The Detection of Lumpy Skin Disease Virus in Samples of Experimentally Infected Cattle Using Different Diagnostic Techniques. *Onderstepoort J. Vet. Res.* **2005**, *72*, 153–164. [CrossRef]
14. Murray, L.; Edwards, L.; Tuppurainen, E.S.M.; Bachanek-Bankowska, K.; Oura, C.A.L.; Mioulet, V.; King, D.P. Detection of Capripoxvirus DNA Using a Novel Loop-Mediated Isothermal Amplification Assay. *BMC Vet. Res.* **2013**, *9*, 90. [CrossRef]
15. Chen, X.; Nie, F.; Xiong, Y.; Lin, L.; Shi, M.; Yang, J.; Wang, Y.; Wang, G.; Li, Y.; Huo, D.; et al. Ultra-Sensitive and Point-of-Care Detection of Capripoxvirus (CaPV) Based on Loop-Mediated Amplification (LAMP) and Trans-Cleavage Activity of CRISPR/Cpf1. *Anal. Chim. Acta* **2022**, *1191*, 339330. [CrossRef]
16. Mwanandota, J.J.; Macharia, M.; Ngeleja, C.M.; Sallu, R.S.; Yongolo, M.G.; Mayenga, C.; Holton, T.A. Validation of a Diagnostic Tool for the Diagnosis of Lumpy Skin Disease. *Vet. Dermatol.* **2018**, *29*, 532-e178. [CrossRef]
17. Möller, J.; Moritz, T.; Schlottau, K.; Krstevski, K.; Hoffmann, D.; Beer, M.; Hoffmann, B. Experimental Lumpy Skin Disease Virus Infection of Cattle: Comparison of a Field Strain and a Vaccine Strain. *Arch. Virol.* **2019**, *164*, 2931–2941. [CrossRef] [PubMed]
18. Wang, H.; Kong, Y.; Mei, L.; Lv, J.; Wu, S.; Lin, X.; Han, X. Multiplex Real-Time PCR Method for Simultaneous Detection and Differentiation of Goat Pox Virus, Sheeppox Virus, and Lumpy Skin Disease Virus. *J. AOAC Int.* **2021**, *104*, 1389–1393. [CrossRef] [PubMed]
19. Haegeman, A.; De Vleeschauwer, A.; De Leeuw, I.; Vidanović, D.; Šekler, M.; Petrović, T.; Demarez, C.; Lefebvre, D.; De Clercq, K. Overview of Diagnostic Tools for Capripox Virus Infections. *Prev. Vet. Med.* **2020**, *181*, 104704. [CrossRef] [PubMed]
20. Amin, D.M.; Shehab, G.; Emran, R.; Hassanien, R.T.; Alagmy, G.N.; Hagag, N.M.; Abd-El-Moniem, M.I.I.; Habashi, A.R.; Ibraheem, E.M.; Shahein, M.A. Diagnosis of Naturally Occurring Lumpy Skin Disease Virus Infection in Cattle Using Virological, Molecular, and Immunohistopathological Assays. *Vet. World* **2021**, *14*, 2230–2237. [CrossRef]
21. Monoclonal Antibodies as Diagnostic Tools for Lumpy Skin Disease—Istituto Zooprofilattico Sperimentale Della Lombardia e Dell'Emilia Romagna. Available online: https://www.izsler.it/pubblicazioni/monoclonal-antibodies-as-diagnostic-tools-for-lumpy-skin-disease/ (accessed on 27 July 2022).
22. Cavalera, S.; Di Nardo, F.; Forte, L.; Marinoni, F.; Chiarello, M.; Baggiani, C.; Anfossi, L. Switching from Multiplex to Multimodal Colorimetric Lateral Flow Immunosensor. *Sensors* **2020**, *20*, 6609. [CrossRef]
23. Cavalera, S.; Russo, A.; Foglia, E.A.; Grazioli, S.; Colitti, B.; Rosati, S.; Nogarol, C.; Di Nardo, F.; Serra, T.; Chiarello, M.; et al. Design of Multiplexing Lateral Flow Immunoassay for Detection and Typing of Foot-and-Mouth Disease Virus Using Pan-Reactive and Serotype-Specific Monoclonal Antibodies: Evidence of a New Hook Effect. *Talanta* **2022**, *240*, 123155. [CrossRef]
24. Zhang, L.; Mazouzi, Y.; Salmain, M.; Liedberg, B.; Boujday, S. Antibody-Gold Nanoparticle Bioconjugates for Biosensors: Synthesis, Characterization and Selected Applications. *Biosens. Bioelectron.* **2020**, *165*, 112370. [CrossRef]
25. Di Nardo, F.; Cavalera, S.; Baggiani, C.; Giovannoli, C.; Anfossi, L. Direct vs Mediated Coupling of Antibodies to Gold Nanoparticles: The Case of Salivary Cortisol Detection by Lateral Flow Immunoassay. *ACS Appl. Mater. Interfaces* **2019**, *11*, 32758–32768. [CrossRef]
26. Turkevich, J.; Stevenson, P.C.; Hillier, J. A Study of the Nucleation and Growth Processes in the Synthesis of Colloidal Gold. *Discuss. Faraday Soc.* **1951**, *11*, 55–75. [CrossRef]
27. Reed, L.J.; Muench, H. A Simple Method of Estimating Fifty per Cent Endpoints. *Am. J. Epidemiol.* **1938**, *27*, 493–497. [CrossRef]
28. Serological Elisas Based on Monoclonal Antibodies as Diagnostic Tools. Available online: https://www.slideshare.net/eufmd1/serological-elisas-based-on-monoclonal-antibodies-as-diagnostic-tools-for-lumpy-skin-disease (accessed on 27 July 2022).
29. Khlebtsov, N.G. Determination of Size and Concentration of Gold Nanoparticles from Extinction Spectra. *Anal. Chem.* **2008**, *80*, 6620–6625. [CrossRef] [PubMed]
30. Zhang, L.; Hu, D.; Salmain, M.; Liedberg, B.; Boujday, S. Direct Quantification of Surface Coverage of Antibody in IgG-Gold Nanoparticles Conjugates. *Talanta* **2019**, *204*, 875–881. [CrossRef] [PubMed]
31. Byzova, N.A.; Safenkova, I.V.; Slutskaya, E.S.; Zherdev, A.V.; Dzantiev, B.B. Less is More: A Comparison of Antibody–Gold Nanoparticle Conjugates of Different Ratios. *Bioconjugate Chem.* **2017**, *28*, 2737–2746. [CrossRef] [PubMed]
32. Cavalera, S.; Di Nardo, F.; Chiarello, M.; Serra, T.; Colitti, B.; Guiotto, C.; Fagioli, F.; Cagnazzo, C.; Denina, M.; Palazzo, A.; et al. Bacterial ligands as flexible and sensitive detectors in rapid tests for antibodies to SARS-CoV-2. *Anal. Bioanal. Chem.* **2022**, 1–10. [CrossRef] [PubMed]

biosensors

MDPI

Article

Model of the SARS-CoV-2 Virus for Development of a DNA-Modified, Surface-Enhanced Raman Spectroscopy Sensor with a Novel Hybrid Plasmonic Platform in Sandwich Mode

Mariia V. Samodelova [1], Olesya O. Kapitanova [1,2,*], Nadezda F. Meshcheryakova [1], Sergey. M. Novikov [2], Nikita R. Yarenkov [1], Oleg A. Streletskii [3], Dmitry I. Yakubovsky [2], Fedor I. Grabovenko [1], Gleb A. Zhdanov [1], Aleksey V. Arsenin [2], Valentyn S. Volkov [2], Elena G. Zavyalova [1], Irina A. Veselova [1] and Maria I. Zvereva [1]

1 Faculty of Chemistry, Lomonosov Moscow State University, Leninskie Gory 1, 119991 Moscow, Russia
2 Center for Photonics and 2D Materials, Moscow Institute of Physics and Technology, 141700 Dolgoprudny, Russia
3 Faculty of Physics, Lomonosov Moscow State University, Leninskie Gory 1, 119991 Moscow, Russia
* Correspondence: kapitanova@inorg.chem.msu.ru

Abstract: The recent severe acute respiratory syndrome coronavirus 2 (SARS-CoV-2) infection has posed a great challenge for the development of ultra-fast methods for virus identification based on sensor principles. We created a structure modeling surface and size of the SARS-CoV-2 virus and used it in comparison with the standard antigen SARS-CoV-2—the receptor-binding domain (RBD) of the S-protein of the envelope of the SARS-CoV-2 virus from the Wuhan strain—for the development of detection of coronaviruses using a DNA-modified, surface-enhanced Raman scattering (SERS)-based aptasensor in sandwich mode: a primary aptamer attached to the plasmonic surface—RBD-covered Ag nanoparticle—the Cy3-labeled secondary aptamer. Fabricated novel hybrid plasmonic structures based on "Ag mirror-SiO_2-nanostructured Ag" demonstrate sensitivity for the detection of investigated analytes due to the combination of localized surface plasmons in nanostructured silver surface and the gap surface plasmons in a thin dielectric layer of SiO_2 between silver layers. A specific SERS signal has been obtained from SERS-active compounds with RBD-specific DNA aptamers that selectively bind to the S protein of synthetic virion (dissociation constants of DNA-aptamer complexes with protein in the range of 10 nM). The purpose of the study is to systematically analyze the combination of components in an aptamer-based sandwich system. A developed virus size simulating silver particles adsorbed on an aptamer-coated sensor provided a signal different from free RBD. The data obtained are consistent with the theory of signal amplification depending on the distance of the active compound from the amplifying surface and the nature of such a compound. The ability to detect the target virus due to specific interaction with such DNA is quantitatively controlled by the degree of the quenching SERS signal from the labeled compound. Developed indicator sandwich-type systems demonstrate high stability. Such a platform does not require special permissions to work with viruses. Therefore, our approach creates the promising basis for fostering the practical application of ultra-fast, amplification-free methods for detecting coronaviruses based on SARS-CoV-2.

Keywords: surface-enhanced Raman scattering (SERS); aptasensor; SARS-CoV-2 virus

Citation: Samodelova, M.V.; Kapitanova, O.O.; Meshcheryakova, N.F.; Novikov, S.M.; Yarenkov, N.R.; Streletskii, O.A.; Yakubovsky, D.I.; Grabovenko, F.I.; Zhdanov, G.A.; Arsenin, A.V.; et al. Model of the SARS-CoV-2 Virus for Development of a DNA-Modified, Surface-Enhanced Raman Spectroscopy Sensor with a Novel Hybrid Plasmonic Platform in Sandwich Mode. *Biosensors* **2022**, *12*, 768. https://doi.org/10.3390/bios12090768

Received: 23 July 2022
Accepted: 13 September 2022
Published: 19 September 2022

1. Introduction

Surface-enhanced Raman scattering (SERS)-based sensors for viral detection are very attractive, combining high recognition ability and sensitivity with a rapidness of the analysis. The ultimate specificity can be achieved for low-molecular compounds, providing identification via the unique Raman spectrum of the compound. In this regard, the development of novel improved SERS-active substrates is of high interest [1]. The development and implementation of reliable methods for the synthesis/fabrication of uniform, reproducible, low-cost SERS substrates with a high enhanced factor and quantitative response within specification limits is being conducted. Thus, the natural reed leaves, without any special pretreatment, coated by silver, have been utilized as a self-assembled plasmonic structure for ultra-sensitive crystal violet trace detection with sensitivity as low as 10^{-13} M using a novel natural surface-enhanced fluorescence [2]. Recently, a scalable design of SERS substrate based on a biopolymer of free-standing chitosan film, silver nanoparticles, and graphene oxide has been demonstrated [3]. The developed plasmon hybrid structures were assessed by their SERS-sensitivity, reproducibility, stability, and quenching capability. The fabricated plasmonic polymer nanocomposite offered a limit of detection for R6G down to 100 pM. Using the 3D jet writing technique, biologically inert poly(lactic-co-glycolic acid) tessellated scaffolds containing fluorescent polymers and/or SERS-encoded gold nanostars have been fabricated [4]. Such a hybrid scaffold provides efficient work with high-molecule volumetric objects such as live cells and offers sufficient optical transparency for both fluorescence and SERS imaging. The novel hybrid metal-dielectric-metal sandwiched structures [5,6] demonstrate high sensitivity and reproducibility of the SERS signal [7,8]. In this geometry, the overall enhancement factor principally benefits from the combination of two main mechanisms. The first one is associated with the excitation of localized surface plasmon resonance (SPR) in gold/silver clusters. The second one is due to the gap surface plasmons' excitation in a thin dielectric layer between the metal mirror and corrugated gold/silver layers. SERS has attracted much attention for its potential in multiplexed sensing. SERS is an optical molecular finger-printing technique that has the ability to resolve analytes from within mixtures [9]. However, the biological macromolecules are composed of the same set of building blocks—amino acids, nucleotides, and sugars, that creates difficulties for SERS detection based on biopolymer nature [10]. The specific determination of one protein in a mixture of dozens of other macromolecules is challenging [11–13]. Speaking of viruses, reported Raman spectra of the S protein of SARS-CoV-2 are not the same [14–17], and Raman spectra of influenza virions are rather different [14,15,18,19]. The differences in spectra could be due to variations of impurities in biological samples or different SERS-active substrates that enhanced Raman spectra of different groups in macromolecules. In both cases, their application for real clinical samples is questionable.

Recognizing molecules, like antibodies or aptamers, can be used to provide the specificity of virus detection along with an increase in the limit of detection [11–13]. The most common approach is to decorate a SERS-active surface with recognizing molecules enhancing the local concentration of the target of interest near the surface [15,20–23]. A complementary approach uses labeled recognizing molecules to trace the spectrum of the artificial molecule that is detectable in the nanomolar range of concentrations [22–25]. The typical Raman active labels used for virus detection are fluorescent dyes like Cyanine 3 (Cy3), Cyanine 5.5 (Cy5.5), Bodipy FL (BDP FL), and Rhodamine Red-X (RRX) [22–25]. Labeled recognizing molecules are either approaching the surface due to sandwich-like ternary complexes [22,23] or retiring from the surface due to binding with viral particles [24,25]. In both cases, the Q-factor and stability in biological fluids are the key parameters that are to be considered. Silver nanoislands and lithography-based substrates were used for the sandwich-like complex aptamer1-virus-aptamer2 [22,23]. The first one provided nice SERS spectra for ternary complexes with Cy3 and BDP FL-labeled aptamers [22]; however, the stability of the nanostructured surface in biological fluids was revealed to be suboptimal (authors' unpublished data). Lithography-based substrates are usually stable in various media; however, the SERS spectra of ternary complexes with Cy5.5-labeled aptamers were not acquired [23]. The detected differences in spectra could be attributed to surface-enhanced fluorescence of utilized model dye.

There is a growing interest in aptamer-based assay methods, without signal amplification, due to their low cost, having high stability, being easy to modify, and, in some cases, they have better specificity and affinity [26]. Among them, an electrochemical aptasensor for the determination of the SARS-CoV-2 receptor-binding domain in human saliva samples demonstrated high sensitivity (down to 7.0 pM) and accuracy [27]. Photonic biosensing technologies that have been explored for possible use in SARS-CoV-2 serology include SPR [28,29] for not only the detection of antibodies but also the profiling of binding kinetics of the complete polyclonal antibody response against the receptor-binding domain. Moreover, the asymmetric Mach–Zehnder Interferometer enables multiplex and ultra-sensitive detection of antibodies that target the viral antigens:spike protein, the receptor-binding domain, and the nucleocapsid protein [30]. SERS-based aptasensors have been used to detect both SARS-CoV-2 spike protein [31] and the whole virus of SARS-CoV-2 [18,23,25,32], providing good accuracy and satisfactory performance for SARS-CoV-2 determination in good agreement with RT-qPCR results [33]. However, the limit of detection of the sensors is still higher than RT-qPCR. Increasing the sensitivity of the method will increase the reliability of the results of the virus detection. The research work requires a good model system for the sensor optimization. The RBD protein per se cannot be used as a model, as the viruses contain dozens or hundreds of surface proteins that bind the solid substrates in multiple points. Therefore, the exemplary limit of detection of the virus is typically lower compared to those for the recombinant protein [18,23,25,31,32]. Production of virus-like particles (VLP) requires genetic engineering, and the stability of these membrane structures is to be traced during storage. The production of VLP requires laboratory space organized according to special strict safety requirements. In our work, we proposed a simplified imitation of the virus surface produced by the adsorption of RBD protein onto silver nanoparticles. We optimized the conditions to make the nanoparticles recognizable by the aptamers to RBD and storable for several weeks. This model of synthetic virion was used to screen SERS-substrates, revealing the mirror silver-dielectric-nanostructured silver sandwiched plasmonic nanostructures as a promising tool for SARS-CoV-2 detection. A developed indicator system for SARS-CoV-2 virions demonstrates applicability of the SERS approach for the selective detection of coronavirus-2 infection.

2. Materials and Methods

2.1. Oligonucleotides and Some Materials

Inorganic salts and buffer solutions, as well as bovine serum albumin, were purchased from Sigma-Aldrich (New York, NY, USA).

The following oligonucleotides were studied. Selected in lab [34], Found-SH (SH-CACCGCTTTTGCCTTTTGGGACGGATATAGGGAAACACGATA GAATCCGAACAGC-ACC), and complementary to Found—CompFound-SH (SH-CACCGCTTTTGCCTTTTGGG-GACGGATAGGTGCTGTTCGGATTCTATCGTGTTTCCCTA)—were kindly synthesized by the group of Timofei Zatsepin (Moscow, Russia). RBD-1C selected in [35] in two forms: RBD-1C-Cy3(Cy3-5′-CAGCACCGACCTTGTGCTTTGGGAGTGCTGGTCCAAGGGCGTTAAT-GGACA-3′) and biotinylated aptamer RBD-1C (biotin-5′-CAGCACCGACCTTGTGCTTT-GGGAGTGCTGGTCCAAGGGCGTTAATGGACA-3′) from Eurogene (Moscow, Russia). The receptor-binding domain of the S-protein of SARS-CoV-2 (further referred to as RBD) was from [36] (for details of aptamer preparation and binding experiments see in supplementary material citation, Figure S1).

2.2. Synthesis of the Model of SARS-CoV-2 Virions

Silver nanoparticles (NP) were obtained by mixing 18 mL of a solution of 3.3 mM NaOH and 2.6 mM $NH_2OH{\cdot}HCl$ with 2 mL of 10 mM $AgNO_3$. The solution of $AgNO_3$ was injected for 10 s with a subsequent stirring for 1 h at room temperature.

RBD, expressed in Chinese hamster ovary cell line [36], was obtained in a concentration of 4.7 mg/mL. An amount of 6.4 μL of RBD protein solution was added to 1 mL of silver nanoparticles, providing a final concentration of the protein of 1 μM. The mixture was incubated at room temperature for 30 min. Unreacted protein was removed by centrifugation for 10 min at 10 rpm. The RBD-coated silver nanoparticles (further referred to as RBD NP) were resuspended in 1 mL of buffer with 10 mM Tris-HCl pH 7.5, 140 mM $NaNO_3$, and 10 mM KNO_3. Bovine serum albumin (further referred to as BSA) coated silver nanoparticles (further referred to as BSA NP) were obtained following the same protocol using 10 μM solution of BSA instead of RBD protein.

2.3. Characterization of Nanoparticles Coated with RBD Protein

A dynamic light-scattering instrument ZetasizerNano ZS (Malvern, Worcestershire, UK) was used to estimate the size and ζ-potential of nanoparticles.

The ability to bind aptamers to RBD was estimated using biolayer interferometry (Blitz equipment from ForteBio, Dallas, TX, USA). The sensor was hydrated in water, and then it was incubated with 1 μM solution of biotinylated aptamer RBD-1C for 2 min. The baseline was acquired in the binding buffer (10 mM Tris-HCl pH 7.5, 140 mM $NaNO_3$, 10 mM KNO_3) for 30 s. Then, the sensor was placed into RBD NP or BSA NP solutions for 200 s. The 2×, 4×, and 8× dilutions of the nanoparticle samples were obtained in the binding buffer. The dissociation was performed in the binding buffer.

2.4. SERS-Active Aptasensor Platform

The idea of the aptasensor platform is a formation of sandwich-like complexes on the silver nanostructured surface. The primary aptamer was modified with a thiol-group providing stable linkage between the plasmonic nanostructured silver surface and the aptamer. Primary aptamers bound RBD NP. Secondary aptamers with a Raman-active label (Cy3) formed complexes with RBD NP immobilized on the surface. The detailed protocol was as following: (1) SERS substrate was incubated in 20 μL of a 20 nM solution of thiol-modified aptamer in PBS for 15 min; (2) the substrate was rinsed with PBS; (3) the substrate was incubated in 20 μL of RBD NP (or BSA NP in the control experiment) solution for 5 min; (4) the substrate was rinsed with PBS; (5) the substrate was incubated in 20 μL of a 200 nM solution of labeled aptamer in PBS for 5 min; (6) the substrate was rinsed with water; (7) the substrate was dried in air.

2.5. SERS-Based Substrates

SERS platforms were prepared by method as previously described [37]. The fabrication of plasmonic substrates was performed in a preliminary evacuated chamber (base pressure 10^{-5} Torr). Then, the chamber was filled with Ar gas. The total pressure in the preparation chamber was 5×10^{-4} Torr and the temperature ~300 K. The 100 nm layer of Ag (99.999%) was deposited on silicon substrate, a 3 nm Ti adhesion interlayer over Ag and 50 nm layer of SiO_2. A thin titanium layer was introduced, as it promotes adhesion, followed by deposited SiO_2 to the silver layer. Then, a layer of 20 nm silver (99.999%) was deposited on the top of SiO_2. These layers were deposited in one regime by electron beam evaporation. Then, the substrates were treated by an Ar ion beam with an energy of 150 eV and dose of 4×10^{16} ion/cm^2. The angle between the Ar beam and substrate was 45°. A Hall-effect ion source with a cold hollow cathode Klan 53-M (Platar Corp.) generated an Ar ion beam. An ion beam modification of silver film allows for the fabrication of plasmonic nanostructured surfaces.

2.6. Characterization of SERS-Based Substrates

The surface morphology of the deposited films was investigated using a scanning electron microscope (SEM) JEOL JSM-7001F (JEOL Ltd., Tokyo, Japan).

Raman spectroscopy and SERS studies were carried out using the Enspectr R532 analyzer (Enhanced Spectrometry, San Jose, CA, USA) with a solid-state laser (wavelength—532 nm, 50 MW). All spectra were collected with the working laser spot diameter of ca. 2 μm through a × 40 objective. The specific power of the laser was set at 0.3 mW, and the accumulation time was 5 s with 10 accumulations. The obtained Raman scattering spectra were fitted using the pseudo-Voigt functions. Measurements were carried out in the range of 200–2000 cm^{-1} using 3rd-order baseline subtraction.

3. Results and Discussion

In this work, we have developed an indicator system for the recognition of SARS-CoV-2 virions in the future. We have determined optimal conditions for obtaining the artificial model of SARS-CoV-2 virions through a specific binding of a Cy3-labeled secondary aptamer and a primary aptamer with RBD-covered Ag nanoparticles (both the primary and secondary aptamers interact with the protein with approximately the same efficiency). In parallel, nonspecific interaction of both aptamers to BSA-covered Ag nanoparticles (Table 1), allows for the creation of a model without SARS-CoV-2 virions. A developed indicator system demonstrates applicability of the SERS approach for the selective detection of coronavirus-2 infection through the control of the distance between the SERS platform and the dye moiety. Figure 1 illustrates the platform of the SERS-active aptasensor using hybrid plasmonic substrates for the quantitative detection of coronaviruses. The self-assembled RBD nanoparticles blocked the direct adsorption of the Cy3-labeled secondary aptamer on a SERS-active nanostructured surface due to the specific interaction between all components in the assay leading to a strong quenching Raman signal of Cy3 (Figure 1).

Table 1. Combinations of aptamers and protein-coated nanoparticles studied.

	Primary Aptamer	Nanoparticles	Secondary Aptamer	Comment
Sample 1	Found-SH	RBD NP	RBD-1C-Cy3	Experiment
Sample 2	CompFound-SH	RBD NP	RBD-1C-Cy3	Control with nonspecific primary oligonucleotide
Sample 3	Found-SH	BSA NP	RBD-1C-Cy3	Control with nonspecific protein instead of RBD

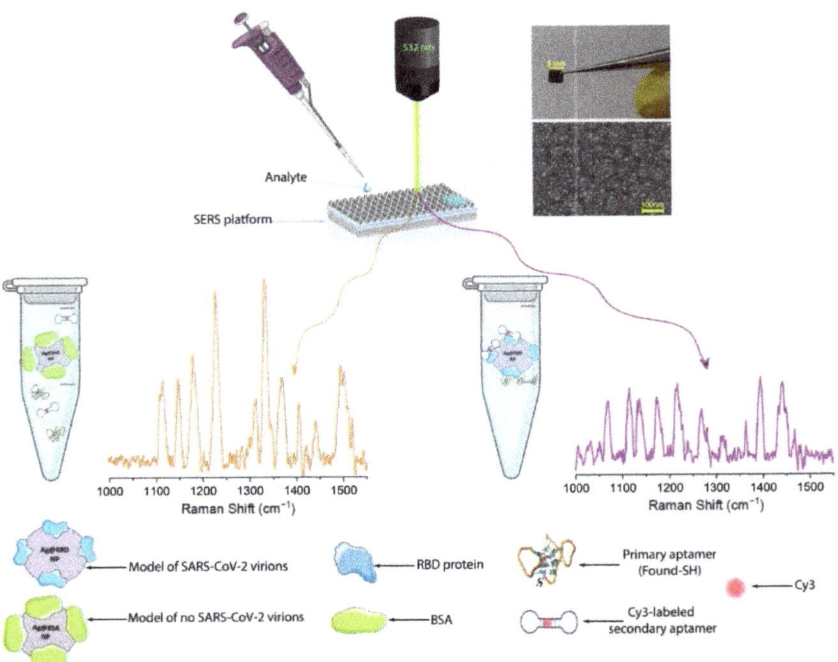

Figure 1. Scheme of the developed indicator system for SARS-CoV-2 virions' detection using the SERS approach.

3.1. Fabrication and Characterization of Model of SARS-CoV-2 Virions

Silver nanoparticles adsorb proteins, readily forming stable coatings [38]. We optimized the conditions for RBD and BSA proteins that were used as coating for silver nanoparticles with a diameter of 80 nm. Stable nanoparticles were obtained after incubation with 1 μM solution of RBD and 10 μM solution of BSA. An amount of 1 μM solution of BSA provided aggregates of nanoparticles during the storage, indicating that the protein coating was not completed. The excess of the proteins was removed by the centrifugation in both cases. The estimated concentration of the protein-coated nanoparticles was 10^{10} particles per mL. The optimized protocol provided nanoparticles that are stable for more than 6 weeks of storage at +4 °C. Unmodified NP and RBD NP were studied with dynamic light scattering. RBD NP was shown to have a significantly increased size compared to uncoated nanoparticles. The diameter of RBD NP was 320 ± 30 nm, whereas the diameter of NP was 80 ± 30 nm (Figure 2a,b). The size of RBD NP is overestimated, as protein coating decreases the mobility of silver nanoparticles. Similarly, ζ-potential was altered significantly from −42 ± 8 mV for NP to −2 ± 5 mV for RBD NP (Figure 2c,d), indicating significant changes of the surface of silver nanoparticles.

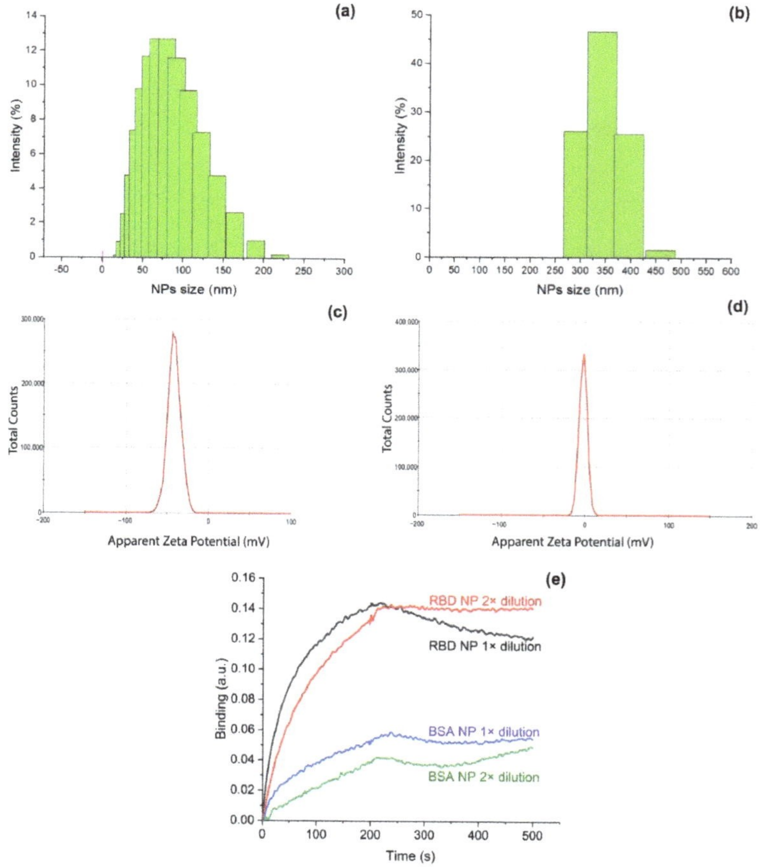

Figure 2. The size distribution for unmodified nanoparticles (**a**) and RBD-coated nanoparticles (**b**) estimated with dynamic light scattering. ζ-potential of unmodified nanoparticles (**c**) and RBD-coated nanoparticles (**d**) estimated with dynamic light scattering. (**e**) Interaction between aptamer RBD-1C and silver nanoparticles coated with RBD or BSA proteins.

We used DNA aptamer RBD-1C selected by Song et al. [35] as a recognition molecule; this aptamer has dissociation constants in the range of 0.8–5.8 nM for the complexes with recombinant RBD protein [39]. Adsorbed RBD retained the ability of binding DNA aptamers as shown by biolayer interferometry (Figure 2e). The binding of BSA NP to DNA was 3–4 times lower, indicating a low level of nonspecific interaction. The overall dataset confirmed the formation of RBD-coated nanoparticles with a size close to the size of SARS-CoV-2 virions; RBD NPs are capable of binding DNA aptamers to RBD protein. This model system was used in further SERS experiments. Proposed particles are simple in preparation, stable during the storage, and composed of one type of protein, providing low non-specific binding.

3.2. Characterization of the Hybrid Plasmonic Sandwiched Structures

First, we obtained an efficient plasmonic sandwiched structure as the SERS-active platform. The surface of the sensor element is represented with a multilayered structure based on subsequently coated silver mirror film, dielectric film, and Ag nanoparticles, further referred to as film-Ag/SiO_2/nano-Ag NPs (Figure 3a). As noted above, the amplification of the Raman signal for sandwiches in comparison with one-contained-layer nanostructured silver surface is associated with the excitation of both gap plasmons in the SiO_2 layer and local surface plasmon resonances in nano-Ag NPs [5–8]. Moreover, the back reflection from the bottom Au layer gives rise to a stronger absorbance and, therefore, the enhancement factor (EF) since the excitation beam passes through the top film twice. The morphology of the obtained substrates was studied by scanning electron microscopy (SEM) (Figure 3b). As it is shown on an SEM image, silver nanoparticles are evenly distributed and the plasmonic surface has a reproducible structure. The average size of a particle was 13 ± 1 nm for 150 particles (Figure 3b, the inset).

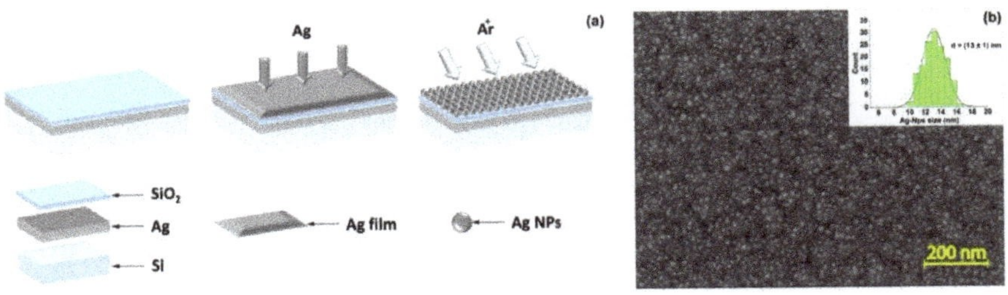

Figure 3. (**a**) Scheme of the fabrication of the hybrid plasmonic sandwiched structure: films of Ag, SiO_2, Ag stepwise formed by electron beam sputtering with the further formation of a plasmonic silver-based structure through the Ar ion beam treatment of the top Ag film. (**b**) A typical SEM image of the sensor element decorated by Ag nanoparticles. The inset shows a size distribution of silver nanoparticles on the sensor element.

The maximum of the plasmonic band of the SERS-active substrate appears at 425 nm (Figure S2). Moreover, a wide plasmonic peak covers the wavelength range from the blue to the IR regions of the spectrum. To evaluate the SERS activity of the designed hybrid substrate, first, we analyzed a model dye Cy3. The EF of the obtained hybrid surfaces was calculated according to the following equation [40]:

$$EF = \frac{I_{SERS}\ C_{ref}}{I_{ref}\ C_{SERS}} \quad (1)$$

where I_{ref} and I_{SERS} are the band's intensities of the Raman scattering and SERS at 1397 cm^{-1} for Cy3; $C_{ref} = 1 \times 10^{-2}$ M and $C_{SERS} = 1 \times 10^{-6}$ M—concentrations of the analytes whose signals correspond to Raman scattering and SERS, respectively. Corresponding SERS and Raman spectra of Cy3 are in Figures S3 and S4 of the supplementary information. The EF of the resulting surface was 9.5×10^4 for Cy3. Moreover, highly reproducible morphology of developed hybrid plasmonic substrates results in a good analytical signal reproducibility from point-to-point of the sample with the relative standard deviation (RSD) for the EF value as low as 7.37%.

3.3. SERS-Active Aptasensor Platform for the Detection of Model of SARS-CoV-2 Virions Using Hybrid Plasmonic Substrates

Figure 4 demonstrates SERS spectra of Cy3 and the Cy3-labeled secondary aptamer. The SERS spectrum of the Cy3-labeled secondary aptamer, as well as the spectrum of pure Cy3, contains well-resolved characteristic signals on 1140, 1175, 1216, 1274, 1397, 1439, 1482, and 1591 cm^{-1}, though there is a slight shift within 1–2 cm^{-1} for some peaks (Figure 4a). However, low-intensity bands in the region of small Raman shifts (588, 693, 744, 938, 1140, 1175 cm^{-1}) disappear after binding an aptamer to the Cy3. Of note, the intensity of characteristic peaks (1216, 1274, 1397, 1439, 1482 cm^{-1}) for the Cy3-aptamer decreased three times in comparison with Cy3. The lateral size of the aptamer is about 2–3 nm [41]. The weakening of the signal can be associated with the increased distance between the dye and the silver nanostructured surface through the primary aptamer, as schematically presented in Figure 4b. The signals at 1272 and 1394 cm^{-1} have been selected as intense characteristic bands for SERS spectra of the platform for aptasensor assays. The peak 1272 cm^{-1} corresponds to the deformation oscillation of the hydroxyl group (COH); 1394 cm^{-1} corresponds to the CN-bond symmetric stretching [42].

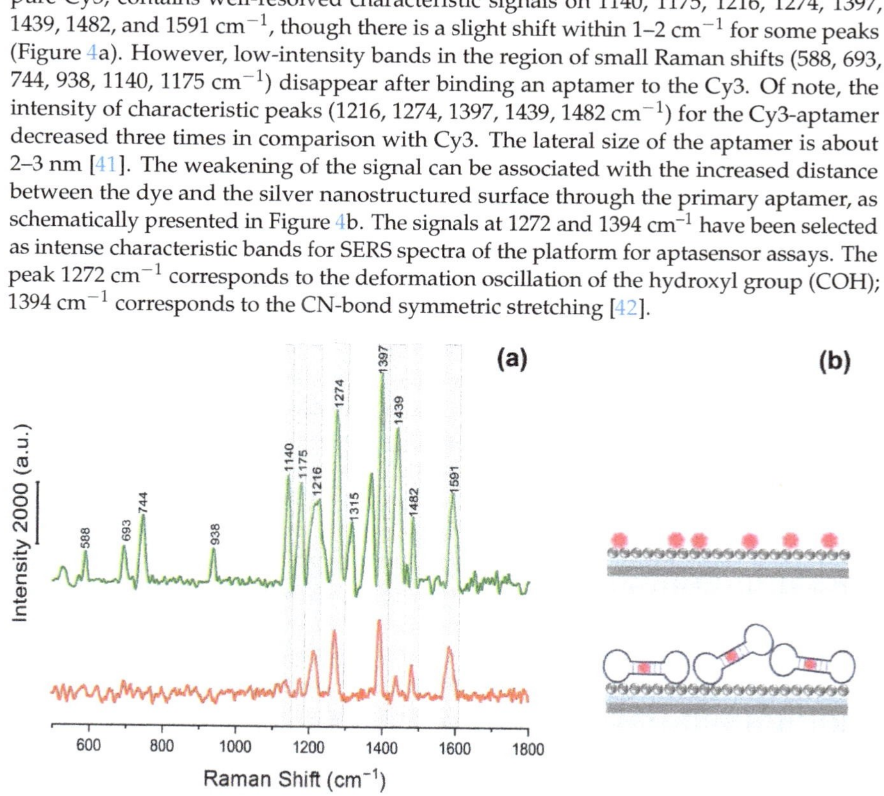

Figure 4. (**a**) SERS spectrum of Cy3 (green) and Cy3-aptamer mixture (red) on the obtained hybrid plasmonic substrate. (**b**) Scheme of the distribution of Cy3 and Cy3-labeled aptamer on a SERS-active hybrid structure.

Further, the SERS spectra of the proposed model for aptasensor assays will be compared with the spectrum of the Cy3-aptamer. The spectrum of Sample 1 has the least similarity with the model mixture Cy3-aptamer spectrum. The spectrum retains signals at 1117, 1137, 1176, 1214, 1272, 1372, 1394, 1439, and 1584 cm^{-1}. At the same time, the signal intensities (1272 and 1394 cm^{-1}) of the dye in Sample 1 fall relative to the other observed peaks. There are also new modes at 1072 and 1315 cm^{-1}, which may be due to a signal from an aptamer or RBD protein fragment. A significant three-fold drop in the signal for Sample 1 is associated with the enhanced distance of the dye from the SERS surface due to the RBD nanoparticle shielding. Consequently, a well-organized, specific interaction of aptasensor components with model RBD protein-covered nanoparticles aids in the reproducible SERS signal quenching from the Cy3-labeled artificial model of SARS-CoV-2 virions.

To test the selectivity of the developed model of aptasensor assay, we performed additional experiments with analytes based on the nonspecific interaction of RDB nanoparticles with the primary aptamer (further referred to as Sample 2) and the nonspecific binding of BSA nanoparticles to the secondary Cy3-labeled aptamer (further referred to as Sample 3). We investigated the dependence of signal intensity on the distance of the SERS-active label from the plasmonic surface. The similarity in the set of signals with a Cy3-aptamer probe has been observed (Figure 5a); all of the main peaks of Cy3 are present in both samples' spectra. The last mode, associated with CH bond vibrations, can be highly intense in the spectra for Samples 2 and 3 due to the contribution of vibrations of both the dye and protein fragments and aptamers. On average, the intensity of the characteristic modes for Samples 2 and 3 is about 2.5 and 2.3 times less than the intensities for the Cy3-aptamer sample, respectively.

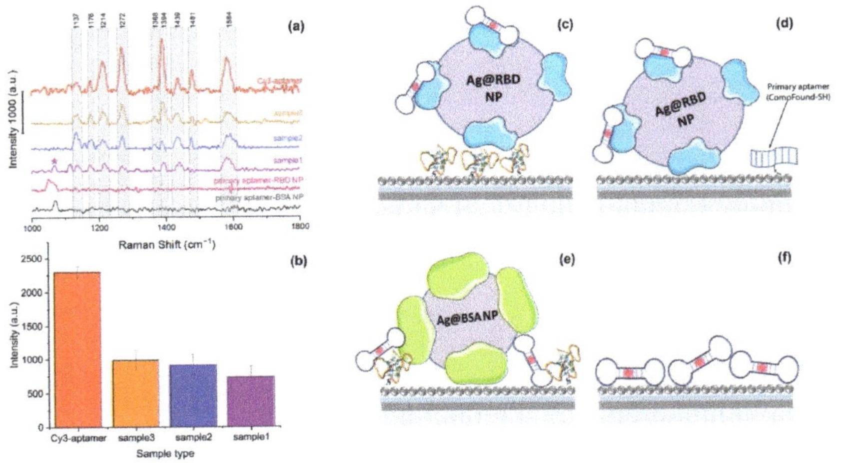

Figure 5. (**a**) SERS spectra of the Cy3-labeled secondary aptamer (red), Sample 3 (yellow), Sample 2 (blue), Sample 1 (violet), primary aptamer-RBD NPs (pink), primary aptamer, and BSA NPs mixture (black) on the obtained hybrid plasmonic substrate. (**b**) SERS signal at 1272 and 1394 cm^{-1} for Cy3-labeled aptamer, Sample 1, Sample 2, and Sample 3. Scheme of the distribution of (**c**) Sample 1, (**d**) Sample 2, (**e**) Sample 3, and (**f**) Cy3-labeled secondary aptamer on an SERS-active hybrid structure.

The signal comparison for all the samples mentioned is shown in Figure 5a,b. All the considered probes are presented in order of decreasing value of EF: from the Cy3-labeled primary aptamer to the full-size probe with specific binding RBD NPs with primary and secondary aptamers. The EF ratio for samples 1, 2, 3 is 1:1.25:1.34, respectively. The self-assembled RBD nanoparticle assay blocked the direct adsorption of the Cy3-labeled secondary aptamer on the SERS-active nanostructured surface due to the specific interaction between all components in the assay leading to strong quenching of the signal of Cy3, which agrees with previously demonstrated SERS data for a real influenza virus [24]. It is important to mention that a developed artificial indicator system gives us the opportunity to investigate simulating viruses as well as their fragments. This confirms the applicability of the proposed artificial indicator system, which does not require special permissions to work with viruses and will speed up the process of developing effective methods for their detection. The increase in signal enhancement from Sample 1 to Sample 3 may be due to the full specific binding of the developed model of SARS-CoV-2 virions with the thiol- aptamer and the Cy3-labeled aptamer (Figure 5c–f). In the case of Sample 2, the interaction of Cy3-labeled RBD NPs is not specific with the thiol-modified aptamer. It results in the direct closer contact of a labeled probe with the SERS-active surface and an increase of EF (Figure 5d). The absence of specific binding in the case of BSA NPs for Sample 3 leads to the uneven distribution of probes including BSA NPs, the labeled secondary aptamer along with the primary one as schematically presented in Figure 5e, and in summary, a further increase of the signal for Cy3 dye in comparison with Samples 1 and

2. Moreover, a comparison of SERS data of RBD protein with the Cy3-labeled secondary aptamer and Sample 1 was performed. As shown in Figure S5, no characteristic peaks of Cy3 for RBD protein with Cy3-labeled aptamer were observed. The absence of the signal can be associated with the possible dense coverage of the SERS substrate by protein fragments. As a result, the dye molecules are away from the SERS-active surface, leading to the absence of the Raman signal of Cy3.

The developed indicator system allowed us to control the distance between the SERS platform and the labeled dye moiety. Moreover, it allows us to distinguish sandwich bioassays from their fragments, which is not possible to achieve with traditional enzyme-linked immunosorbent assay (ELISA) test. The application of a fabricated efficient hybrid plasmonic structure provides detection with a sensitivity to the target virus down to 10 nM (10^{10} particles/mL). This value is slightly above the upper necessary bound of a virus in real samples. The typical SARS-CoV-2 load in the clinical samples from the patient is in the range from 10^6 to 10^9 virus particles/mL [43]. Thus, the analytical sensitivity of the obtained platform in our work should be further enhanced to be competitive with the limit of detection (LOD) of 10^2–10^3 viral particles/mL for SARS-CoV-2 for polymerase chain reaction [44,45], the LOD of 2×10^4 viral particles/mL for loop-mediated isothermal amplification [46], and the LOD of 1×10^6–4×10^8 viral particles/mL for rapid antibody-based assays [47,48]. The developed novel plasmonic structure through varying the deposition parameters of the films and post-coating treatment paves a promising avenue to improve SERS performance of the sensor element, including the LOD, and fosters the practical application of the ultra-fast, amplification-free determination of coronaviruses.

4. Conclusions

Thus, we have proposed a novel SERS sensing platform based on protein-coated silver nanoparticles with a size close to the viral size as a highly stable structure for studying the detection of SARS-CoV-2 virions. Such a platform does not require special permissions to work with viruses and will speed up the process of developing effective methods for their detection. The optimal conditions for obtaining an artificial SERS-active system due to both specific and nonspecific binding of the primary aptamer and the Cy3-labeled secondary aptamer to synthetic virions were determined. Utilizing the receptor-binding domain of the S-protein of the envelope of the SARS-CoV-2 virus as a shell for model nanoparticles simulating the virus results in specific interactions by sandwich formation of it and both aptamers while bovine serum albumin results in no specific way, respectively. The developed indicator system allows us to control the distance between the fabricated efficient SERS hybrid platform and the labeled dye moiety. The ability to detect the target virus due to specific interactions with such aptameric DNA is quantitatively controlled by the degree of the quenching SERS signal from the labeled compound linked to an aptamer and paves a promising avenue to foster the practical application of the ultra-fast, amplification-free determination of coronaviruses.

Supplementary Materials: The following supporting information can be downloaded at: https://www.mdpi.com/article/10.3390/bios12090768/s1, Figure S1: Sensorgrams of 100 nM found aptamer sequence (Bt-Found), aptamer from the literature and cmpFound-Bt (nonspecific control), all carrying Bt-T10—linker at 5′(or 3′)-end; Figure S2: Reflection spectrum of SERS-active substrate; Figure S3: SERS spectrum of Cy3 (1×10^{-6} M). (532 nm, 0.3 mW, 5 s, 10 acc., ×40) on the obtained hybrid plasmonic substrate; Figure S4: Raman spectrum of Cy3 (1×10^{-2} M). (532 nm, 0.3 mW, 5 s, 10 acc., ×40) on the obtained hybrid plasmonic substrate. Figure S5: SERS spectra of Sample 1 (violet), primary aptamer-RBD protein with Cy3-labeled aptamer (magenta) (532 nm, 0.3 mW, 5 s, 10 acc., ×40).

Author Contributions: Conceptualization, M.V.S., O.O.K., E.G.Z., I.A.V. and M.I.Z.; methodology, M.V.S., O.O.K., N.F.M., S.M.N., O.A.S. and E.G.Z.; validation, M.V.S., O.O.K., S.M.N. and E.G.Z.; formal analysis, M.V.S., N.F.M., N.R.Y., F.I.G. and G.A.Z.; investigation, M.V.S., O.O.K., N.F.M., S.M.N., N.R.Y., O.A.S., D.I.Y., F.I.G. and G.A.Z.; resources, A.V.A., V.S.V., I.A.V. and M.I.Z.; data curation, O.O.K., E.G.Z., I.A.V. and M.I.Z.; writing—original draft preparation, M.V.S., O.O.K. and

E.G.Z.; writing—review and editing, all authors; visualization, M.V.S., O.O.K., S.M.N. and G.A.Z.; supervision, O.O.K., E.G.Z., I.A.V. and M.I.Z.; project administration, M.I.Z.; funding acquisition, I.A.V. and M.I.Z. All authors have read and agreed to the published version of the manuscript.

Funding: This research was funded by Russian Foundation for Basic Research, grant number 20-04-60477.

Institutional Review Board Statement: Not applicable.

Informed Consent Statement: Not applicable.

Data Availability Statement: Not applicable.

Acknowledgments: This research has been conducted in the frame of the Interdisciplinary Scientific and Educational School of Moscow University "Molecular Technologies of the Living Systems and Synthetic Biology". We are grateful to Ivan Vorobiev for the sample of RBD protein expressed in CHO, Timofey Zatsepin for thiol-modified oligonucleotides, and PhD student Mikhail Tatmyshevskiy for the measurement of the reflection spectra for hybrid plasmonic structure.

Conflicts of Interest: The authors declare no conflict of interest.

References

1. Langer, J.; de Aberasturi, D.J.; Aizpurua, J.; Alvarez-Puebla, R.A.; Auguié, B.; Baumberg, J.J.; Bazan, G.C.; Bell, S.E.J.; Boisen, A.; Brolo, A.G.; et al. Present and Future of Surface-Enhanced Raman Scattering. *ACS Nano* **2020**, *14*, 28–117. [CrossRef] [PubMed]
2. Cao, H.; Cao, H.; Li, Y.; Sun, Z.; Yang, Y.; Jiao, T. A Novel Natural Surface-Enhanced Fluorescence System Based on Reed Leaf as Substrate for Crystal Violet Trace Detection. *Chinese Phys. B* **2022**. [CrossRef]
3. Samodelova, M.V.; Kapitanova, O.O.; Evdokimov, P.V.; Eremina, O.E.; Goodilin, E.A.; Veselova, I.A. Plasmonic Features of Free-Standing Chitosan Nanocomposite Film with Silver and Graphene Oxide for SERS Applications. *Nanotechnology* **2022**, *33*, 1–10. [CrossRef] [PubMed]
4. Lenzi, E.; De Jimenez Aberasturi, D.; Henriksen-Lacey, M.; Piñeiro, P.; Muniz, A.J.; Lahann, J.; Liz-Marzán, L.M. SERS and Fluorescence-Active Multimodal Tessellated Scaffolds for Three-Dimensional Bioimaging. *ACS Appl. Mater. Interfaces* **2022**, *14*, 20708–20719. [CrossRef] [PubMed]
5. Lu, Y.; Dong, W.; Chen, Z.; Pors, A.; Wang, Z.; Bozhevolnyi, S.I. Gap-Plasmon Based Broadband Absorbers for Enhanced Hot-Electron and Photocurrent Generation. *Sci. Rep.* **2016**, *6*, 30650. [CrossRef]
6. Nielsen, M.G.; Gramotnev, D.K.; Pors, A.; Albrektsen, O.; Bozhevolnyi, S.I. Continuous Layer Gap Plasmon Resonators. *Opt. Express* **2011**, *19*, 19310. [CrossRef]
7. Tatmyshevskiy, M.K.; Yakubovsky, D.I.; Kapitanova, O.O.; Solovey, V.R.; Vyshnevyy, A.A.; Ermolaev, G.A.; Klishin, Y.A.; Mironov, M.S.; Voronov, A.A.; Arsenin, A.V.; et al. Hybrid Metal-Dielectric-Metal Sandwiches for SERS Applications. *Nanomaterials* **2021**, *11*, 3205. [CrossRef]
8. Novikov, S.M.; Boroviks, S.; Evlyukhin, A.B.; Tatarkin, D.E.; Arsenin, A.V.; Volkov, V.S.; Bozhevolnyi, S.I. Fractal Shaped Periodic Metal Nanostructures Atop Dielectric-Metal Substrates for SERS Applications. *ACS Photonics* **2020**, *7*, 1708–1715. [CrossRef]
9. Kasera, S.; Herrmann, L.O.; Del Barrio, J.; Baumberg, J.J.; Scherman, O.A. Quantitative Multiplexing with Nano-Self-Assemblies in SERS. *Sci. Rep.* **2014**, *4*, 6785. [CrossRef]
10. Pisarev, E.K.; Kapitanova, O.O.; Vesolova, I.A.; Zvereva, M.I. Amplification-Free Identification and Determination of Nucleic Acids by Surface Plasmon Resonance and Surface-Enhanced Raman Spectroscopy. *Mosc. Univ. Chem. Bull.* **2021**, *76*, 353–360. [CrossRef]
11. Tadesse, L.F.; Safir, F.; Ho, C.S.; Hasbach, X.; Khuri-Yakub, B.P.; Jeffrey, S.S.; Saleh, A.A.E.; Dionne, J. Toward Rapid Infectious Disease Diagnosis with Advances in Surface-Enhanced Raman Spectroscopy. *J. Chem. Phys.* **2020**, *152*, 240902. [CrossRef] [PubMed]
12. Saviñon-Flores, F.; Méndez, E.; López-Castaños, M.; Carabarin-Lima, A.; López-Castaños, K.A.; González-Fuentes, M.A.; Méndez-Albores, A. A Review on Sers-Based Detection of Human Virus Infections: Influenza and Coronavirus. *Biosensors* **2021**, *11*, 66. [CrossRef] [PubMed]
13. Ambartsumyan, O.; Gribanyov, D.; Kukushkin, V.; Kopylov, A.; Zavyalova, E. SERS-Based Biosensors for Virus Determination with Oligonucleotides as Recognition Elements. *Int. J. Mol. Sci.* **2020**, *21*, 3373. [CrossRef] [PubMed]
14. Paria, D.; Kwok, K.S.; Raj, P.; Zheng, P.; Gracias, D.H.; Barman, I. Label-Free Spectroscopic SARS-CoV-2 Detection on Versatile Nanoimprinted Substrates. *Nano Lett.* **2022**, *22*, 3620–3627. [CrossRef]
15. Payne, T.D.; Klawa, S.J.; Jian, T.; Kim, S.H.; Papanikolas, M.J.; Freeman, R.; Schultz, Z.D. Catching COVID: Engineering Peptide-Modified Surface-Enhanced Raman Spectroscopy Sensors for SARS-CoV-2. *ACS Sensors* **2021**, *6*, 3436–3444. [CrossRef]
16. Abdullah, M.B.; Dab, C.; Almalki, M.; Alnaim, A.; Abuzir, A.; Awada, C. Ultrafast Detection of SARS-CoV-2 Spike Protein (S) and Receptor-Binding Domain (RBD) in Saliva Using Surface-Enhanced Raman Spectroscopy. *Appl. Sci.* **2022**, *12*, 5039. [CrossRef]

17. Sarychev, A.K.; Sukhanova, A.; Ivanov, A.V.; Bykov, I.V.; Bakholdin, N.V.; Vasina, D.V.; Gushchin, V.A.; Tkachuk, A.P.; Nifontova, G.; Samokhvalov, P.S.; et al. Label-Free Detection of the Receptor-Binding Domain of the SARS-CoV-2 Spike Glycoprotein at Physiologically Relevant Concentrations Using Surface-Enhanced Raman Spectroscopy. *Biosensors* **2022**, *12*, 300. [CrossRef]
18. Ye, J.; Yeh, Y.-T.; Xue, Y.; Wang, Z.; Zhang, N.; Liu, H.; Zhang, K.; Ricker, R.; Yu, Z.; Roder, A.; et al. Accurate Virus Identification with Interpretable Raman Signatures by Machine Learning. *Proc. Natl. Acad. Sci. USA* **2022**, *119*, e2118836119. [CrossRef]
19. Shanmukh, S.; Jones, L.; Driskell, J.; Zhao, Y.; Dluhy, R.; Tripp, R.A. Rapid and Sensitive Detection of Respiratory Virus Molecular Signatures Using a Silver Nanorod Array SERS Substrate. *Nano Lett.* **2006**, *6*, 2630–2636. [CrossRef]
20. Lee, J.H.; Kim, B.C.; Oh, B.K.; Choi, J.W. Rapid and Sensitive Determination of HIV-1 Virus Based on Surface Enhanced Raman Spectroscopy. *J. Biomed. Nanotechnol.* **2015**, *11*, 2223–2230. [CrossRef]
21. Jen Lin, Y. A Rapid and Sensitive Early Diagnosis of Influenza Virus Subtype via Surface Enhanced Raman Scattering. *J. Biosens. Bioelectron.* **2014**, *5*, 1–5. [CrossRef]
22. Kukushkin, V.I.; Ivanov, N.M.; Novoseltseva, A.A.; Gambaryan, A.S.; Yaminsky, I.V.; Kopylov, A.M.; Zavyalova, E.G. Highly Sensitive Detection of Influenza Virus with SERS Aptasensor. *PLoS ONE* **2019**, *14*, e0216247. [CrossRef] [PubMed]
23. Ebrem Bilgin, B.; Torun, H.; Ilgü, M.; Yanik, C.; Batur, S.N.; Çelik, S.; Öztürk, M.; Dogan, Ö.; Ergönül, Ö.; Solaroglu, I.; et al. *Clinical Validation of SERS Metasurface SARS-CoV-2 Biosensor*; SPIE: Bellingham, WA, USA, 2022; Volume 1195708, p. 36. [CrossRef]
24. Chen, H.; Park, S.G.; Choi, N.; Moon, J.I.; Dang, H.; Das, A.; Lee, S.; Kim, D.G.; Chen, L.; Choo, J. SERS Imaging-Based Aptasensor for Ultrasensitive and Reproducible Detection of Influenza Virus A. *Biosens. Bioelectron.* **2020**, *167*, 112496. [CrossRef] [PubMed]
25. Chen, H.; Park, S.K.; Joung, Y.; Kang, T.; Lee, M.K.; Choo, J. SERS-Based Dual-Mode DNA Aptasensors for Rapid Classification of SARS-CoV-2 and Influenza A/H1N1 Infection. *Sens. Actuators B Chem.* **2022**, *355*, 131324. [CrossRef] [PubMed]
26. Lim, W.Y.; Lan, B.L.; Ramakrishnan, N. Emerging Biosensors to Detect Severe Acute Respiratory Syndrome Coronavirus 2 (SARS-CoV-2): A Review. *Biosensors* **2021**, *11*, 434. [CrossRef]
27. Tabrizi, M.A.; Acedo, P. An Electrochemical Impedance Spectroscopy-Based Aptasensor for the Determination of SARS-CoV-2-RBD Using a Carbon Nanofiber–Gold Nanocomposite Modified Screen-Printed Electrode. *Biosensors* **2022**, *12*, 142. [CrossRef]
28. Qu, J.H.; Leirs, K.; Maes, W.; Imbrechts, M.; Callewaert, N.; Lagrou, K.; Geukens, N.; Lammertyn, J.; Spasic, D. Innovative FO-SPR Label-Free Strategy for Detecting Anti-RBD Antibodies in COVID-19 Patient Serum and Whole Blood. *ACS Sens.* **2022**, *7*, 477–487. [CrossRef]
29. Calvo-Lozano, O.; Sierra, M.; Soler, M.; Estévez, M.C.; Chiscano-Camón, L.; Ruiz-Sanmartin, A.; Ruiz-Rodriguez, J.C.; Ferrer, R.; González-López, J.J.; Esperalba, J.; et al. Label-Free Plasmonic Biosensor for Rapid, Quantitative, and Highly Sensitive COVID-19 Serology: Implementation and Clinical Validation. *Anal. Chem.* **2022**, *94*, 975–984. [CrossRef]
30. Besselink, G.; Schütz-trilling, A.; Veerbeek, J.; Verbruggen, M.; Van Der Meer, A. Asymmetric Mach—Zehnder Interferometric Biosensing for Quantitative and Sensitive Multiplex Detection of Anti-SARS-CoV-2 Antibodies in Human Plasma. *Biosensors* **2022**, *12*, 553. [CrossRef]
31. Svobodova, M.; Skouridou, V.; Jauset-Rubio, M.; Viéitez, I.; Fernández-Villar, A.; Cabrera Alvargonzalez, J.J.; Poveda, E.; Bofill, C.B.; Sans, T.; Bashammakh, A.; et al. Aptamer Sandwich Assay for the Detection of SARS-CoV-2 Spike Protein Antigen. *ACS Omega* **2021**, *6*, 35657–35666. [CrossRef]
32. Chen, H.; Park, S.G.; Choi, N.; Kwon, H.J.; Kang, T.; Lee, M.K.; Choo, J. Sensitive Detection of SARS-CoV-2 Using a SERS-Based Aptasensor. *ACS Sens.* **2021**, *6*, 2378–2385. [CrossRef] [PubMed]
33. Zhang, D.; Zhang, X.; Ma, R.; Deng, S.; Wang, X.; Wang, X.; Zhang, X.; Huang, X.; Liu, Y.; Li, G.; et al. Ultra-Fast and Onsite Interrogation of Severe Acute Respiratory Syndrome Coronavirus 2 (SARS-CoV-2) in Waters via Surface Enhanced Raman Scattering (SERS). *Water Res.* **2021**, *200*, 117243. [CrossRef] [PubMed]
34. Khrenova, M.G.; Nikiforova, L.; Grabovenko, F.; Orlova, N.; Sinegubova, M. In Vitro Selection of an Aptamer Targeting SARS-CoV-2 Spike Protein with Nanopore Sequence Identification Reveals Discrimination Between the Authentic Strain and Omicron. *ChemRxiv* **2022**, 1–23. [CrossRef]
35. Song, Y.; Song, J.; Wei, X.; Huang, M.; Sun, M.; Zhu, L.; Lin, B.; Shen, H.; Zhu, Z.; Yang, C. Discovery of Aptamers Targeting the Receptor-Binding Domain of the SARS-CoV-2 Spike Glycoprotein. *Anal. Chem.* **2020**, *92*, 9895–9900. [CrossRef]
36. Sinegubova, M.V.; Orlova, N.A.; Kovnir, S.V.; Dayanova, L.K.; Vorobiev, I.I. High-Level Expression of the Monomeric SARS-CoV-2 S Protein RBD 320-537 in Stably Transfected CHO Cells by the EEF1A1-Based Plasmid Vector. *PLoS ONE* **2021**, *16*, e0242890. [CrossRef]
37. Novikov, S.M.; Streletskiy, O.A.; Doroshina, N.V.; Yakubovsky, D.I.; Mironov, M.S.; Sychev, V.V.; Voronov, A.A.; Arsenin, A.V.; Volkov, V.S. Long-Term Stable Structures Formed by Ion-Beam Modification of Silver Film for SERS Applications. *J. Phys. Conf. Ser.* **2021**, *2015*, 8–12. [CrossRef]
38. Waghmare, M.; Khade, B.; Chaudhari, P.; Dongre, P. Multiple Layer Formation of Bovine Serum Albumin on Silver Nanoparticles Revealed by Dynamic Light Scattering and Spectroscopic Techniques. *J. Nanoparticle Res.* **2018**, *20*, 185. [CrossRef]
39. Grabovenko, F.; Nikiforova, L.; Yanenko, B.; Ulitin, A.; Loktyushov, E.; Zatsepin, T.; Zavyalova, E.; Zvereva, M. Glycosylation of Receptor Binding Domain of SARS-CoV-2 S-Protein Influences on Binding to Immobilized DNA Aptamers. *Int. J. Mol. Sci.* **2022**, *23*, 557. [CrossRef]
40. Le Ru, E.C.; Blackie, E.; Meyer, M.; Etchegoint, P.G. Surface Enhanced Raman Scattering Enhancement Factors: A Comprehensive Study. *J. Phys. Chem. C* **2007**, *111*, 13794–13803. [CrossRef]

41. Eremina, O.E.; Zatsepin, T.S.; Farzan, V.M.; Veselova, I.A.; Zvereva, M.I. DNA Detection by Dye Labeled Oligonucleotides Using Surface Enhanced Raman Spectroscopy. *Mendeleev Commun.* **2020**, *30*, 18–21. [CrossRef]
42. Jaworska, A.; Pyrak, E.; Kudelski, A. Comparison of the Efficiency of Generation of Raman Radiation by Various Raman Reporters Connected via DNA Linkers to Different Plasmonic Nano-Structures. *Vib. Spectrosc.* **2019**, *101*, 34–39. [CrossRef]
43. Despres, H.W.; Mills, M.G.; Shirley, D.J.; Schmidt, M.M.; Huang, M.L.; Roychoudhury, P.; Jerome, K.R.; Greninger, A.L.; Bruce, E.A. Measuring Infectious SARS-CoV-2 in Clinical Samples Reveals a Higher Viral Titer:RNA Ratio for Delta and Epsilon vs. Alpha Variants. *Proc. Natl. Acad. Sci. USA* **2022**, *119*, 6–8. [CrossRef] [PubMed]
44. Kim, H.N.; Yoon, S.Y.; Lim, C.S.; Yoon, J. Comparison of Three Molecular Diagnostic Assays for SARS-CoV-2 Detection: Evaluation of Analytical Sensitivity and Clinical Performance. *J. Clin. Lab. Anal.* **2022**, *36*, e24242. [CrossRef] [PubMed]
45. Yang, J.; Han, Y.; Zhang, R.; Zhang, R.; Li, J. Comparison of Analytical Sensitivity of SARS-CoV-2 Molecular Detection Kits. *Int. J. Infect. Dis.* **2021**, *111*, 233–241. [CrossRef] [PubMed]
46. Yan, C.; Cui, J.; Huang, L.; Du, B.; Chen, L.; Xue, G.; Li, S.; Zhang, W.; Zhao, L.; Sun, Y.; et al. Rapid and Visual Detection of 2019 Novel Coronavirus (SARS-CoV-2) by a Reverse Transcription Loop-Mediated Isothermal Amplification Assay. *Clin. Microbiol. Infect.* **2020**, *26*, 773–779. [CrossRef] [PubMed]
47. Chan, K.H.; To, K.K.W.; Chan, J.F.W.; Li, C.P.Y.; Chen, H.; Yuen, K.Y. Analytical Sensitivity of Seven Point-of-Care Influenza Virus Detection Tests and Two Molecular Tests for Detection of Avian Origin H7N9 and Swine Origin H3N2 Variant Influenza a Viruses. *J. Clin. Microbiol.* **2013**, *51*, 3160–3161. [CrossRef]
48. Peters, T.R.; Blakeney, E.; Vannoy, L.; Poehling, K.A. Evaluation of the Limit of Detection of the BD VeritorTM System Flu A+B Test and Two Rapid Influenza Detection Tests for Influenza Virus. *Diagn. Microbiol. Infect. Dis.* **2013**, *75*, 200–202. [CrossRef]

biosensors

Article

Detection of A and B Influenza Viruses by Surface-Enhanced Raman Scattering Spectroscopy and Machine Learning

Artem Tabarov [1,*], Vladimir Vitkin [1,*], Olga Andreeva [1], Arina Shemanaeva [1], Evgeniy Popov [1], Alexander Dobroslavin [1], Valeria Kurikova [1], Olga Kuznetsova [1], Konstantin Grigorenko [1], Ivan Tzibizov [1], Anton Kovalev [1], Vitaliy Savchenko [2], Alyona Zheltuhina [3], Andrey Gorshkov [3] and Daria Danilenko [3]

[1] Institute of Advanced Data Transfer Systems, ITMO University, Birzhevaya Liniya 14, 199034 Saint Petersburg, Russia
[2] National Research Center "Kurchatov Institute", Akademika Kurchatova Sq. 1, 123182 Moscow, Russia
[3] Smorodintsev Research Institute of Influenza, Prof. Popova Str. 15/17, 197376 Saint Petersburg, Russia
* Correspondence: a_t_tabarov@itmo.ru (A.T.); vitkin@itmo.ru (V.V.)

Abstract: We demonstrate the possibility of applying surface-enhanced Raman spectroscopy (SERS) combined with machine learning technology to detect and differentiate influenza type A and B viruses in a buffer environment. The SERS spectra of the influenza viruses do not possess specific peaks that allow for their straight classification and detection. Machine learning technologies (particularly, the support vector machine method) enabled the differentiation of samples containing influenza A and B viruses using SERS with an accuracy of 93% at a concentration of 200 μg/mL. The minimum detectable concentration of the virus in the sample using the proposed approach was ~0.05 μg/mL of protein (according to the Lowry protein assay), and the detection accuracy of a sample with this pathogen concentration was 84%.

Keywords: surface-enhanced Raman spectroscopy; SERS; influenza A virus; influenza B virus; detection; machine learning

Citation: Tabarov, A.; Vitkin, V.; Andreeva, O.; Shemanaeva, A.; Popov, E.; Dobroslavin, A.; Kurikova, V.; Kuznetsova, O.; Grigorenko, K.; Tzibizov, I.; et al. Detection of A and B Influenza Viruses by Surface-Enhanced Raman Scattering Spectroscopy and Machine Learning. *Biosensors* **2022**, *12*, 1065. https://doi.org/10.3390/bios12121065

Received: 21 October 2022
Accepted: 20 November 2022
Published: 23 November 2022

1. Introduction

Outbreaks and epidemics of viral diseases in recent years have raised the active search for new methods for differential diagnostics and virus detection. Modern methods of virus detection, such as polymerase chain reactions (PCR) [1] and enzyme-linked immunosorbent assays (ELISA) [2], have high sensitivity in determining the presence of influenza viruses in a sample [3,4] but have a number of disadvantages: laboriousness, time duration of the assays, low versatility, and a high percentage of false positive results [5,6].

Surface-enhanced Raman spectroscopy (SERS) is considered a promising alternative for the differential detection of viral infections [7–11]. The viral envelope contains surface proteins that differ in their amino acid composition and conformation which leads to the difference in Raman spectra of various viruses. Since the Raman scattering signal is rather weak, the SERS is used to enhance it by implementing special substrates with a surface layer containing nanoparticles. The latter allows amplification of the Raman scattering signal of the test sample due to the plasmon resonance effect, by 10^3–10^9 times, and the subsequent detection of the pathogen in low concentrations (5×10^8 PFU/mL) [7]. Identification of viral particles by SERS spectroscopy has a number of potential advantages over classical diagnostic methods, such as the speed of testing and the possibility of diagnosing without the need to use specific antibodies or aptamers [7,9–11]. On the other hand, there are developments that are aimed at the selective detection of a specific pathogen based on SERS substrates made from precious metal nanoparticles and immobilized antibodies or aptamers [12–14]. This approach, however, deprives versatility, which is the SERS technology's main advantage, since each antigen has its own specific features in the spectrum [15].

The label-free SERS platform, in combination with machine learning, can become a powerful tool for diagnosing viral diseases not only due to antigen detection, but also due to the detection of biomarkers in the sample [16,17].

Many research efforts are underway to create a universal SERS substrate made from metal nanoparticles and nanorods which will allow us to determine the presence of viruses (including their type) in a sample [18–22]. The optimal parameters of nanostructures in these studies are selected to amplify the Raman signal. Thus, the differences in spectra could be noticeable visually [22] or be differentiated by classical methods of mathematical analysis, such as multivariate calibration and partial least squares regression (PLS/PSR) [23]. It should be noted, however, that these studies were carried out with purified samples of viral particles. In field work dealing with clinical samples, virus detection using this approach can be complicated.

Spectral patterns of different influenza strains have similarities [24,25] and to make their robust classification possible, mathematical processing methods can be used. For example, machine learning technologies improve the accuracy of differentiation and classification of SERS spectra [26]. Huang J. et al. managed to achieve an accuracy of 87.7% using the technology of recurrent neural networks (RNNs) in identifying the spectra of SARS-CoV-2 spike proteins [27]. The experiments by Paria D. et al. demonstrated how the random forest algorithm provides an accuracy of 83% when differentiating Zika, coronavirus (SARS-CoV-2), influenza A (H1N1), and Marburg viruses [11]. Yeh Y.T. et al. described a technique for recognizing differences in the SERS spectra of rhinovirus, the influenza A virus, and the type 3 parainfluenza virus with an accuracy of 93% [24]. Their scientific team also conducted a study on samples containing avian influenza viruses and confirmed the data processing efficiency by machine learning using the logistic regression algorithm [24,28]. The work of Lim J.Y. et al. [25] demonstrates the possibility of classifying the cells infected with wild and mutant influenza A viruses by their SERS spectra using the principal component analysis (PCA) [29].

The paper by Zhang Z. et al. described an experiment on modifying the structures of silver nanosubstrates by adding acetonitrile, bromine, and calcium ions; the minimum number of viruses for successful detection, in this case, was 100 particles per test [30]. In their work, the PCA was used to classify the samples containing and not containing viruses; as a result, differentiation analysis showed the spectra separation with a 95% accuracy (yet the researchers noted that the cell proteins interfere with the detection of the viral proteins themselves). In continuing experiments with SERS substrates, Zhang Z. et al. used sodium borohydride to change the conformation of silver nanostructures in order to detect adenovirus, coronavirus (SARS-CoV-2), and the influenza A (H1N1) virus [31]. The analysis time was 2 minutes and the minimum number of viral particles for detection was reduced to 10 units. Their work used the Latent Dirichlet allocation algorithm [32] with a differentiation accuracy of 95%.

The work of Durmanov et al. describes the usage of a fabricated SERS substrate, which was composed of nanoporous mica with the addition of a thin silver layer by electron beam physical vapor deposition method [33]. Four types of viruses were selected to demonstrate the practical application of the new SERS substrate: the myxoma virus (MYXV), the canine distemper virus (CDV), the tobacco mosaic virus (TMV), and the potato virus X (PVX). The SERS substrate performance was tested in its ability to obtain the spectra of viral particles of different sizes, morphology, structural composition, and physicochemical properties. Various methods of data analysis were used to identify the viruses by their spectra. Data spectral analyses were carried out by the method of machine learning, the principal components, and the linear discriminant analysis (PCA-LDA), in particular. The classification model accuracy was double-checked by 5-fold repetition of the 3-fold cross-validation, resulting in an average accuracy of 99.4%.

Paria D. et al. reported on the creation of a label-free SERS platform with a metal-insulator-metal nanostructure [34]. Combined with machine learning methods, this structure made it possible to differentiate four viruses: the influenza A virus, coronavirus,

Zika virus, and Marburg virus. The PCA method was used for visual analysis of the data obtained from the SERS spectra collection, and the random forest algorithm was used to classify the spectral data set. The resulting classification accuracy for labeling an unknown virus sample was ranging from 83 to 95%.

Song C. et al. were able to demonstrate the possibility of using the portable Raman spectrometer to detect the influenza A virus [35]. In their work, SERS spectra of the three influenza virus types (A/Mute Swan/MI/06/451072-2/2006, A/chicken/Pennsylvania/13609/1993, and A/chicken/TX/167280-4/02) and the control sample spectra were obtained, with the PCA method used to classify the spectra. Visualization of the results on the principal component axes indicated 100% accuracy (n = 10 for each sample). Despite the small number of data set samples, this study demonstrated the portability and versatility of the SERS virus particle detection technology.

The support vector machine (SVM) method has been previously applied to the detection of varied biomarkers using spectroscopic data [36]. A recent study by Yang Y. et al. demonstrates highly accurate differentiation of respiratory disease virus agents by implementing custom-fabricated SERS substrates with silver nanorods and the SVM classification procedure with data preprocessing [37]. The authors were able to differentiate a variety of viruses, including a coronavirus, influenza A and B viruses, and adenovirus. In their work, influenza viruses were contained in chick embryo allantoic fluid that could influence spectral uniformity and produce fluorescence. The inclusion of organic components may influence pattern recognition and defining fingerprints characteristic for viral particles.

In this paper, we propose implementing SERS and subsequent processing of the spectra to detect and differentiate the A and B influenza viruses in an STE buffer medium and investigate the limits of virus detection using this approach. Here, to the best of our knowledge, we, for the first time, demonstrate differentiating the A and B influenza viruses in an STE buffer based on SERS and SVM. We demonstrate successful detection and 93% accuracy in the differentiation of the viruses with low-cost commercial substrates, a simple STE buffer solution, and a standard machine learning algorithm that does not require time-consuming preprocessing steps of the spectral data. This makes our approach affordable and effective for use in real-life applications.

2. Materials and Methods

2.1. Viruses

Influenza A and influenza B viruses are typical agents causing acute respiratory infections. The antigenic structure of the hemagglutinin of influenza A and B viruses differs by ~70% [38], which made these pathogens a representative model for research.

Influenza A (A/California/07/2009, A(H1N1)pdm09) and influenza B (B/Hong Kong/269/2017) viruses were grown in chick embryos; purified virus concentrate was then obtained by differential centrifugation of virus-containing allantoic fluid followed by ultracentrifugation reprecipitation. The Pierce BCA Protein Assay Kit (Thermo Fisher Scientific, Rockford, IL, USA) was used to measure the protein concentration in purified virus concentrates. The result of determining the concentration of protein, according to Lowry, in concentrated viral suspensions, was <200 μg/mL. The hemagglutination titer of the concentrate was 1:256. The specificity of the obtained viral concentrates was confirmed by the ELISA test using specific monoclonal antibodies to the hemagglutinating protein of influenza A and B viruses.

The suspension of purified viral particles was placed in an STE buffer medium for further storage and use.

2.2. Buffer Medium

The STE containing NaCl, Tris, and EDTA with pH = 8.0 was used as a buffer medium. The choice of this medium was justified by the absence of a fluorescence signal in the range required. The absorption spectrum of the STE buffer medium (in the visible and IR ranges) is shown in Figure 1. It absorbs, in the IR region, near 980 nm lying far from the SERS signal.

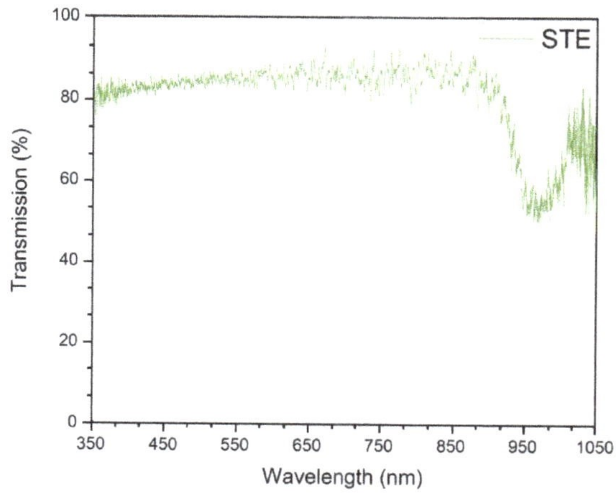

Figure 1. Absorption spectra of the STE buffer medium.

2.3. *SERS Substrates*

Commercial SERS substrates from SERSitive [39] were used in this work. This choice was made due to the substrate's hydrophilicity and tenfold Raman signal amplification at a 633 nm wavelength. The substrates consisted of glass, indium tin oxide, and contained nanostructured electrodeposited silver nanoparticles. The average size of substrate nanoparticles was 100–150 nm and the distance between nanoparticles was 100–200 nm. This substrate structure makes it possible to adsorb viral particles and provide high signal reproducibility over the entire surface. The image of the substrate obtained by scanning electron microscopy (the accelerating voltage was 10 kV) is shown in Figure 2.

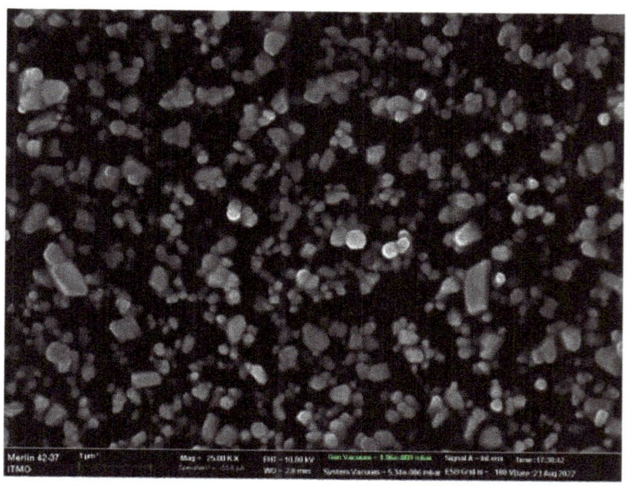

Figure 2. Scanning electron microscopy image of SERS substrate.

2.4. Raman Spectroscopy Setup

The measurements were carried out using a Horiba LabRam Raman spectrometer (Horiba Jobin Yvon S.A.S., Longjumeau, France). The setup included a laser source (a 632.8 nm wavelength, 0.01 mW of power), a spectrometer (600 lines/mm grating, a 500 nm blaze wavelength), a Mitutoyo Apo Plan 50x VIS lens, and a CCD camera (a 2000 × 800-pixel matrix in the receiving area, a 15 × 15 μm size of the pixel). The design of the spectrometer is shown in Figure 3.

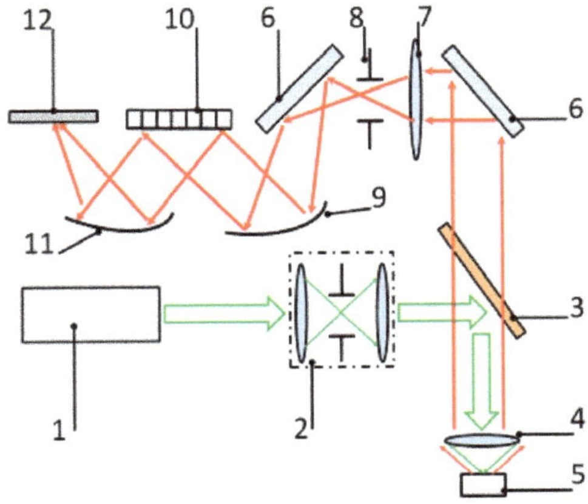

Figure 3. Design of the Raman spectroscopy setup: 1—laser source; 2—spatial filter; 3—edge-filter; 4—lens; 5—sample; 6—mirror; 7—condenser; 8—entrance slit of the spectrometer; 9, 11—aspherical mirrors; 10—diffraction grating; 12—CCD camera.

The excitation radiation of the laser source was put through an aperture diaphragm serving as a spatial filter. The edge filter reflecting the excitation light and transmitting the Raman signal directs the excitation beam to the sample fixed at the object stage. The Raman scattering signal was then collected by the lens, passed through the edge filter, and directed to the spectrometer. The spectrally resolved signal was projected onto a CCD camera to obtain a Raman spectrum.

It was experimentally found that the selected laser's power and wavelength make possible the fluorescence signal reduction and allow for minimizing the probability of destroying the biological structures due to the nanoparticle heating.

3. Results

3.1. Spectra of A Pure Buffer Medium and Viruses in A Buffer Medium

During the experiments, the first dark spectra and spectra of a pure substrate without a sample were obtained, with no Raman scattering peaks or artifacts visible during the process. Then, 1 μL volume samples of a clean STE buffer medium with influenza A and B viral particles (having concentrations of 500 μg/mL and 200 μg/mL according to the Lowry method, correspondingly), were applied to the SERS substrates surface. Further, the samples were dried at room temperature and placed in the spectrometer. SERS spectra were obtained in the range from 550 to 2000 cm^{-1}, which includes the main vibrational modes of organic compounds; the exposure time was 60 s. Several spectra were taken from five to six randomly selected spatial points on the substrate. As a result, 25–30 spectra were obtained for each sample, some of which are given in Figure 4.

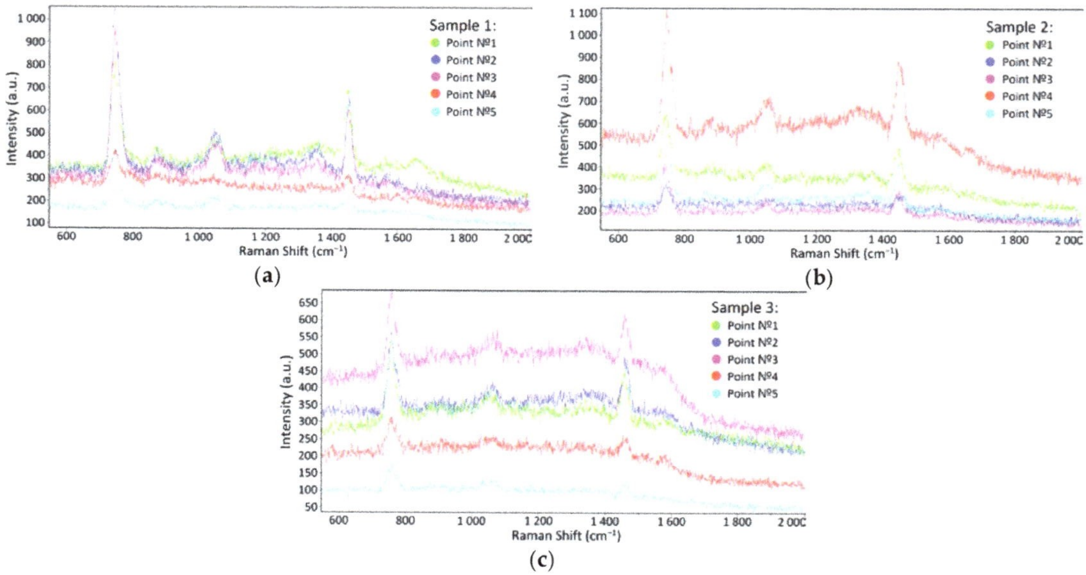

Figure 4. SERS spectra examples at different spatial points on the substrate having: (**a**) a pure buffer medium; (**b**) an influenza A virus in the buffer medium; (**c**) an influenza B virus in the buffer medium. The spectra of one sample taken at different points are indicated by different line colors.

The Raman spectra of the influenza A and B containing samples are visually similar to the buffer medium spectra due to the strong signal from the latter. In this case, the differentiation of spectra and virus detection become complicated (see Figure 5), prompting the use of mathematical processing methods (see Section 3.3).

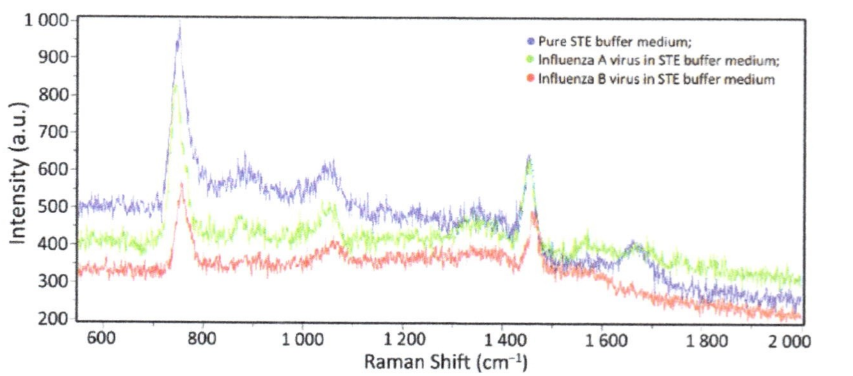

Figure 5. Relative comparison of the spectra: blue line—the pure STE buffer medium; green line—influenza A virus in the STE buffer medium; red line—influenza B virus in the STE buffer medium.

3.2. Spectra of Influenza A Virus in A Buffer Medium at Different Concentrations

To determine the method sensitivity limits of the SERS spectra and classification using mathematical processing, the suspension with the influenza A virus was diluted with the STE buffer medium in various ratios: 1:10, 1:100, 1:1000, and 1:10,000. The initial protein concentration of the sample with the viral suspension was 500 μg/mL, according to the Lowry method [40]. For this purpose, the STE buffer medium was poured into four test tubes in portions of 10 μL. Then, 1 μL of the influenza A virus concentrated suspension was added to the first tube and thoroughly mixed. Next, 1 μL was taken from the resulting suspension and put into the next tube, etc. The protein concentration in the resulting diluted samples was 50 μg/mL, 5 μg/mL, 0.5 μg/mL, and 0.05 μg/mL, respectively. The samples were then placed on a SERS substrate and dried. The process of obtaining Raman scattering spectra was the same as the method described in Section 3.1; ten samples were obtained for each concentration. The spectra did not differ visually, as shown in Figure 6.

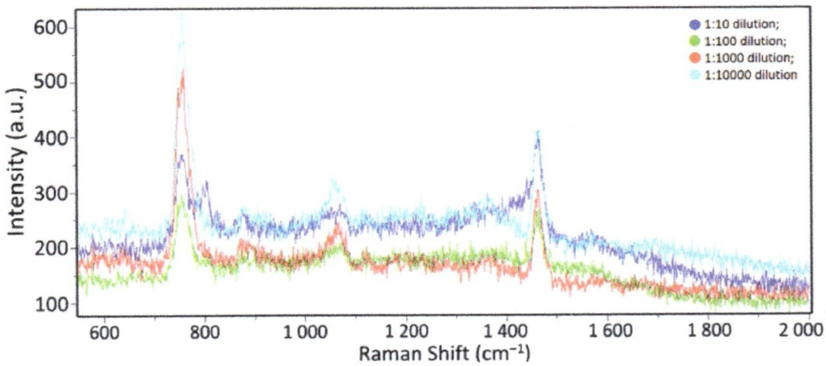

Figure 6. Influenza A virus in different dilutions: dark blue line—1:10 dilution; green line—1:100 dilution; red line—1:1000 dilution; light blue line—1:10,000 dilution.

3.3. *Mathematical Processing of Spectra*

3.3.1. Detection of Influenza A Virus in a Buffer Medium

The experimental spectra obtained were combined into a table in which the columns corresponded to wavelengths and the rows corresponded to intensity values. The table that is used as a sample set had the dimensions of 57 × 1332 cells, where 29 lines corresponded to the spectra of the STE buffer medium and 28 lines corresponded to the spectra of the influenza A virus. For the mathematical processing, we used the package Scikit-learn v. 0.24.2 for Python v. 3.10.7 [41].

The detection of the viral particles was carried out by binarily classifying the spectra of the pure buffer medium and the spectra of virus A in a buffer medium (with a concentration of 500 μg/mL, according to Lowry). For preliminary processing of the data, we applied standardization (StandardScaler) and normalization (Normalizer) to the sample set. For classification, the support vector machine (SVM) method was used. The SVM is one of the most commonly used machine learning algorithms, the purpose of which is to solve the classification problem by constructing an optimal separating hyperplane [42].

Samples that are closest to the hyperplane are the support vectors. The hyperplane is constructed so that the distance between it and the support vectors is maximized. This distance is called a margin. Accordingly, the remaining objects must have a distance to the hyperplane greater than the margin to perform the classification. This strict rule corresponds to the linear kernel.

Cross-validation was carried out using StratifiedKFold (stratification of the data set into two subsets at a ratio of 9 to 1, a training set of 51–52 samples, and a test set of 5–6 samples, with equal distribution of objects belonging to the different classes).

The average virus detection accuracy was 95.5% (see Table 1). Visualization of the classification is shown in Figure 7 (in this case, the values of the spectra were separated by a hyperplane).

Table 1. Spectrum Classification Accuracy Results.

Iteration	1	2	3	4	5	6	7	8	9	10
Test set size	6	6	6	6	6	6	6	5	5	5
Accuracy	100%	94%	94%	100%	94%	94%	94%	88%	100%	94%

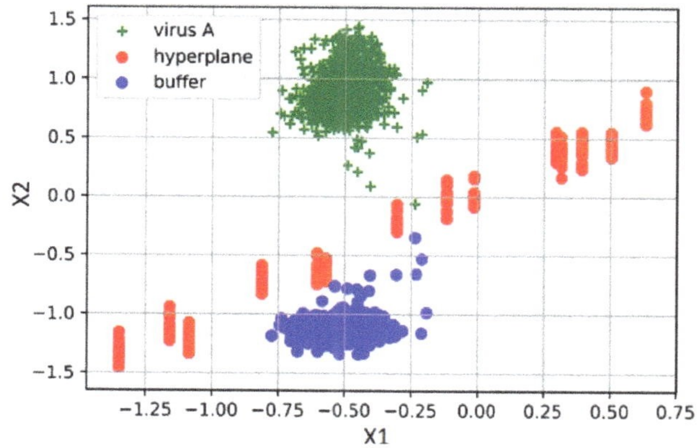

Figure 7. Visualization of spectra clustering of the pure STE buffer medium and influenza A virus in the STE buffer medium using projection onto support vectors X1 and X2. The red dots mark the projection of hyperplane points, which allows to classification of the spectra.

3.3.2. Differentiation of Influenza A, Influenza B, and Pure Buffer Medium Spectra

The spectra of influenza A and B viruses were combined into a table and the total size was 90 × 1332 cells (a total of 90 spectrum samples). An additional class-labeled column was introduced for classification purposes (the blank buffer medium was labeled 0, the influenza A virus was labeled 1, and the influenza B virus was labeled 2). Each class of spectra was represented by an equal number of samples (30 for each). Similarly, the methods of standardization (StandardScaler) and normalization (Normalizer) of the samples were used for preprocessing. Here, we used the SVM method with the hinge loss kernel function for classification that introduces soft boundaries of a margin.

Cross-validation was carried out using the StratifiedKFold method (the total data set was stratified into two subsets at a ratio of 9 to 1, where the size of the training set was 81 and the size of the test set was 9 samples. The training (test) set included per 27 (3) spectrum samples of the virus A, the virus B, and the pure buffer medium. Thus, a balanced division of the sample set into subsets with respect to the classes was obtained). In the full dataset, the mean cross-validation accuracy of classification was 93% (Table 2). For different subsamples, the prediction accuracy varied within 85–100% and the standard deviation (SD) was 4.8% (0.04835) at 10 cross-validation iterations. The visualization results are shown in Figure 8.

Table 2. Accuracy of the pure STE buffer medium, the influenza A virus in the STE buffer medium, and the influenza B virus in the STE buffer medium spectrum differentiation.

Iteration	1	2	3	4	5	6	7	8	9	10
Accuracy	93%	89%	93%	100%	93%	96%	89%	85%	100%	96%
SD					4.8%					

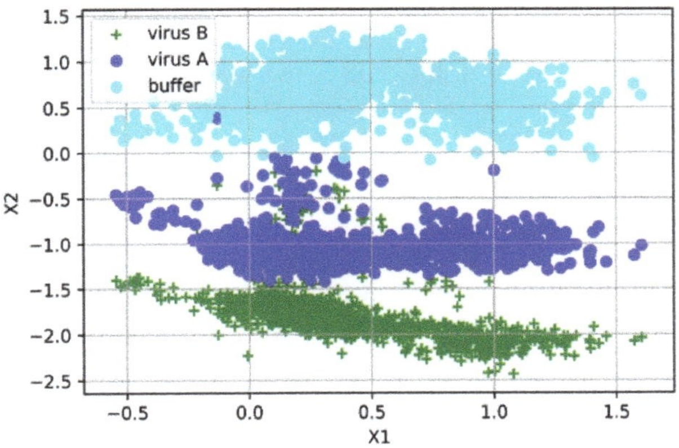

Figure 8. Visual representation of the spectra classification by support vector machine showing the projection of the samples onto support vectors X1 and X2: light blue—the pure STE buffer medium; dark blue—the influenza A virus in the STE buffer medium; green—the influenza B virus in the STE buffer medium.

3.3.3. Determination of the Minimum Allowable Concentration of Viral Particles for The Detection and Classification of the Influenza A Virus

In order to solve the binary classification problem of the pure buffer medium spectra from the influenza A virus spectra at different dilutions, the support vector machine with a linear classification kernel was used.

For samples with different influenza A virus concentrations in the buffer medium (1:10, 1:100, 1:1000, and 1:10,000), a data set was formed, which consisted of 10 measurements of the virus A spectra (for each concentration) and 10 measurements of the buffer spectra. Five subsets were formed for cross-validation (cross-validation was carried out using the stratification method to maintain the balance of virus and buffer classes). Thus, the test set for cross-validation had a dimension of 4 $\times$ 1332 (two spectra of the influenza A virus and two spectra of the pure buffer medium), and the training set was 16 $\times$ 1332 (eight spectra of the influenza A virus and eight spectra of the pure buffer medium).

The model always accurately differentiated the virus and the buffer at high virus concentrations (the classification accuracy was 100%), but the classification accuracy decreased while lowering the concentration. The results of differentiation are presented in Table 3. The samples with a virus concentration of 1:10,000 are of the greatest practical interest, and the average accuracy of the virus/buffer binary classification for the test sets was 0.84 (84%) in this case.

Table 3. Results of spectra differentiation by the algorithm at different influenza A virus dilutions.

Sample Dilution	Average Accuracy	SD of Accuracy	Average Spectrum RSD	Total Number of Samples	Training Set Size	Test Set Size
1:10	100%	0%	24%	20	15	5
1:100	100%	0%	42%	20	15	5
1:1000	94.5%	6.9%	26%	24	20	4
1:10,000	84%	15.2%	28%	20	15	5

We determined the spectrum relative SD (RSD) to investigate the intensity signal deviation of the spectrum samples and their potential influence on the classification accuracy. We calculated the normalized intensity SD at each spectral wavenumber, normalized it by a mean value to obtain an RSD at each point, and then averaged it over the entire wavenumber range. The average RSD was 24% for 1:10, 42% for 1:100, 26% for 1:100, and 28% for 1:10,000. It should be noted that such relatively high values may be caused by using a label-free SERS substrate, and that no special techniques for ensuring uniformity of a sample distribution were undertaken. Despite these RSD values, no apparent influence of signal deviation was found during the classification process.

4. Discussion

Through a combination with machine learning, photonics technology has flourished in biosensing by the implementation of extremely sophisticated methods to prepare biosamples, tailored substrates, and complicated preprocessing of spectral data for numerical analysis. In this work, we were focusing on design approaches to easy-to-use and affordable biosensing applications, and demonstrated the possibility of detecting and differentiating viral particles in a sample containing the STE buffer medium. The use of surface-enhanced Raman spectroscopy technology in combination with machine learning algorithms made it possible to differentiate samples with the influenza A virus, the influenza B virus, and without the virus, even for the limited data set sizes. The spectra were classified using the support vector machine. The accuracy of virus detection was 93%. To approach clinical studies, the concentration of the influenza A virus was reduced by 10^4 times. At low concentrations, the method used was able to differentiate between virus-free and virus-containing samples with an accuracy of 84%. The results of the study will help to develop fast, cheap but reliable diagnostic methods for real life working with clinical samples.

Author Contributions: Conceptualization, A.T. and V.V.; methodology, A.T. and D.D.; software, A.D.; validation, A.K. and E.P.; investigation, A.T., D.D., A.Z. and A.G.; writing—original draft preparation, A.T.; writing—review and editing, O.A., A.S., V.K., O.K.; engineer, I.T.; visualization, K.G.; supervision, A.K.; project administration, V.V.; funding acquisition, V.S. All authors have read and agreed to the published version of the manuscript.

Funding: The work was supported by the Ministry of Science and Higher Education of the Russian Federation (Project 075-15-2021-1349).

Institutional Review Board Statement: The study was approved by the Commission on Bioethics of the Smorodintsev Research Institute of Influenza, the meeting protocol #48/1 dated by 20 January 2022.

Informed Consent Statement: Not applicable.

Data Availability Statement: The data presented in this study are available on reasonable request from the corresponding authors.

Conflicts of Interest: The authors declare no conflict of interest.

References

1. Sedlak, R.H.; Jerome, K.R. Viral diagnostics in the era of digital polymerase chain reaction. *Diagn. Microbiol. Infect. Dis.* **2013**, *75*, 1–4. [CrossRef] [PubMed]
2. Atmar, R.L. *Viral Infections of Humans*; Kaslow, R., Stanberry, L., Le Duc, J., Eds.; Springer: Boston, MA, USA, 2014; Immunological detection and characterization; pp. 47–62. ISBN 978-1-4899-7448-8.
3. Yamada, A.; Imanishi, J.; Nakajima, E.; Nakajima, K.; Nakajima, S. Detection of influenza viruses in throat swab by using polymerase chain reaction. *Microbiol. Immunol.* **1991**, *35*, 259–265. [CrossRef] [PubMed]
4. Akylbayeva, K.K.; Shynybekova, G.O.; Tlenchiyeva, T.M.; Sultankulova, K.T.; Sandybaev, N.T. Лабораторная диагностика гриппа типов А и Б методом ПЦР. *Вестник КазНУ. Серия биологическая* **2017**, *70*, 100–108.
5. Wikramaratna, P.S.; Paton, R.S.; Ghafari, M.; Lourenço, J. Estimating the false-negative test probability of SARS-CoV-2 by RT-PCR. *Eurosurveillance* **2020**, *25*, 2000568. [CrossRef] [PubMed]
6. Boonham, N.; Kreuze, J.; Winter, S.; van der Vlugt, R.; Bergervoet, J.; Tomlinson, J.; Mumford, R. Methods in virus diagnostics: From ELISA to next generation sequencing. *Virus Res.* **2014**, *186*, 20–31. [CrossRef]
7. Shanmukh, S.; Jones, L.; Driskell, J.; Zhao, Y.; Dluhy, R.; Tripp, R.A. Rapid and sensitive detection of respiratory virus molecular signatures using a silver nanorod array SERS substrate. *Nano Lett.* **2006**, *6*, 2630–2636. [CrossRef]

8. Savinon-Flores, F.; Mendez, E.; Lopez-Castanos, M.; Carabarin-Lima, A.; Lopez-Castanos, K.A.; Gonzalez-Fuentes, M.A.; Mendez-Albores, A. A review on SERS-based detection of human virus infections: Influenza and coronavirus. *Biosensors* **2021**, *11*, 66. [CrossRef] [PubMed]
9. Tabarov, A.T.; Vitkin, V.V.; Danilenko, D.M.; Andreeva, O.V.; Shemanaeva, A.A.; Popov, E.E.; Dobroslavin, A.A.; Kurikova, V.V.; Kuznetsova, O.B. Using the SERS method and machine learning technology to detect the influenza A virus. In Proceedings of the 2022 International Conference Laser Optics (ICLO), Saint Petersburg, Russia, 20–24 June 2022.
10. Tabarov, A.T.; Vitkin, V.V.; Danilenko, D.M.; Andreeva, O.V.; Shemanaeva, A.A.; Popov, E.E.; Dobroslavin, A.A.; Kurikova, V.V.; Kuznetsova, O.B. Detection of Influenza A and B viruses by Raman spectroscopy. In Proceedings of the 2022 International Conference Laser Optics (ICLO), Saint Petersburg, Russia, 20–24 June 2022.
11. Paria, D.; Kwok, K.S.; Raj, P.; Zheng, P.; Gracias, D.; Barman, I. Label-Free SARS-CoV-2 Detection on Flexible Substrates. *medRxiv* **2021**. [CrossRef]
12. Eom, G.; Hwang, A.; Lee, D.K.; Guk, K.; Moon, J.; Jeong, J.; Jung, J.; Kim, B.; Lim, E.K.; Kang, T. Superb specific, ultrasensitive, and rapid identification of the Oseltamivir-resistant H1N1 virus: Naked-eye and SERS dual-mode assay using functional gold nanoparticles. *ACS Appl. Bio Mater.* **2019**, *2*, 1233–1240. [CrossRef] [PubMed]
13. Wang, C.; Wang, C.; Wang, X.; Wang, K.; Zhu, Y.; Rong, Z.; Wang, W.; Xiao, R.; Wang, S. Magnetic SERS strip for sensitive and simultaneous detection of respiratory viruses. *ACS Appl. Mater. Interfaces* **2019**, *11*, 19495–19505. [CrossRef] [PubMed]
14. Wang, J.; Drelich, A.J.; Hopkins, C.M.; Mecozzi, S.; Li, L.; Kwon, G.; Hong, S. Gold nanoparticles in virus detection: Recent advances and potential considerations for SARS-CoV-2 testing development. *Wiley Interdiscip. Rev. Nanomed. Nanobiotechnol.* **2022**, *14*, e1754. [CrossRef] [PubMed]
15. Alvarez-Puebla, R.A.; Liz-Marzán, L.M. SERS-based diagnosis and biodetection. *Small* **2010**, *6*, 604–610. [CrossRef] [PubMed]
16. Othman, N.H.; Lee, K.Y.; Radzol, A.R.M.; Mansor, W.; Ramlan, N.N.M. Linear discriminant analysis for detection of salivary NS1 from SERS spectra. In Proceedings of the TENCON 2017–2017 IEEE Region 10 Conference, Penang, Malaysia, 5–8 November 2017; pp. 2876–2879.
17. Ralbovsky, N.M.; Lednev, I.K. Towards development of a novel universal medical diagnostic method: Raman spectroscopy and machine learning. *Chem. Soc. Rev.* **2020**, *49*, 7428–7453. [CrossRef] [PubMed]
18. Sivashanmugan, K.; Liao, J.D.; You, J.W.; Wu, C.L. Focused-ion-beam-fabricated Au/Ag multilayered nanorod array as SERS-active substrate for virus strain detection. *Sens. Actuators B Chem.* **2013**, *181*, 361–367. [CrossRef]
19. Chang, C.W.; Liao, J.D.; Shiau, A.L.; Yao, C.K. Non-labeled virus detection using inverted triangular Au nano-cavities arrayed as SERS-active substrate. *Sens. Actuators B Chem.* **2011**, *156*, 471–478. [CrossRef]
20. Lin, Y.Y.; Liao, J.D.; Yang, M.L.; Wu, C.L. Target-size embracing dimension for sensitive detection of viruses with various sizes and influenza virus strains. *Biosens. Bioelectron.* **2012**, *35*, 447–451. [CrossRef]
21. Lin, Y.Y.; Liao, J.D.; Ju, Y.H.; Chang, C.W.; Shiau, A.L. Focused ion beam-fabricated Au micro/nanostructures used as a surface enhanced Raman scattering-active substrate for trace detection of molecules and influenza virus. *Nanotechnology* **2011**, *22*, 185308. [CrossRef]
22. Lim, J.Y.; Nam, J.S.; Yang, S.E.; Shin, H.; Jang, Y.H.; Bae, G.U.; Kang, T.; Lim, K.I.; Choi, Y. Identification of newly emerging influenza viruses by surface-enhanced Raman spectroscopy. *Anal. Chem.* **2015**, *87*, 11652–11659. [CrossRef]
23. Alexander, T.A. Development of methodology based on commercialized SERS-active substrates for rapid discrimination of Poxviridae virions. *Anal. Chem.* **2008**, *80*, 2817–2825. [CrossRef]
24. Yeh, Y.T.; Gulino, K.; Zhang, Y.; Sabestien, A.; Chou, T.W.; Zhou, B.; Lin, Z.; Albert, I.; Lu, H.; Swaminathan, V.; et al. A rapid and label-free platform for virus capture and identification from clinical samples. *Proc. Natl. Acad. Sci. USA* **2020**, *117*, 895–901. [CrossRef] [PubMed]
25. Lim, J.Y.; Nam, J.S.; Shin, H.; Park, J.; Song, H.I.; Kang, M.; Lim, K.I.; Choi, Y. Identification of newly emerging influenza viruses by detecting the virally infected cells based on surface enhanced Raman spectroscopy and principal component analysis. *Anal. Chem.* **2019**, *91*, 5677–5684. [CrossRef]
26. Lussier, F.; Thibault, V.; Charron, B.; Wallace, G.Q.; Masson, J.F. Deep learning and artificial intelligence methods for Raman and surface-enhanced Raman scattering. *TrAC Trends Anal. Chem.* **2020**, *124*, 115796. [CrossRef]
27. Huang, J.; Wen, J.; Zhou, M.; Ni, S.; Le, W.; Chen, G.; Wei, L.; Zeng, Y.; Qi, D.; Pan, M.; et al. On-site detection of SARS-CoV-2 antigen by deep learning-based surface-enhanced raman spectroscopy and its biochemical foundations. *Anal. Chem.* **2021**, *93*, 9174–9182. [CrossRef]
28. LaValley, M.P. Logistic regression. *Circulation* **2008**, *117*, 2395–2399. [CrossRef]
29. Abdi, H.; Williams, L.J. Principal component analysis. *Wiley Interdiscip. Rev. Comput. Stat.* **2010**, *2*, 433–459. [CrossRef]
30. Zhang, Z.; Jiang, S.; Wang, X.; Dong, T.; Wang, Y.; Li, D.; Gao, X.; Qu, Z.; Li, Y. A novel enhanced substrate for label-free detection of SARS-CoV-2 based on surface-enhanced Raman scattering. *Sens. Actuators B Chem.* **2022**, *359*, 131568. [CrossRef]
31. Zhang, Z.; Li, D.; Wang, X.; Wang, Y.; Lin, J.; Jiang, S.; Wu, Z.; He, Y.; Gao, X.; Zhu, Z.; et al. Rapid detection of viruses: Based on silver nanoparticles modified with bromine ions and acetonitrile. *Chem. Eng. J.* **2022**, *438*, 135589. [CrossRef]
32. Blei, D.M.; Ng, A.Y.; Jordan, M.I. Latent dirichlet allocation. *J. Mach. Learn. Res.* **2003**, *3*, 993–1022. [CrossRef]
33. Durmanov, N.N.; Guliev, R.R.; Eremenko, A.V.; Boginskaya, I.A.; Ryzhikov, I.A.; Trifonova, E.A.; Putlyaev, E.V.; Mukhin, A.N.; Kalnov, S.L.; Balandina, M.V.; et al. Non-labeled selective virus detection with novel SERS-active porous silver nanofilms fabricated by Electron Beam Physical Vapor Deposition. *Sens. Actuators B Chem.* **2018**, *257*, 37–47. [CrossRef]

34. Paria, D.; Kwok, K.S.; Raj, P.; Zheng, P.; Gracias, D.H.; Barman, I. Label-free spectroscopic SARS-CoV-2 detection on versatile nanoimprinted substrates. *Nano Lett.* **2022**, *22*, 3620–3627. [CrossRef]
35. Song, C.; Driskell, J.D.; Tripp, R.A.; Cui, Y.; Zhao, Y. The use of a handheld Raman system for virus detection. In Proceedings of the Chemical, Biological, Radiological, Nuclear, and Explosives (CBRNE) Sensing XIII, Baltimore, MD, USA, 24–27 April 2012; Volume 8358, pp. 109–115.
36. Modak, S.; Mehta, S.; Sehgal, D.; Valadi, J. Application of support vector machines in viral biology. In *Global Virology III: Virology in the 21st Century*; Shapshak, P., Balaji, S., Eds.; Springer: Cham, Switzerland, 2019; pp. 361–403. ISBN 978-3-030-29021-4.
37. Yang, Y.; Xu, B.; Murray, J.; Haverstick, J.; Chen, X.; Tripp, R.A.; Zhao, Y. Rapid and quantitative detection of respiratory viruses using surface-enhanced Raman spectroscopy and machine learning. *Biosens. Bioelectron.* **2022**, *217*, 114721. [CrossRef]
38. Johansson, B.E.; Brett, I.C. Recombinant influenza B virus HA and NA antigens administered in equivalent amounts are immunogenically equivalent and induce equivalent homotypic and broader heterovariant protection in mice than conventional and live influenza vaccines. *Hum. Vaccines* **2008**, *4*, 420–424. [CrossRef] [PubMed]
39. SERSitive. User Guide and Datasheed. Available online: https://sersitive.eu/wp-content/uploads/Manual_product-sheet.pdf (accessed on 19 October 2022).
40. Waterborg, J.H.; Matthews, H.R. *Basic Protein and Peptide Protocols*; Walker, J.M., Ed.; Humana Press: Totowa, NJ, USA, 1994; The Lowry method for protein quantitation; Volume 32, pp. 1–4. ISBN 978-1-59259-519-8.
41. Pedregosa, F.; Varoquaux, G.; Gramfort, A.; Michel, V.; Thirion, B.; Grisel, O.; Blondel, M.; Prettenhofer, P.; Weiss, R.; Dubourg, V.; et al. Scikit-learn: Machine Learning in Python. *J. Mach. Learn. Res.* **2011**, *12*, 2825–2830.
42. Griffel, L.M.; Delparte, D.; Edwards, J. Using Support Vector Machines classification to differentiate spectral signatures of potato plants infected with Potato Virus Y. *Comput. Electron. Agric.* **2018**, *153*, 318–324. [CrossRef]

biosensors

MDPI

Article

Plier Ligands for Trapping Neurotransmitters into Complexes for Sensitive Analysis by SERS Spectroscopy

Olga E. Eremina [1], Olesya O. Kapitanova [1], Alexei V. Medved'ko [2], Alexandra S. Zelenetskaya [1], Bayirta V. Egorova [1], Tatyana N. Shekhovtsova [1,*], Sergey Z. Vatsadze [2] and Irina A. Veselova [1]

1 Chemistry Department, Moscow State University, Moscow 119991, Russia
2 N.D. Zelinsky Institute of Organic Chemistry, Russian Academy of Sciences, Moscow 119991, Russia
* Correspondence: shekhovtsovatn@my.msu.ru

Abstract: Catecholamines–dopamine, noradrenaline and adrenaline are important biomarkers of neurotransmitter metabolism, indicating neuroendocrine tumors and neurodegenerative diseases. Surface-enhanced Raman spectroscopy (SERS) is a promising analytical technique with unprecedented multiplexing capabilities. However, not all important analytes exhibit strong SERS signals on stable and robust nanostructured substrates. In this work, we propose a novel indicator system based on the formation of mixed ligand complexes with bispidine-based bis-azole ligands which can serve as pliers to trap Cu(II) ions and stabilize its complexes with catecholamines. Four synthesized ligands with different functional groups: carboxyl, amino, benzyl, and methoxybenzyl, were applied for forming stable complexes to shift maximum absorbance of catecholamines from the ultraviolet region to 570–600 nm. A new absorbance band in the visible range resonates with the local surface plasmon resonance (LSPR) band of metal nanoparticles and most used laser wavelengths. This match allowed use of Molecular Immobilization and Resonant Raman Amplification by Complex-Loaded Enhancers (MIRRACLE) methodology to measure intense Raman signals on a nanostructured silver-based SERS-active substrate. The synthesized plier-like ligands fixed and stabilized catecholamine complexes with Cu(II) on the SERS sensor surface, which facilitated the determination of dopamine in a 3.2×10^{-12}–1×10^{-8} M concentration range.

Keywords: catecholamines; dopamine; neurotransmitter; resonance Raman spectroscopy; silver nanoparticles; surface-enhanced Raman spectroscopy (SERS)

Citation: Eremina, O.E.; Kapitanova, O.O.; Medved'ko, A.V.; Zelenetskaya, A.S.; Egorova, B.V.; Shekhovtsova, T.N.; Vatsadze, S.Z.; Veselova, I.A. Plier Ligands for Trapping Neurotransmitters into Complexes for Sensitive Analysis by SERS Spectroscopy. *Biosensors* **2023**, *13*, 124. https://doi.org/10.3390/bios13010124

Received: 13 December 2022
Revised: 5 January 2023
Accepted: 6 January 2023
Published: 11 January 2023

1. Introduction

Neurotransmitter metabolism (or metabolism of biogenic amines) forms the basis of neural mediation both in the peripheral and central nervous systems. The concentration of biogenic amines varies in normal and pathological conditions, which allows them to be used as molecular diagnostic markers both in basic research and in clinical medical practice. Catecholamines (CAs), such as dopamine (DA), noradrenaline (NA), adrenaline (AD), their precursor–l-dioxyphenylalanine (L-DOPA), and metabolites–homovanillic (HVA) and vanillylmandelic acids (VMA) are biologically active substances involved in the regulation of nervous activity, the cardiovascular system, the lipid peroxidation system, energy metabolism and myocardial contractility, in microcirculation and tissue oxygen supply, and in the regulation of embryogenesis [1]. NA in blood plasma serves as a prognostic biomarker for disorders in the functioning of the cardiovascular system [2]. The ratios of CAs and their metabolites in the urine are used to diagnose an adrenal tumor–pheochromocytoma [3]. Deviations in the synthesis and metabolism of CAs lie in the etiology of depression and related conditions [1]. However, there are two main groups of diseases associated with impaired neurotransmitter metabolism. The first group includes neurodegenerative disorders characterized by the death of neuronal cells in various areas of the brain and spinal cord, which corresponds to a decrease in the content of CA and their metabolites in the body.

The main types of such diseases include Alzheimer's (AD), Parkinson's (PD), Huntington's disease and amyotrophic lateral sclerosis [4]. The second group of diseases includes neuroendocrine disorders–neuroblastoma, pheochromocytoma, paraganglioma, carcinoid tumors, which are caused by excessive CA secretion and irritation of the autonomic nervous system [5]. Biochemical low-molecular markers of various diseases are associated with disorders of neurotransmitter metabolism in living organisms in clinical diagnostics are the precursor (L-DOPA), catecholamines (DA, AD, and NA) and their final metabolites (HMA and HVA).

The first major challenge for detecting and quantifying CAs in blood plasma or other biological fluids is that they are normally present at very low concentrations (at the level of 1 nM and lower). Moreover, with various pathological disorders, such as AD and PD, the concentration of CAs in biological samples decreases by an order of magnitude or more. The second important aspect of reliable diagnosis is the possibility of simultaneous selective determination of several analytes in a single sample [1]. Notably, in the blood, CAs can be quickly oxidized by platelet monoamine oxidases; thus, the determination of CAs in the body should be ideally carried out rather quickly, within 15–30 min [6]. Several analytical methods have been described to measure CAs and related biogenic amines in biological fluids, e.g., blood, urine, and cerebrospinal fluid. The most common methods are electrochemical [7–10], enzyme immunoassay [11–13], fluorometric [6,14,15], high-performance liquid chromatography (HPLC) [16–18] and capillary electrophoresis (CE) [19–21]. Even though these methods have good selectivity and low limits of detection (LODs), they cannot always be used for the multiplex determination of CA. Furthermore, simultaneous quantification of catecholamines and related biogenic amines remains an analytical challenge because of their low concentrations in plasma, the oxidation-prone catechol moiety, potential chromatographic interferences, and poor fragmentation characteristics in the mass spectrometer [22,23]. De facto, most of these methods are laborious, time-consuming, relatively imprecise, require bulk instrumentation and sample pretreatment steps with oxidation-sensitive CAs, and use large sample volumes [22]. Chemical sensors and biosensors have been among the most demanded tools of modern analytical chemistry over the past decades compared to relatively bulky, expensive and complex analytical instruments [24]. Currently, the most widely used in clinical diagnostics are electrochemical sensors, due to their high efficiency, portability, simplicity, and relatively low cost. However, as a rule, electrochemical sensors do not allow simultaneous, performance, i.e., multiplexed detection of several target analytes in one sample.

Surface-enhanced Raman spectroscopy (SERS) is widely used in the fields of chemical and biological analysis as a fast, sensitive, highly selective, and informative technique for the qualitative and semiquantitative determination of amino acids, nucleotides, water-soluble and membrane proteins, and nucleic acids [25–30]. In one of the first works on SERS determination of CAs, the multiplex determination of DA and NA was carried out in model solutions with non-modified silver nanoparticles (AgNPs) [31]. Despite the similarity of the SERS spectra of DA and NA, clear differences in the SERS spectra in the region of 1271–1325 cm^{-1}, associated with the adsorption behavior of CAs on the surface of AgNPs, were reported. The achieved LODs were 5 nM for DA and NA, and addition of 0.5 wt% of albumin into the analyzed sample had no effect on the SERS spectra of CA. Moody and Sharma demonstrated multiplexed determinations of seven neurotransmitters: melatonin, serotonin, glutamate, DA, GABA, NA, and AD, by SERS with gold and silver nanoparticles in combination using 532, 633, and 785 nm laser wavelengths [32]. The reported LODs were 0.2, 1, and 10 μM for DA, NA, and AD, respectively. To increase the sensitivity of CA determination and match required for diagnosis LODS, one can vary the size and shape of nanoparticles, as well as modify chemical design of nanoparticle surface. For instance, ultrasensitive SERS detection of DA and serotonin down to 0.1 nM was demonstrated on graphene-gold heterostructural nanopyramids [33]. The additional 10-fold increase in the SERS intensity supposedly occurred as a result of chemical amplification originating from π-π interactions between graphene and aromatic ring structures in DA and serotonin. Thus,

SERS is a promising technique for the multiplexed sensitive determination of CAs and related biogenic amines, but few quantitative methods have been proposed so far, most of which are based on the determination of only one compound. Therefore, the development and search for new sensing approaches to the determination of neurotransmitters is one of important tasks for chemical analysis.

Herein, we propose a novel indicator system based on plier-like ligands for trapping catecholamines into stable, colored complexes with Cu(II) ions. We synthesized four plier-like bispidine-based bis-azole ligands with different substituents: carboxylic, amino, benzyl, and methoxybenzyl groups. The stability of the obtained complexes with Cu(II) was studied with spectrophotometry and potentiometry. Importantly, the formation of ternary complexes allowed measurement of strong SERS signals on the silver nanostructured substrate. High sensitivity was achieved as a result of applying Molecular Immobilization and Resonant Raman Amplification by Complex-Loaded Enhancers (MIRRACLE) methodology, as the ligands and Cu(II) allowed to shift the absorbance of catecholamines from the ultraviolet into visible range which provided additional signal gain.

2. Materials and Methods

2.1. Materials

Dopamine hydrochloride (≥98%), norepinephrine (or noradrenaline, ≥98%), epinephrine (or adrenaline, ≥98%), chitosan (CS, with Mw∼150 kDa and a degree of deacetylation of 85%), and copper sulfate ($CuSO_4 \cdot 5H_2O$ were purchased from Sigma-Aldrich, USA. Hydrochloric, nitric, and acetic acids were obtained from Khimmed, Russia. ACS grade solvents were purchased from Khimmed, Russia, and used as received. All aqueous solutions were prepared using deionized water with a specific resistance of at least 18.2 MΩ·cm (Millipore, Martillac, France).

2.2. Synthesis of Bispidine-Based Bis-Azole Ligands

General procedure for [3+2]-cycloaddition. To a solution of 1,5-dimethyl-3,7-di(prop-2-yn-1-yl)-3,7-diazabicyclo[3.3.1]nonan-9-one (1 mmol) and respective azide (2 mmol) in 10 mL of tert-butanol the solutions of sodium ascorbate (0.2 mmol) in 2.5 mL of water and $CuSO_4 \cdot 5H_2O$ (0.1 mmol) in 2.5 mL of water were added sequentially. The reaction mixture was stirred for 36 h at room temperature under argon. The solution was evaporated to dryness and quenched with 30 mL of dichloromethane (DCM) and 10 mL of water. The organic layer was separated and washed with water until discoloration of water layer. The organic phase was dried with sodium sulfate and evaporated to dryness.

tBu2L^1. White foam. Yield 80%. ^{1}H-NMR (400 MHz, $CDCl_3$, δ, ppm) 0.93 (s, 6H, CH_3); 1.48 (s, 18H), 2.42 (d, 4H, 10.8 Hz), 3.05 (d, 4H, 10.8 Hz), 3.69 (s, 4H), 5.08 (s, 4H), 7.66 (s, 2H). ^{13}C-NMR (100 MHz, $CDCl_3$, δ, ppm) 19.68, 27.91, 46.57, 51.43, 51.83, 65.10, 83.55, 124.24, 145.09, 165.37, 215.21. HRMS-ESI. Calculated for [$C_{27}H_{42}N_8O_5+H^+$]: 559.3351. Found: 559.3356

Boc2L^2. White foam. Yield 56%. Purified with gradient column chromatography eluting from DCM:MeOH 10:1 to DCM:MeOH 3:1 and to DCM:MeOH:25% $NH_3 \cdot H_2O$ 1:1:0.01. ^{1}H-NMR (400 MHz, DMSO-d_6, δ, ppm) 0.82 (s, 6H, CH_3), 1.30 (s, 18H, Boc), 2.32 (d, 4H, 10.5 Hz), 3.00 (d, 4H, 10.6 Hz), 3.33–3.36 (t, 4H, 5.7 Hz), 3.59–3.64 (m, overlapped with H_2O), 4.36 (t, 4H, 5.7 Hz), 7.86 (s, 2H). ^{13}C-NMR (100 MHz, DMSO-d_6, δ, ppm) 20.12, 28.44, 40.37, 46.31, 49.48, 51.38, 64.70, 78.46, 124.20, 143.52, 155.92, 217.88. HRMS-ESI. Calculated for [$C_{31}H_{38}N_8O_8+H^+$]: 571.3145. Found: 571.3145.

L^3. Yield 83%. All data are in accordance with our previous work [34].

L^4. White powder. Yield 71%. ^{1}H-NMR (400 MHz, $CDCl_3$, δ, ppm) 0.89 (s, 6H, CH_3); 2.37 (d, 4H, 10.9 Hz), 2.98 (d, 4H, 10.9 Hz), 3.62 (s, 4H), 3.79 (s, 6H), 5.46 (s, 4H), 6.88 (d, 4H, 8.9 Hz), 7.23 (d, 4H, 8.9 Hz), 7.44 (s, 2H). ^{13}C-NMR (100 MHz, $CDCl_3$, δ, ppm) 19.68, 46.56, 51.96, 53.59, 55.29, 65.14, 114.15, 114.37, 122.45, 126.83, 129.53, 129.71, 145.23, 159.81. HRMS-ESI. Calculated for [$C_{29}H_{48}N_{10}O_5+H^+$]: 617.3887. Found: 617.3882.

General procedure for Boc-deprotection. Cooled TFA (56.0 mmol) was added to cooled Boc-protected compound (0.56 mmol) in dry DCM (5 mL) and the resulting solution was stirred overnight at room temperature. The solution was evaporated to dryness, the residue was dissolved in 5 mL of DCM and evaporated to dryness again. The dissolving-evaporation procedure was repeated 3 times. The resulting oil was triturated with dry diethyl ether till formation of powder. The latter was filtered, washed with dry diethyl ether, and dried in vacuo over P_2O_5.

L1·2TFA. White hygroscopic powder. Yield 97%. ^{1}H-NMR (400 MHz, D_2O, δ, ppm, mixture of ketone and gem-diol) 0.77, 0.82 (s, 6H, CH_3); 2.85–2.97 (m, 6.5H), 3.46 (d, 1.5H, 11.2 Hz), 3.92 (s, 2.5H), 4.03 (s, 1.5 H), 5.24 (s, 4H), 8.03, 8.05 (s, 2H). ^{13}C-NMR (D_2O, mixture of ketone and gem-diol) 13.83, 14.52, 39.23, 45.40, 48.96, 49.32, 50.44, 59.00, 62.60, 92.99, 111.34, 114.24, 117.13, 119.67 (JC-F 291 Hz), 127.07, 127.12, 139.01, 139.23, 161.87, 162.22 (JC-F 35 Hz), 169.62, 169.69, 210.86. Calculated for $C_{19}H_{26}N_8O_5 \cdot 2C_2F_3O_2H$: C 40.96, H 4.18, N 16.61. Found: C 41.26 H 4.61 N 16.55.

L2·3.3TFA. White hygroscopic powder. Yield 95%. ^{1}H-NMR (400 MHz, D_2O, δ, ppm, mixture of ketone and gem-diol) 0.82, 0.89 (s, 6H, CH_3), 2.88–3.05 (m, 6.85H), 3.51–3.60 (m, 5.50H), 4.02 (s, 2.75H), 4.16 (s, 1.25H), 8.11, 8.13 (s, 2H). ^{13}C-NMR (D_2O, mixture of ketone and gem-diol) 13.59, 13.87, 38.37, 39.20, 45.47, 46.90, 48.84, 49.07, 58.81, 62.42, 93.04, 111.49, 114.37, 117.27, 120.17 (JC-F 292 Hz), 126.30, 126.38, 138.90, 139.17, 162.22, 162.57 (JC-F 35 Hz), 210.72. Calculated for $C_{19}H_{32}N_{10}O \cdot 3.3C_2F_3O_2H$: C 38.78, H 4.49, N 17.67. Found: C 38.86 H 4.92 N 17.66.

2.3. SERS Sensor Fabrication

A SERS-active AgNP-based planar substrate was prepared by a universal approach as previously reported [35,36]. Briefly, to 100 mL of freshly prepared 0.17 mM aqueous silver nitrate solution (Carl-Roth, Karlsruhe, Germany) 30 mL of 2.5 M sodium hydroxide solution (Sigma-Aldrich, St. Louis, MO, USA) was added dropwise for complete precipitation of a dark brown silver(I) oxide. The prepared oxide then was washed 5 times with water and completely dissolved in 35 mL of 30% aqueous ammonia solution (Sigma-Aldrich, USA). To obtain a 0.1 M solution of $[Ag(NH_3)_2]OH$ we added water to 170 mL and then diluted it to prepare final 12.5 mM solution of silver(I) complex. In the ultrasonic silver rain deposition process, this initial ammonia solution of $[Ag(NH_3)_2]OH$ was nebulized into mist. AgNPs were deposited on a preheated surface of a glass slide (290–320 °C) in the form of droplets of 1–5 mm in size, which then evaporated, forming "coffee rings". Silver nanoparticles formed aggregates with average size 185 ± 6 nm (see Supplementary Information, Figure S12). In addition, to immobilize the target analytes–catecholamines, for their preconcentration and uniform distribution, the silver surface was additionally modified with an optically transparent polymer film–chitosan (CS) 1.7 ± 0.1 μm thick. This feature of chitosan film was demonstrated by our group previously [36–38].

2.4. Immobilization of Cu(II) Ions and Bispidine-Based Bis-Azole Ligands into Chitosan Film

The chitosan film was prepared by drop-casting 10 μL of a 1 wt% CS solution in 1vol% acetic acid. After drying at room temperature for at least 2 h, 10 μL of a mixture of Cu(II) sulfate with one of the studied bispidine-based bis-azole ligands: L^1, L^2, L^3 or L^4, was applied onto a CS-coated AgNPs surface. Then, the prepared modified sensor was left drying at room temperature for at least 1 h.

2.5. Material Characterization and Raman Measurements

The obtained substrates were examined by XRD measurements using a Rigaku D/MAX 2500 machine (Tokyo, Japan) with a rotating copper anode (Cu Kα irradiation, 5–90° 2θ range, 0.02° step). Diffraction maxima were indexed using the PDF-2 database. UV−vis absorption spectra were recorded using the Lambda 950 (PerkinElmer, Waltham, MA, USA) UV−vis spectrophotometer with an attached diffuse reflectance accessory. Measurements were performed in the spectral range of 250–850 nm with a scan step of 1 nm. The obtained nanostructured substrates were characterized by scanning electron microscopy (SEM) Carl Zeiss NVision 40 and transmission electron microscopy (TEM) combined with electron diffraction (ED) on a 215 LEO 912 AB OMEGA, Carl Zeiss setup at 100 kV accelerating voltage. To perform Raman measurements, an InVia Raman confocal microscope (Renishaw, Wotton-under-Edge, UK) was used. All SERS spectra were acquired using a 17 mW 633 nm argon laser with a power neutral density filter of 10%. The spectra were collected using a 50× objective lens and with 10 s of acquisition time. A silicon wafer was used for instrument calibration.

3. Results and Discussion

3.1. Complexes of Catecholamines with Cu(II) Ions

Catecholamines are known to form complexes with transition metal ions [39–41]. Some of these complexes have a band with a high molar absorption coefficient in the 500–750 nm region, which is characteristic of the intramolecular charge transfer. The formation of complexes that absorb light in the visible region allows additional resonance enhancement in SERS [42]. First, we studied the possibility of the formation of complexes of dopamine, norepinephrine, or adrenaline with metal ions: Mg(II), Al(III), Cu(II), Co(II), Ni(II), Fe(II), and Fe (III) in a concentration ratio of 1:1 (see Supplementary Information file, Figure S1). The obtained UV-vis spectra of aqueous solutions of CAs with the metal ions showed that high molar absorption coefficients in the region of 750–900 nm are characteristic of CA complexes with Cu(II) ions. Moreover, compared to Fe(III), whose complexes showed similar molar absorption coefficients in this region, Cu(II) ions in aqueous solutions do not have such pronounced oxidizing ability and react only with strong reducing agents, which as a rule are absent in the biological fluids. In addition, copper-containing reagents are widely available, and the resulting complex compounds are quite stable.

First, we studied the possibility of measuring SERS spectra of catecholamine complexes. A previously developed and characterized SERS-active platform consisted of AgNPs adhered onto a glass slide and coated with a chitosan layer [36]. However, the application of a DA solution onto the SERS substrate did not result in high-intensity characteristic signals in SERS spectra [43]. This could be explained by the mismatch of the maximum absorption of CAs in UV region (230–300 nm) and the local surface plasmon resonance (LSPR) position of the obtained stable AgNP-based substrate. Thus, for the "visualization" of CA characteristic bands, we proposed to use catecholamine complexes that are capable of intensively absorbing in the visible spectrum (450–800 nm). We called this approach as Molecular Immobilization and Resonant Raman Amplification by Complex-Loaded Enhancers (MIRRACLE) [43]. Therefore, the SERS sensor surface was modified with a Cu(II) salt solution, then a model DA, AD or NA solution was applied onto the sensor and the Cu(II)–CA complex was formed in the CS layer (Figure 1). The SERS spectra for DA, AD, and NA in the form of their respective complexes with Cu(II) demonstrated intense characteristic bands: 1596 cm^{-1} (ν $(C–C)_{ring}$), 1530 cm^{-1} (ν (C–C)), 1382 cm^{-1} (ν (C1–C2) + ν (C=O)), 1209 cm^{-1} (ν (C–H) + ν (C=O)), 948 cm^{-1} (ν (C–C) + ν (C=O)), and 497 cm^{-1} (ν (Cu_{ring})). The obtained SERS spectra had vibrations characteristic of the $(Cu(semiquinone)_3)^-$ complexes formed on the AgNP surface. The peaks shown in Figure 1 agree with theoretically calculated and previously published bands [44], representing a "molecular fingerprint". Furthermore, we observed lower signal enhancement and signal-to-noise ratio for AD and NA compared to DA.

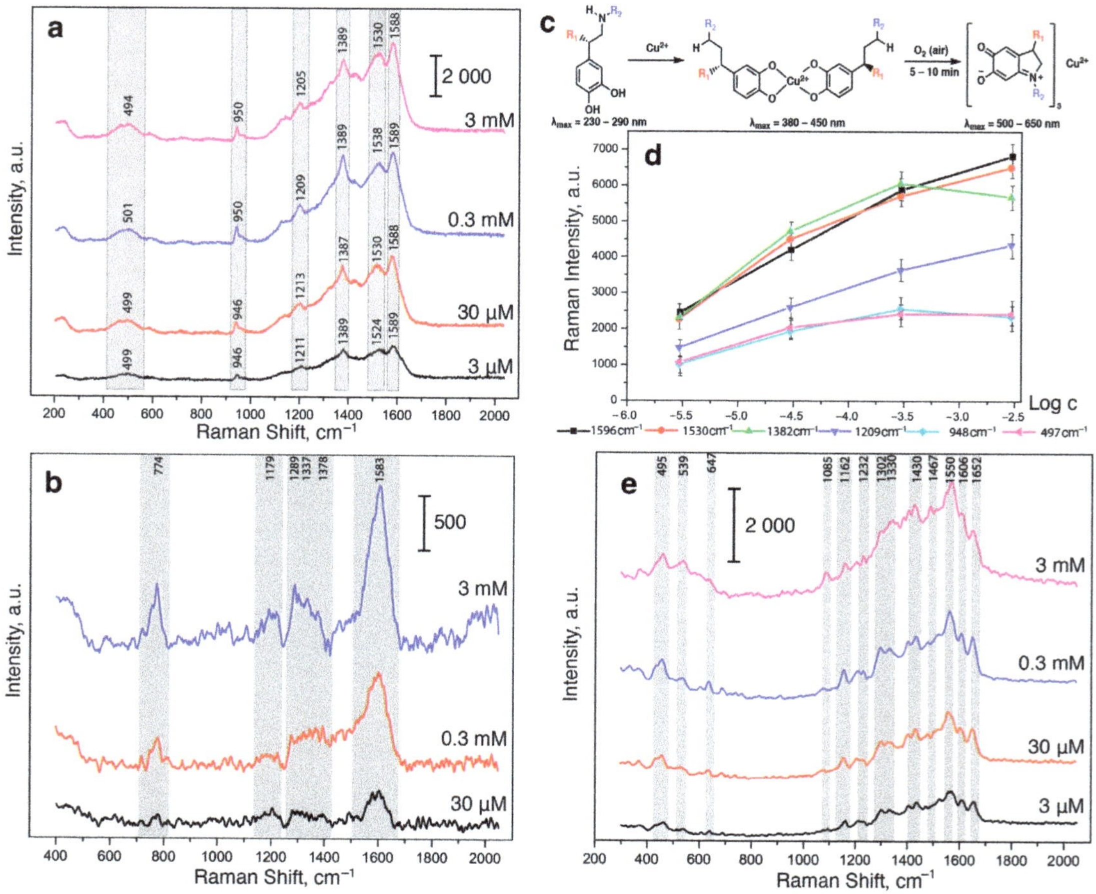

Figure 1. (**a**) SERS spectra of DA at 3 mM, 0.3 mM, 30 μM, and 3 μM concentrations with 3 mM Cu(II). (**b**) SERS spectra of NA at 3 mM, 0.3 mM, and 30 μM concentrations with 3 mM Cu(II). (**c**) Scheme of (Cu(semiquinone)$_3$)$^-$ formation [45]. (**d**) Dependence of the intensity of SERS signals (at 1596, 1530, 1382, 1209, 948 and 497 cm^{-1}) on the logarithm of the DA concentration (M) in the presence of Cu^{2+} (3 mM). Error bars are shown for $n = 20$, $p = 0.95$. (**e**) SERS spectra of AD at 3 mM, 0.3 mM, 30 μM, and 3 μM concentrations with 3 mM Cu(II). Spectra were measured on the silver nanostructured surface coated with the chitosan layer at 633 nm laser wavelength.

When the concentration of CAs in the solution applied to the SERS-sensor element was changed, the peak positions did not shift but the intensities of each of the observed signals changed. However, due to the instability of the resulting complexes and the oxidation of CAs on the AgNP surface, the dependences of the intensities of the observed signals could not be linearized even in logarithmic coordinates (Figure 1d). This could be connected to the formation of CA complexes with copper ions of various stoichiometry. Moreover, UV-vis spectra of the CA complexes with Cu(II) demonstrated a gradual increase in optical density at a ca. 450 nm and a decrease at ca. 750 nm during the formation of complexes of the (Cu(semiquinone)$_3$)$^-$ type (see Supplementary Information File, Figure S2). The described change can be explained by the oxidation of catecholamines to the corresponding catecholates (Figure 1c). It should be noted that the oxidation of DA, NA, and AD with atmospheric oxygen, additionally catalyzed by copper ions, was completed within 15 min after mixing the solutions at mM concentrations.

The formation of such semiquinone Cu(II) complexes, on the one hand, is of interest since it is accompanied by a bathochromic shift of the absorption maxima for CAs. Thus, the resonance SERS effect can be achieved because of the overlap of the absorption maximum of the CA and the LSPR band of AgNPs and laser wavelength, which allow measurement of intense SERS signals. However, on the other hand, for future applications, when analyzing a real biological sample of a biological fluid, the selectively of CA determination in the presence of their metabolites (formed in the body due to oxidative stress) would be compromised. Therefore, in order to develop an appropriate indicator system for the determination of CAs, it is important that the proposed approach prevents the oxidation of the catechol fragment of DA, NA, and AD, i.e., ensures the stabilization of their complexes with copper ions.

3.2. Complexes of Cu(II) Ions and Bispidine-Based Bis-Azole Ligands

In accordance with the classification of vibrational transitions and the selection rules for SERS spectra of molecules with sufficiently high symmetry, the most enhanced lines are the lines due to totally symmetric vibrations that transform according to the unit irreducible representation. In addition, the appearance of forbidden lines associated with antisymmetric vibrations for centrally symmetric molecules can be observed in the SERS spectra. Thus, we proposed the determination of CAs with symmetric polydentate ligands of the bispidine-based bis-azole class (Figure 2) in the presence of Cu(II) ions. As such ligands, we synthesized a set of bispidine-based bis-azoles: L^1, L^2, L^3, and L^4, which possess similar moiety, differing only in the substituents.

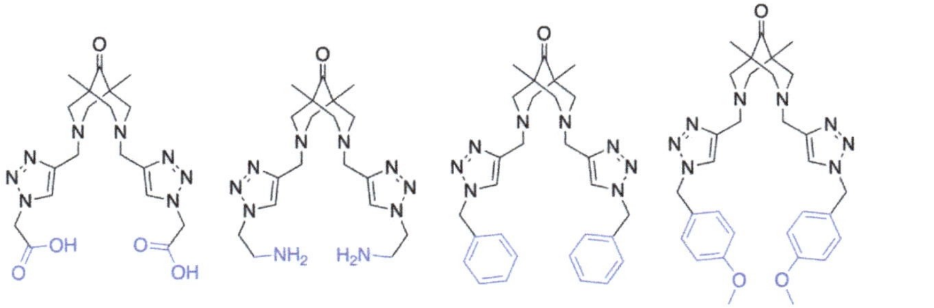

Figure 2. Structures of the bispidine-based bis-azole ligands used in the work. L^1·2TFA–2,2′-(((1,5-dimethyl-9-oxo-3,7-diazabicyclo[3.3.1]nonane-3,7-diyl)bis(methylene))bis(1H-1,2,3-triazole-4,1-diyl))diacetic acid. L^2·3.3TFA–3,7-bis((1-(2-aminoethyl)-1H-1,2,3-triazol-4-yl)methyl)-1,5-dimethyl-3,7-diazabicyclo-[3.3.1]nonan-9-one. L^3–3,7-bis((1-benzyl-1H-1,2,3-triazol-4-yl)methyl)-1,5-dimethyl-3,7-diazabicyclo[3.3.1]nonan-9-one. L^4–3,7-bis((1-(4-methoxybenzyl)-1H-1,2,3-triazol-4-yl)methyl)-1,5-dimethyl-3,7-diazabicyclo-[3.3.1]nonan-9-one.

We studied the formation of complexes between the Cu(II) cation and the bispidine-based bis-azole ligands used, investigated their stability and established the stoichiometry of the resulting complexes (see Supplementary Information File). To do this, each ligand was spectrophotometrically titrated (1 mM) with a $Cu(ClO_4)_2$ solution at 7.5 mM concentration (Figure 3d). The stability constants were calculated using the Bouguer-Lambert-Beer law and the balance of acting masses. To determine the molar absorption coefficients of the resulting complexes, we assumed that at the initial stage of complex formation, the equilibrium concentration of the complex was equal to the concentration of copper cations in the solution. Then, according to the constructed graph of the dependence of the optical density of the solution on the concentration of Cu(II) cations, the molar absorption coefficient was calculated. Based on the data obtained, the true equilibrium concentrations of the ligand and Cu^{2+} and the stability constant of each complex were determined using the relation $K = \frac{[Cu_xL_y]}{[Cu]^x[L]^y}$. As shown in Table 1, the complexes had a preferrable stoichiometry of 1:1.

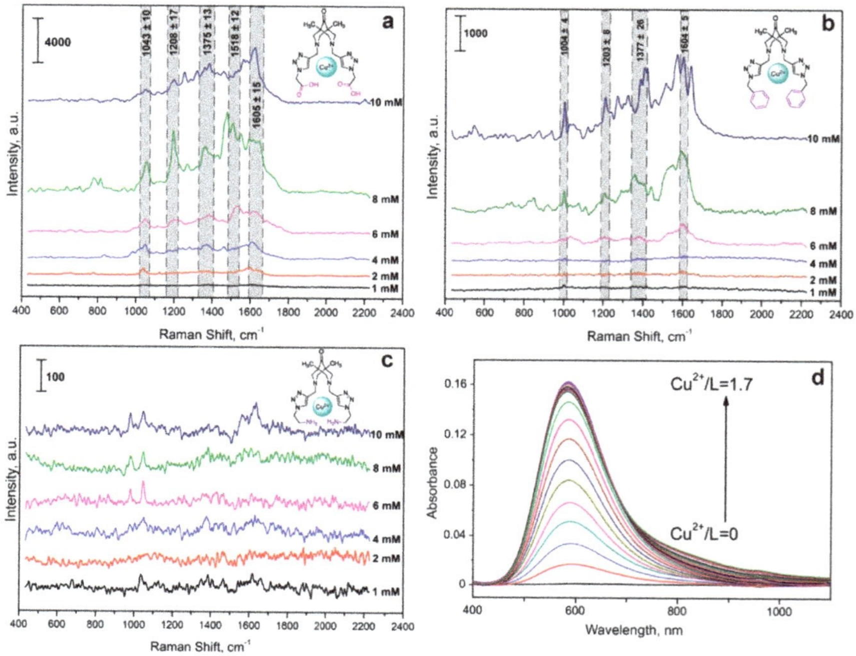

Figure 3. SERS spectra of the bispidine-based bis-azole ligands L^1 (**a**), L^2 (**b**), and L^3 (**d**) with Cu(II) ions at different concentrations. Spectra were measured on the silver nanostructured surface coated with the chitosan layer at 633 nm laser wavelength. (**c**) Typical UV-vis spectra of the triazole ligand L^3 with Cu(II) ions at the increasing L^3 concentration.

Table 1. Stability constants of complexes of bispidine-based bis-azoles with Cu(II) with various stoichiometry. Stability constants were calculated based on the UV-vis measurements using the Bouguer-Lambert-Beer law and the mass action law (see Supplementary Information file).

Ligand, Solvent	L^1, H_2O	L^2, EtOH	L^3, MeCN	L^4, MeCN
lgK(CuL_2)	–*	–*	8.3 ± 0.5	–*
lgK(CuL)	–*	4.0 ± 0.1	6.0 ± 0.5	5.2 ± 0.3
lgK(Cu_2L_3)	23.0 ± 0.3 (pH 4.0)	–*	–*	–*
λ_{max}, nm	577	595	583	577

* the complexes of this stoichiometry did not form.

Importantly, λ_{max} of the obtained complexes were in the visible wavelength range and closer to 633 nm laser (Table 1) compared to absorbance of CAs [43]. The measured stability constants demonstrated that more stable CuL complexes were formed between the Cu(II) cation and L^3, as with the most nonpolar compound. For the most stable complex, CuL^3, it was possible to obtain a solid sample and take an X-ray spectrum, confirming the 1:1 structure stoichiometry. For L^1, a 1:1 complex could not be obtained, which was presumably due to the excessive polarity of the substituent. Based on the obtained data, a 1:1:1 complex was chosen for further determination of CA with various bispidine-based bis-azoles.

We first studied the intrinsic spectra of Cu(II) with the synthesized bispidine-based bis-azole ligands, which could cause a background in SERS spectra for CA detection. Thus, the optimal concentrations of the indicator system were selected by varying the Cu(II) and each of the considered ligands concentrations in the corresponding solvents from Table 1 in the range of 1–10 mM. The complex formation for L^1 with copper ions on the SERS surface resulted in the significant interfering background signals (Figure 3a). This made determination of catecholamines in the corresponding ternary complexes challenging. Most likely, this was related to the additional interaction of –COOH groups of L^1 with $-NH_2$ groups of chitosan deposited onto silver nanoparticles. Thus, the indicator system with Cu(II) and L^1 was not further considered for sensitive determination of CAs. For L^3, its complex formation with copper ions also contributed to the additional spectral background (Figure 3b) due to the intrinsic intense characteristic vibrations of the phenyl groups in L^3 structure. The SERS spectra shown in Figure 3 demonstrated that L^2 caused the least background, and the ligands could be used at 1, 4, and 10 mM concentrations for L^1, L^3, and L^2, respectively. L^3 was advantageous as it showed higher SERS intensities, while L^2 provided lower spectral background. Thus, L^3 and L^2 were chosen for further development of approach for CA detection.

3.3. Complexes of Catecholamines with Cu(II) Ions and Bispidine-Based Bis-Azole Ligands

To identify CA by SERS, the optimal concentration of the indicator system was selected by varying the Cu(II) and each of the considered ligands concentrations in the corresponding solvents from Table 1 in the range of 1–10 mM at a constant concentration of DA, as a model catecholamine. The SERS spectra were recorded with a 633 nm laser wavelength and an extinction time of 10 s. Raman spectra of CAs have been preliminary modeled with density functional theory (DFT) (see Supplementary Information file). Strong modes of CA agreed well with experimental SERS of CA complexes. The SERS spectra in Figure 4a showed that even at an indicator system concentration of 1 mM, intense signals were observed in the SERS spectra for the DA complex with L^3 and Cu(II) ions. The indicator system with Cu(II) and L^2 demonstrated stronger SERS signals of DA at 8 mM concentration based on the majority of Raman peaks (Figure 4b). The difference in the optimal concentrations might be connected to the differences in chemical structure of the studied ligands. From the SERS spectra shown in Figure 3, carboxyl and phenyl groups of L^1 and L^3, respectively, demonstrated more effective coordination to the AgNPs surface, compared to amino groups of L^2.

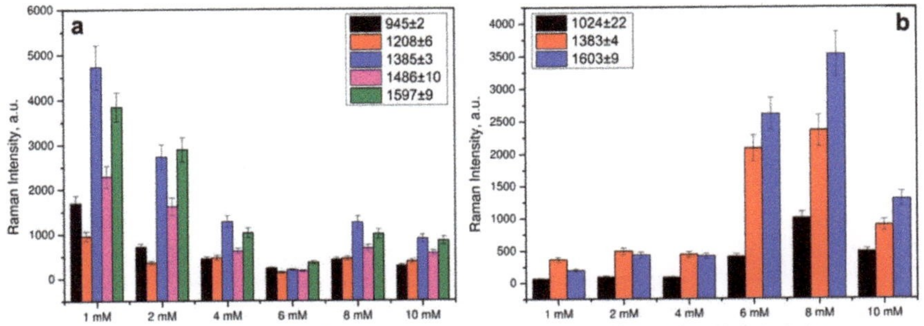

Figure 4. Raman intensities of characteristic DA bands in the SERS spectra at different concentrations of Cu(II) and ligands for: (**a**) L^3 and (**b**) L^2. Error bars are shown for $n = 20$, $p = 0.95$.

With using identical experimental conditions, areas of the SERS-sensing surface, concentrations and volumes of solutions of the Cu(II):bispidine-based bis-azole components applied to the surface, and Raman microscope settings, we obtained the calibration curves for CAs (Figure 5). Notably, the L^3 ligand with phenyl groups demonstrated sharper SERS peaks as well as higher sensitivity compared to the indicator system with amino groups in L^2. This could be explained by NH_2-groups in L^2 structure also interacting with the

AgNP-based surface resulting in the presence of differently oriented DA:Cu(II):L^2 complexes on the SERS substrate surface. Figure 5c shows the SERS spectra of DA obtained on the developed SERS-active surface at various concentrations of DA in the reaction mixture (the indicator system is L^3:Cu(II) ions at 1 mM concentration). The linear trend for DA determination was observed at a 3.2×10^{-12}–1×10^{-8} M concentration range. Normal values of catecholamines in blood plasma were reported to be 4×10^{-11}–4.5×10^{-9} M [46], 4.5×10^{-10}–2.49×10^{-9} M [47], and 2×10^{-11}–4.67×10^{-10} M [47] for DA, NA, and AD, respectively. Thus, the proposed approach demonstrated promise for rapid multiplex determination of catecholamines: dopamine, adrenaline, and noradrenaline which could be applied for "point-of-care" analysis (Figure 6).

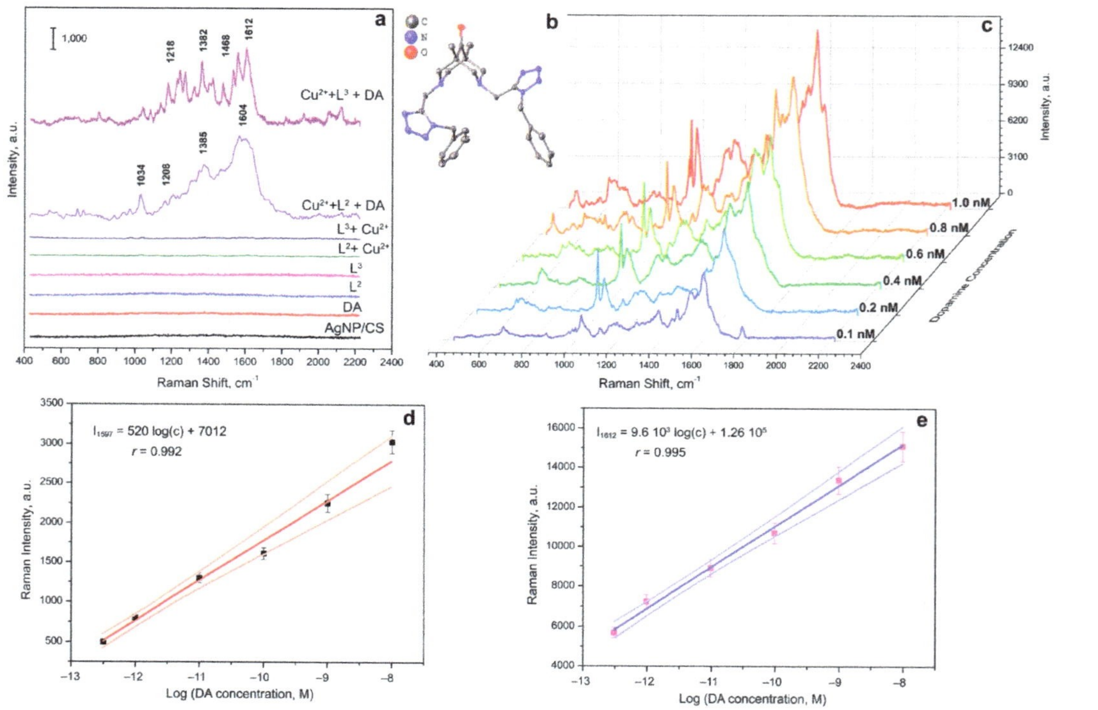

Figure 5. (**a**) Typical SERS spectra of the blank non-modified SERS substrate, DA (0.1 mM), L^2 (8 mM), L^3 (1 mM), the indicator system: Cu(II) ions with L^2 (8 mM), the indicator system: Cu(II) ions with L^3 (1 mM), DA (0.1 mM) applied onto the SERS substrate modified with Cu(II) ions and L^2 (8 mM), and DA (0.1 mM) applied onto the SERS substrate modified with Cu(II) ions and L^3 (1 mM). (**b**) Single-crystal structure of the bispidine-based bis-azole ligand L^3. (**c**) SERS spectra of various DA concentrations applied onto the SERS substrate modified with Cu(II) ions and L^3 (1 mM). All SERS spectra were measured on the silver nanostructured surface coated with the chitosan layer at the resonant 633 nm laser wavelength. (**d**) Calibration curve for DA determination with Cu(II) ions and L^2 (8 mM) based on Raman intensity of 1597 cm^{-1} peak. Pale orange lines demonstrate confidence corridor with $p = 0.95$. (**e**) Calibration curve for DA determination with Cu(II) ions and L^3 (1 mM) based on Raman intensity of 1612 cm^{-1} peak. Pale blue lines demonstrate confidence corridor with mboxemphp = 0.95.

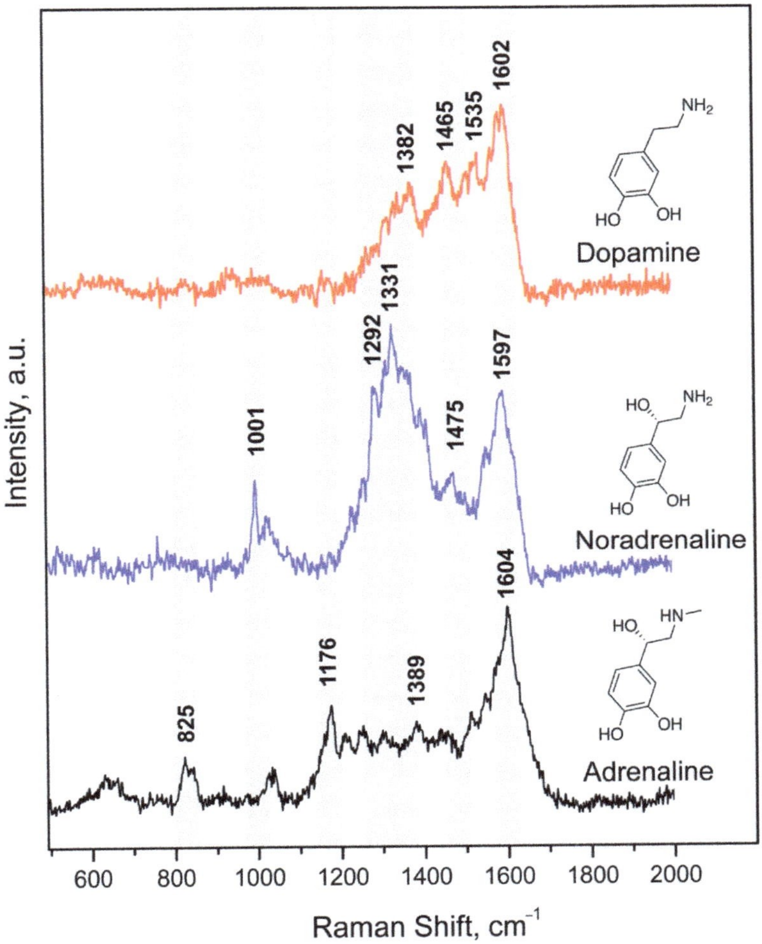

Figure 6. SERS spectra of the DA (0.1 mM), NA (0.1 mM), and AD (0.1 mM) applied onto the SERS substrate modified with Cu(II) ions and L^3 (1 mM) at 633 nm excitation laser wavelength.

4. Conclusions

Herein, we propose an approach for catecholamine detection based on trapping them into colored stable complexes on the SERS-active surface. The MIRRACLE (Molecular Immobilization and Resonant Raman Amplification by Complex-Loaded Enhancers) methodology allowed for sensitive determination of dopamine in 3.2×10^{-12}–1×10^{-8} M concentration range. The determination of catecholamines in complexes with Cu(II) and bispidine-based bis-azole ligands on a nanostructured silver surface coated with a thin layer of chitosan by SERS allowed the selective determination of dopamine, adrenaline, and noradrenaline. Four synthesized bispidine-based bis-azole ligands served as pliers in the ternary complex with Cu(II) ions and catecholamines. The limits of detection of catecholamines using the developed method make it possible to determine them at the level of picomolar concentrations, which corresponds to the reference values in the blood plasma of healthy people. Therefore, the use of the proposed technique provides a potential opportunity to determine catecholamine neurotransmitters in biological fluids using SERS spectroscopy, specifically for diagnosis of neuroendocrine disorders. Further prospects in this area lie in the improvement and optimization of methods for the multiplexed determination of the largest possible set of molecules–markers of neurotransmitter metabolism disorders, as well as the subsequent detection of the correlation of neuroendocrine and neurodegenerative diseases with concentrations and their ratios in biological fluids with the possibility of non-invasive diagnostics.

Supplementary Materials: The following supporting information can be downloaded at: https://www.mdpi.com/article/10.3390/bios13010124/s1. Figure S1: UV-vis spectra of 5 mM solutions of 8a–AD, 8b–DA and their corresponding complexes with: 1–Cu(II), 2–Al(III), 3–Fe(II), 4–Co(II), 5–Fe(III), 6–Ni(II), 7–Mg(II). Figure S2: UV-vis spectra of DA and Cu(II) equimolar mixture in time 0–15 min. Table S1: Raman bands theoretically calculated, published in the literature, and experimentally observed on the AgNP-based substrate for $[Cu(semiquinone)_3]^-$ (λ_{ex} 632.8 nm). Figure S3: UV-vis spectra for spectrophotometric titration of L^4 with Cu^{2+} solution in (a) 100% MeCN; (b) 50% MeCN–50% H_2O. Figure S4: Spectrophotometric titration of L^4 with Cu^{2+} solution in (a) 100% MeCN (λ_{abs} = 577 nm); (b) 50% MeCN–50% H_2O (λ_{abs} = 583 nm). Figure S5: (a) UV-vis spectra for spectrophotometric titration of Cu^{2+} with L^4 solution in 50% MeCN–50% H_2O. (b) Spectrophotometric titration of Cu^{2+} with L^4 solution in 50% MeCN–50% H_2O (λ_{abs} = 583 nm). Figure S6: UV-vis spectra for spectrophotometric titration of L^3 with Cu^{2+} solution in (a) 100% MeCN; (b) 50% MeCN–50% H_2O. Spectrophotometric titration of L^3 with Cu^{2+} solution in (c) 100% MeCN (λ_{abs} = 577 nm and λ_{abs} = 697 nm); (b) 50% MeCN–50% H_2O (λ_{abs} = 583 nm). Figure S7: (a) UV-vis spectra for spectrophotometric titration of Cu^{2+} with L^3 solution in 50% MeCN–50% H_2O. (b) Spectrophotometric titration of Cu^{2+} with L^3 solution in 50% MeCN–50% H_2O (λ_{abs} = 583 nm). Figure S8: (a) UV-vis spectra for spectrophotometric titration of Cu^{2+} with L^4 solution in 100% MeCN. Figure S9: Spectrophotometric titration of L^1 with Cu^{2+} solution in H_2O (λ_{abs} = 577 nm). Table S2: The calculated values of the constants of complex formation of copper with water-insoluble bispidine-type ligands. Table S3: The protonation constants for L^1 in H_2O (μ = 0.1 M) at 25 °C. Figure S10: (a) Potentiometric titration of L^1 (1 mM) in H_2O in the presence of $HClO_4$ (4 mM), μ = 0.1 M KNO_3 at 25 °C with NaOH solution (0.1 M). (b) Distribution of protonated and deprotonated forms of L^1 (1 mM) in solution depending on pH. Table S4: The stability constants of complexes of L^1 with cations Cu^{2+} and hydroxocomplexes Cu^{2+} in an aqueous solution (μ = 0.1 M) at 25 °C. Figure S11: (a) Potentiometric titration of L^1 and L^1 in the presence of Cu^{2+} (Cu^{2+} + L) in H_2O (μ = 0.1 M KNO_3) at 25 °C. (b) Distribution of different forms of L^1 (1 mM) in the presence of Cu(II) perchlorate (1 mM) depending on pH. Figure S12: Typical SEM images of a proposed silver nanoparticle sensor surface (a) coated with a silver chitosan film of a silver nanoparticle surface (d) and its cross section (c); optical photograph of a silver nanostructured surface consisting of overlapping "coffee rings" (b). Figure S13: SERS spectra of various AD concentrations applied onto the SERS substrate modified with Cu(II) ions and L^3 (1 mM). All SERS spectra were measured on the silver nanostructured surface coated with the chitosan layer at the resonant 633 nm laser wavelength. Figure S14: SERS spectra of various NA concentrations applied onto the SERS substrate modified with Cu(II) ions and L^3 (1 mM). All SERS spectra were measured on the silver nanostructured surface coated with the chitosan layer at the resonant 633 nm laser wavelength. Figure S15: Optimized geometry for dopamine (DA) in "Q-Chem" with PBE0 method and 6-31G* basis. Figure S16: Calculated normalized Raman spectrum for dopamine (DA) in "Q-Chem" with PBE0 method and 6-31G* basis (Lorentzian). Table S5: Calculated normalized Raman shifts with intensities for dopamine (DA) in "Q-Chem" with PBE0 method and 6-31G* basis (Lorentzian). Figure S17: Optimized geometry for noradrenaline (NA) in "Q-Chem" with PBE0 method and 6-31G* basis. Figure S18: Calculated normalized Raman spectrum for noradrenaline (NA) in "Q-Chem" with PBE0 method and 6-31G* basis (Lorentzian). Table S6: Calculated normalized Raman shifts with intensities for noradrenaline (NA) in "Q-Chem" with PBE0 method and 6-31G* basis (Lorentzian). Figure S19: Optimized geometry for adrenaline (AD) in "Q-Chem" with PBE0 method and 6-31G* basis. Figure S20: Calculated normalized Raman spectrum for adrenaline (AD) in "Q-Chem" with PBE0 method and 6-31G* basis (Lorentzian). Table S7: Calculated normalized Raman shifts with intensities for adrenaline (AD) in "Q-Chem" with PBE0 method and 6-31G* basis (Lorentzian).

Author Contributions: Conceptualization, O.E.E., T.N.S. and I.A.V.; methodology, O.E.E., O.O.K., A.V.M. and B.V.E.; software, O.E.E., A.S.Z. and A.V.M.; validation, O.E.E., O.O.K. and A.S.Z.; formal analysis, A.S.Z., A.V.M. and B.V.E.; investigation, O.E.E., O.O.K., A.S.Z., A.V.M. and B.V.E.; resources, S.Z.V., T.N.S. and I.A.V.; data curation, O.E.E., O.O.K., A.S.Z., A.V.M. and B.V.E.; writing—original draft preparation, O.E.E.; writing—review and editing, all authors; visualization, O.E.E. and A.S.Z.;

supervision, O.E.E., I.A.V. and S.Z.V.; project administration, T.N.S. and I.A.V.; funding acquisition, T.N.S. and I.A.V. All authors have read and agreed to the published version of the manuscript.

Funding: This research was supported by the Russian Science Foundation (grant No. 20-13-00330).

Institutional Review Board Statement: Not applicable.

Informed Consent Statement: Not applicable.

Data Availability Statement: Not applicable.

Acknowledgments: This research was supported by the Interdisciplinary Scientific and Educational School of Moscow University "Future Planet: Global Environmental Change." The research was carried out using the equipment of MSU Shared Research Equipment Center "Technologies for obtaining new nanostructured materials and their complex study" and purchased by MSU in the frame of the Equipment Renovation Program (National Project "Science") and in the frame of the MSU Program of Development.

Conflicts of Interest: The authors declare no conflict of interest.

References

1. Veselova, I.A.; Sergeeva, E.A.; Makedonskaya, M.I.; Eremina, O.E.; Kalmykov, S.N.; Shekhovtsova, T.N. Methods for determining neurotransmitter metabolism markers for clinical diagnostics. *J. Anal. Chem.* **2017**, *71*, 1155–1168. [CrossRef]
2. Schroeder, C.; Jordan, J. Norepinephrine transporter function and human cardiovascular disease. *Am. J. Physiol. Heart Circ. Physiol.* **2012**, *303*, H1273–H1282. [CrossRef]
3. Miyamoto, S.; Yoshida, Y.; Ozeki, Y.; Okamoto, M.; Gotoh, K.; Masaki, T.; Nishida, H.; Shibuya, T.; Shin, T.; Daa, T.; et al. Dopamine-Secreting Pheochromocytoma and Paraganglioma. *J. Endocr. Soc.* **2021**, *5*, bvab163. [CrossRef]
4. Dwivedi, N.; Shah, J.; Mishra, V.; Tambuwala, M.; Kesharwani, P. Nanoneuromedicine for management of neurodegenerative disorder. *J. Drug Deliv. Sci. Technol.* **2019**, *49*, 477–490. [CrossRef]
5. Liu, S.Q.; Li, B.; Li, J.J.; Sun, S.; Sun, S.R.; Wu, Q. Neuroendocrine regulations in tissue-specific immunity: From mechanism to applications in tumor. *Front. Cell Dev. Biol.* **2022**, *10*, 896147. [CrossRef] [PubMed]
6. Makedonskaya, M.I.; Veselova, I.A.; Kalmykov, S.N.; Shekhovtsova, T.N. Novel biosensing system for the simultaneous multiplex fluorescent determination of catecholamines and their metabolites in biological liquids. *J. Pharm. Biomed. Anal.* **2018**, *156*, 133–141. [CrossRef] [PubMed]
7. Ma, X.; Chao, M.; Wang, Z. Electrochemical detection of dopamine in the presence of epinephrine, uric acid and ascorbic acid using a graphene-modified electrode. *Anal. Methods* **2012**, *4*, 1687–1692. [CrossRef]
8. Quan, D.; Shin, W. Amperometric detection of catechol and catecholamines by immobilized laccase from DeniLite. *Electroanalysis* **2004**, *16*, 1576–1582. [CrossRef]
9. Yogeswaran, U.; Chen, S.M. Multi-walled carbon nanotubes with poly(methylene blue) composite film for the enhancement and separation of electroanalytical responses of catecholamine and ascorbic acid. *Sens. Actuators B Chem.* **2008**, *130*, 739–749. [CrossRef]
10. Kang, W.J.; Niu, L.M.; Ma, L. 2,3-Dimercaptosuccinic acid self-assembled gold electrode for the simultaneous determination of epinephrine and dopamine. *Chin. Chem. Lett.* **2009**, *20*, 221–224. [CrossRef]
11. Murphy, J.F.; Davies, D.H.; Smith, C.J. The development of enzyme-linked immunosorbent assays (ELISA) for the catecholamines adrenalin and noradrenalin. *J. Immunol. Methods* **1992**, *154*, 89–98. [CrossRef] [PubMed]
12. Da Prada, M.; Zurcher, G. Simultaneous radioenzymatic determination of plasma and tissue adrenaline, noradrenaline and dopamine within the femtomole range. *Life Sci.* **1976**, *19*, 1161–1174. [CrossRef] [PubMed]
13. Kim, J.; Park, H.; Ryu, J.; Jeon, O.; Paeng, I.R. Competitive enzyme-linked immunosorbent assay for a selective and sensitive determination of dopamine in the presence of ascorbic acid and uric acid. *J. Immunoass. Immunochem.* **2010**, *31*, 33–44. [CrossRef]
14. Ghasemi, F.; Hormozi-Nezhad, M.R.; Mahmoudi, M. Identification of catecholamine neurotransmitters using fluorescence sensor array. *Anal. Chim. Acta* **2016**, *917*, 85–92. [CrossRef] [PubMed]
15. Zhang, Y.; Wang, B.; Xiong, H.; Wen, W.; Cheng, N. A ratiometric fluorometric epinephrine and norepinephrine assay based on carbon dot and CdTe quantum dots nanocomposites. *Microchem. J.* **2019**, *146*, 66–72. [CrossRef]
16. Fotopoulou, M.A.; Ioannou, P.C. Post-column terbium complexation and sensitized fluorescence detection for the determination of norepinephrine, epinephrine and dopamine using high-performance liquid chromatography. *Anal. Chim. Acta* **2002**, *462*, 179–185. [CrossRef]
17. Zheng, J.; Mandal, R.; Wishart, D.S. A sensitive, high-throughput LC-MS/MS method for measuring catecholamines in low volume serum. *Anal. Chim. Acta* **2018**, *1037*, 159–167. [CrossRef]
18. Carrera, V.; Sabater, E.; Vilanova, E.; Sogorb, M.A. A simple and rapid HPLC-MS method for the simultaneous determination of epinephrine, norepinephrine, dopamine and 5-hydroxytryptamine: Application to the secretion of bovine chromaffin cell cultures. *J. Chromatogr. B* **2007**, *847*, 88–94. [CrossRef]

19. Espina-Benitez, M.B.; Marconi, F.; Randon, J.; Demesmay, C.; Dugas, V. Evaluation of boronate affinity solid-phase extraction coupled in-line to capillary isoelectric focusing for the analysis of catecholamines in urine. *Anal. Chim. Acta* **2018**, *1034*, 195–203. [CrossRef]
20. Xu, X.; Zhang, H.; Shi, H.; Ma, C.; Cong, B.; Kang, W. Determination of three major catecholamines in human urine by capillary zone electrophoresis with chemiluminescence detection. *Anal. Biochem.* **2012**, *427*, 10–17. [CrossRef]
21. Li, X.; Jin, W.; Weng, Q. Separation and determination of homovanillic acid and vanillylmandelic acid by capillary electrophoresis with electrochemical detection. *Anal. Chim. Acta* **2002**, *461*, 123–130. [CrossRef]
22. Tesoro, C.; Lelario, F.; Ciriello, R.; Bianco, G.; Di Capua, A.; Acquavia, M.A. An Overview of Methods for L-Dopa Extraction and Analytical Determination in Plant Matrices. *Separations* **2022**, *9*, 224. [CrossRef]
23. Yu, C.; Zhang, Q.; Wang, L.; Zhang, Y.; Cui, M.; Bi, K.; Li, D.; Li, Q. Nanoconfined liquid phase nanoextraction combined with in-fiber derivatization for simultaneous quantification of seventy amino-containing metabolites in plasma by LC-MS/MS: Exploration of lung cancer screening model. *Talanta* **2022**, *245*, 123452. [CrossRef] [PubMed]
24. Bandodkar, A.J.; Wang, J. Non-invasive wearable electrochemical sensors: A review. *Trends Biotechnol.* **2014**, *32*, 363–371. [CrossRef]
25. Satheeshkumar, E.; Karuppaiya, P.; Sivashanmugan, K.; Chao, W.T.; Tsay, H.S.; Yoshimura, M. Biocompatible 3D SERS substrate for trace detection of amino acids and melamine. *Spectrochim. Acta A Mol. Biomol. Spectrosc.* **2017**, *181*, 91–97. [CrossRef]
26. Rajapandiyan, P.; Yang, J. Sensitive cylindrical SERS substrate array for rapid microanalysis of nucleobases. *Anal. Chem.* **2012**, *84*, 10277–10282. [CrossRef]
27. Shafer-Peltier, K.E.; Haynes, C.L.; Glucksberg, M.R.; van Duyne, R.P. Toward a glucose biosensor based on surface-enhanced Raman scattering. *J. Am. Chem. Soc.* **2003**, *125*, 588–593. [CrossRef]
28. Raju, N.R.C. Hemoglobin detection on AgO surface enhanced Raman scattering (SERS)-substrates. *Mater. Lett.* **2014**, *130*, 274–276. [CrossRef]
29. Lv, Y.; Qin, Y.; Svec, F.; Tan, T. Molecularly imprinted plasmonic nanosensor for selective SERS detection of protein biomarkers. *Biosens. Bioelectron.* **2016**, *80*, 433–441. [CrossRef]
30. Eremina, O.E.; Semenova, A.A.; Sergeeva, E.A.; Brazhe, N.A.; Maksimov, G.V.; Shekhovtsova, T.N.; Goodilin, E.A.; Veselova, I.A. Surface-enhanced Raman spectroscopy in modern chemical analysis: Advances and prospects. *Russ. Chem. Rev.* **2018**, *87*, 741–770. [CrossRef]
31. Kneipp, K.; Wang, Y.; Dasari, R.R.; Feld, M.S. Near-infrared surface-enhanced Raman scattering (NIR-SERS) of neurotransmitters in colloidal silver solutions. *Spectrochim. Acta Part A Mol. Biomol. Spectrosc.* **1995**, *51*, 481–487. [CrossRef]
32. Moody, A.S.; Sharma, B. Multi-metal, Multi-wavelength Surface-Enhanced Raman Spectroscopy Detection of Neurotransmitters. *ACS Chem. Neurosci.* **2018**, *9*, 1380–1387. [CrossRef] [PubMed]
33. Wang, P.; Xia, M.; Liang, O.; Sun, K.; Cipriano, A.F.; Schroeder, T.; Liu, H.; Xie, Y.H. Label-Free SERS Selective Detection of Dopamine and Serotonin Using Graphene-Au Nanopyramid Heterostructure. *Anal. Chem.* **2015**, *87*, 10255–10261. [CrossRef] [PubMed]
34. Vatsadze, S.Z.; Medved'ko, A.V.; Bodunov, A.A.; Lyssenko, K.A. Bispidine-based bis-azoles as a new family of supramolecular receptors: The theoretical approach. *Mendeleev Commun.* **2020**, *30*, 344–346. [CrossRef]
35. Semenova, A.A.; Goodilin, E.A.; Brazhe, N.A.; Ivanov, V.K.; Baranchikov, A.E.; Lebedev, V.A.; Goldt, A.E.; Sosnovtseva, O.V.; Savilov, S.V.; Egorov, A.V.; et al. Planar SERS nanostructures with stochastic silver ring morphology for biosensor chips. *J. Mater. Chem.* **2012**, *22*, 24530–24544. [CrossRef]
36. Eremina, O.E.; Sidorov, A.V.; Shekhovtsova, T.N.; Goodilin, E.A.; Veselova, I.A. Novel multilayer nanostructured materials for recognition of polycyclic aromatic sulfur pollutants and express analysis of fuel quality and environmental health by surface enhanced Raman spectroscopy. *ACS Appl. Mater. Interfaces* **2017**, *9*, 15058–15067. [CrossRef]
37. Sidorov, A.V.; Eremina, O.E.; Veselova, I.A.; Goodilin, E.A. Polymer-coated substrates for surface enhanced Raman spectroscopy. *Mendeleev Commun.* **2015**, *25*, 460–462. [CrossRef]
38. Eremina, O.E.; Kapitanova, O.O.; Ferree, M.V.; Lemesh, I.A.; Eremin, D.B.; Goodilin, E.A.; Veselova, I.A. Ultrasensitive and multiplex SERS determination of anthropogenic phenols in oil fuel and environmental samples. *Environ. Sci. Nano* **2022**, *9*, 964–974. [CrossRef]
39. Suzuki, Y. Design and synthesis of fluorescent reagents for selective detection of dopamine. *Sens. Actuators B* **2017**, *239*, 383–389. [CrossRef]
40. Kareem, M.J.; Al-Hamdani, A.A.S.; Jirjees, V.Y.; Khan, M.E.; Allaf, A.W.; Al Zoubi, W. Preparation, spectroscopic study of Schiff base derived from dopamine and metal Ni(II), Pd(II), and Pt(IV) complexes, and activity determination as antioxidants. *J. Phys. Org. Chem.* **2020**, *34*, e4156. [CrossRef]
41. Nagles, E.; Ibarra, L.; Llanos, J.P.; Hurtado, J.; Garcia-Beltrán, O. Development of a novel electrochemical sensor based on cobalt(II) complex useful in the detection of dopamine in presence of ascorbic acid and uric acid. *J. Electroanal. Chem.* **2017**, *788*, 38–43. [CrossRef]
42. Sidorov, A.; Vashkinskaya, O.; Grigorieva, A.; Shekhovtsova, T.; Veselova, I.; Goodilin, E. Entrapment into charge transfer complexes for resonant Raman scattering enhancement. *Chem. Commun.* **2014**, *50*, 6468–6470. [CrossRef] [PubMed]
43. Eremina, O.E.; Yarenkov, N.R.; Kapitanova, O.O.; Zelenetskaya, A.S.; Smirnov, E.A.; Shekhovtsova, T.N.; Goodilin, E.A.; Veselova, I.A. Molecular Immobilization and Resonant Raman Amplification by Complex-Loaded Enhancers (MIRRACLE)

on copper (II)-chitosan-modified SERS-active metallic nanostructured substrates for multiplex determination of dopamine, norepinephrine, and epinephrine. *Mikrochim. Acta* **2022**, *189*, 211. [CrossRef] [PubMed]
44. Barreto, W.J.; Barreto, S.R.; Ando, R.A.; Santos, P.S.; DiMauro, E.; Jorge, T. Raman, IR, UV-vis and EPR characterization of two copper dioxolene complexes derived from L-dopa and dopamine. *Spectrochim. Acta Part A* **2008**, *71*, 1419–1424. [CrossRef] [PubMed]
45. Pierpont, C.G.; Buchanan, R.M. Transition metal complexes of o-benzoquinone, o-semiquinone, and catecholate ligands. *Coord. Chem. Rev.* **1981**, *38*, 45–87. [CrossRef]
46. Pluto, R.; Burger, P. Normal values of catecholamines in blood plasma determined by high-performance liquid chromatography with amperometric detection. *Int. J. Sports Med.* **1988**, *9*, 75–78. [CrossRef]
47. Peaston, R.T.; Weinkove, C. Measurement of catecholamines and their metabolites. *Ann. Clin. Biochem.* **2004**, *41*, 17–38. [CrossRef]

MDPI
St. Alban-Anlage 66
4052 Basel
Switzerland
Tel. +41 61 683 77 34
Fax +41 61 302 89 18
www.mdpi.com

Biosensors Editorial Office
E-mail: biosensors@mdpi.com
www.mdpi.com/journal/biosensors

www.ingramcontent.com/pod-product-compliance
Lightning Source LLC
Chambersburg PA
CBHW061630210926
51850CB00009B/2493
9783036583167